Oxidative Stress in Aging

From Model Systems to Human Diseases

Edited by

Satomi Miwa, PhD
Institute for Ageing and Health, Newcastle University,
Newcastle upon Tyne, United Kingdom

Kenneth B. Beckman, PhD
Functional Genomics Core,
Children's Hospital Oakland Research Institute, Oakland, California

Florian L. Muller, PhD
Department of Cellular and Structural Biology,
University of Texas Health Science Center at San Antonio, San Antonio, Texas

Humana Press

Editors
Satomi Miwa, Ph.D.
Institute for Ageing and Health
Newcastle University
Newcastle upon Tyne
UK
satomi.miwa@ncl.ac.uk

Kenneth B. Beckman, Ph.D.
Functional Genomics Core
Children's Hospital Oakland Research
  Institute
Oakland, CA
USA
kbeckman@chori.org

Florian L. Muller, Ph.D.
Department of Cellular and Structural
  Biology
University of Texas Health Science
  Center at San Antonio
San Antonio, TX
USA
aettius@aol.com

Series Editors
Robert J. Pignolo
Division of Geriatric Medicine
University of Pennsylvania School
  of Medicine
Philadelphia, PA

Mary Ann Forciea
Division of Geriatric Medicine
University of Pennsylvania Health System
Philadelphia, PA

Jerry C. Johnson
Division of Geriatric Medicine
University of Pennsylvania School
  of Medicine
Philadelphia, PA

ISBN: 978-1-58829-991-8   e-ISBN: 978-1-59745-420-9
DOI: 10.1007/978-1-59745-420-9

Library of Congress Control Number: 2007942719

Cover illustration: Cover art provided by Conor Lawless

Printed on acid-free paper

9 8 7 6 5 4 3 2 1

springer.com

Preface

Aging remains one of the biggest unsolved problems in biology. More than 50 years ago, Denham Harman proposed the free radical theory of aging, arguing that cumulative damage from oxygen free radicals was causal to the process of aging. The fundamental idea is simple, yet scientifically plausible. Although at first it received limited attention, a recent explosion of investigative interest and endeavors has made the free radical theory the most extensively tested of all aging theories. The body of literature is now so vast and encompasses so many different techniques and model systems that an integrated evaluation of the evidence is difficult to say the least. Our goal in writing *Oxidative Stress in Aging: From Model Systems to Human Diseases* was to provide an easily accessible assessment of the free radical theory. We believe that the reader will find our "model system by model system" approach helpful, for it greatly simplifies a sometimes contradictory body of evidence.

A further development for the free radical theory is the realization that oxidative stress may contribute not only to aging per se, but also to the pathogenesis of a variety of age-related diseases. No assessment of the free radical theory of aging would be complete without an up-to-date account of major trends on the role of oxygen free radicals in the pathology of age-related diseases.

We thank all the contributors for taking time to produce outstanding chapters that not only summarize key data but also provide critical evaluation, discussion, and interpretation. This book is only possible because of their efforts.

We are also grateful to Dr. Matthew Ford for help with proofreading, and to Springer, particularly Richard Lansing, for exceptionally efficient help and advice throughout, and for accepting our demanding requests, one after another.

Satomi Miwa
Kenneth Bruce Beckman
Florian L. Muller

Contents

Section IV Future

Contributors

Muhammad A. Abdul-Ghani, MD, PhD
Division of Diabetes, University of Texas Health Science Center at San Antonio, San Antonio, TX

Scott W. Ballinger, PhD
Division of Molecular and Cellular Pathology, University of Alabama at Birmingham, Birmingham, AL

Gustavo Barja, PhD
Department of Animal Physiology II, Faculty of Biological Sciences, Complutense University, Madrid, Spain

Kenneth B. Beckman, PhD
Functional Genomics Core, Children's Hospital Oakland Research Institute, Oakland, CA

Paul S. Brookes, PhD
Department of Anesthesiology, University of Rochester Medical Center, Rochester, NY

Sophocles Chrissobolis, PhD
Department of Internal Medicine, Carver College of Medicine, University of Iowa, Iowa City, IA

Ralph A. DeFronzo, MD
Division of Diabetes, University of Texas Health Science Center at San Antonio, San Antonio, TX

Sean P. Didion, PhD
Department of Internal Medicine, Carver College of Medicine, University of Iowa, Iowa City, IA

Ryan Doonan, PhD
Department of Biology, University College London, London, UK

Frank M. Faraci, PhD
Departments of Internal Medicine and Pharmacology, Carver College
of Medicine, University of Iowa, Iowa City, IA

Toren Finkel, MD, PhD
Cardiology Branch, National Heart, Lung, and Blood Institute,
National Institutes of Health, Bethesda, MD

David Gems, PhD
Department of Biology, University College London, London, UK

Stavros Gonidakis, BA
Integrative and Evolutionary Biology and Andrus Gerontology Center,
University of Southern California, Los Angeles, CA

David L. Hoffman, PhD
Department of Biochemistry and Biophysics, University of Rochester Medical
Center, Rochester, NY

Valter D. Longo, PhD
Andrus Gerontology Center and Department of Biological Sciences,
University of Southern California, Los Angeles, CA

Mónica López-Torres, PhD
Department of Animal Physiology II, Faculty of Biological Sciences,
Complutense University, Madrid, Spain

Satomi Miwa, PhD
Institute for Ageing and Health, Newcastle University, Newcastle upon Tyne, UK

Robin J. Mockett, PhD
Department of Biomedical Sciences, College of Allied Health Professions,
University of South Alabama, Mobile, AL

Florian L. Muller, PhD
Department of Cellular and Structural Biology, University of Texas Health
Science Center at San Antonio, San Antonio, TX

William C. Orr, PhD
Department of Biological Sciences, Southern Methodist University, Dallas, TX

Heinz D. Osiewacz, PhD
Johann Wolfgang Goethe University, Institute of Molecular Biosciences,
Frankfurt, Germany

Melissa M. Page, MSc
Department of Biological Sciences, Brock University, St. Catharines, Ontario, Canada

João F. Passos, PhD
Institute for Ageing and Health, Newcastle University, Newcastle upon Tyne, UK

Ilsa I. Rovira, PhD
Cardiology Branch, National Heart, Lung, and Blood Institute,
National Institutes of Health, Bethesda, MD

Christian Q. Scheckhuber, PhD
Johann Wolfgang Goethe University, Institute of Molecular Biosciences,
Frankfurt, Germany

Mauro Serafini, PhD
Antioxidant Research Laboratory at the Unit of Human Nutrition,
National Institute for Food and Nutrition Research, Rome, Italy

Rajindar S. Sohal, PhD
School of Pharmacy, University of Southern California, Los Angeles, CA

Jeffrey A. Stuart, PhD
Department of Biological Sciences, Brock University, St. Catharines, Ontario, Canada

Thomas von Zglinicki, PhD
Institute for Ageing and Health, Newcastle University, Newcastle upon Tyne, UK

Section I
Introduction

1
Introduction

Satomi Miwa, Florian L. Muller, and Kenneth B. Beckman

1 Living in the Presence of Oxygen

The earliest life—simple cells—appeared 4 billion years ago, and rocks 3.5 billion years old contain microfossils of primitive one-celled organisms, the prokaryotes. At that time, the atmosphere contained little oxygen, and all life was anaerobic. About 3 billion years ago, photosynthetic blue-green algae (cyanobacteria) appeared, which slowly and eventually (800 million years later) raised atmospheric oxygen concentration [1, 2]. The rise in oxygen tension in the atmosphere coincides with development of complex eukaryotic life [3–5], and the Cambrian explosion, a profound diversification of animal life approximately 540 million years ago, occurred when oxygen reached concentrations close to those of today (21%) [6, 7]. Indeed, molecular and genetic analyses suggest a correlation, during the past 2.3 billion years, between the development of aerobic metabolism and increased organism complexity [4]. Aerobic metabolism using oxygen (oxidative phosphorylation) in mitochondria is highly efficient in generating energy from organic compounds, and mathematical analysis has shown that the presence of molecular oxygen in metabolic pathways allows far more complex reactions to occur than in those without oxygen [8]. According to one hypothesis, eukaryotes emerged after the engulfment of respiring bacteria (symbionts) by ancient anaerobic bacteria (hosts) – the symbionts being the origin of mitochondria [9]. Thus, the ability to perform oxidative metabolism is seen as a crucial factor in the emergence of complex multicellular life and evolution of animals. Oxidative metabolism remains the major form of energy production in virtually all animals today. Another significance of the appearance of oxygen in the atmosphere was the formation of the ozone layer, which protected the land from lethal levels of UV radiation.

However, despite its many beneficial effects, the emergence of oxygen as a major constituent of the atmosphere has presented all life forms living in its presence with a fundamental problem: how to efficiently protect themselves from its toxicity. In fact, many of the primitive anaerobic organisms are thought to have had died out when atmospheric oxygen levels rose. The only survivors were those species that evolved efficient mechanisms to detoxify oxygen, and those that colonized anaerobic or microaerobic environments [10, 11].

From: *Aging Medicine: Oxidative Stress in Aging: From Model Systems to Human Diseases*
Edited by: S. Miwa, K.B. Beckman, and F.L. Muller © Humana Press, Totowa, NJ

Indeed, it was the study of oxygen toxicity that first led Gerschman and her co-workers to the realization that oxygen free radicals can exist in vivo, and that these free radicals can be highly toxic [12]. The free radical theory of aging ultimately arose from studies of the basic toxicity of oxygen.

1.1   History of Free Radical Theory from Lavoisier to Harman

Oxygen was discovered independently by the Swedish apothecary Karl W. Scheele, in 1772, and by the English amateur chemist Joseph Priestley, in 1774. Oxygen was originally described as a gas that promotes combustion, but Priestley (who called it "dephlogisticated air") also realized that it was produced by plants (i.e., photosynthesis) and that it was essential for survival of animals (i.e., respiration).

Antoine Lavoisier, a chemist who corresponded with Priestley, repeated Priestley's experiments, and he further characterized the gas and gave it its current name *oxygen*, from the Greek root meaning "acid producer," because it was wrongly thought that acid contained oxygen. Remarkably, Lavoisier is said to have soon realized the harmful effects of oxygen inhalation [13]. Ever since, excessive oxygen has been known to be harmful to a wide range of life forms.

About 100 years after Lavoisier, the toxicity of oxygen was experimentally established by Paul Bert (1878) [14] and J. Lorain Smith (1899) [15] in the course of studies on hyperbaric oxygen. The effects of oxygen on the central nervous system are known as the *Bert effect* and on the lungs as the *Lorain Smith effect* or *Smith effect*, which were named after these pioneers in each subject, respectively.

However, the mechanism of such toxic effects of oxygen remained unknown for another 50 years until Rebecca Gerschman and colleagues postulated that it involved oxygen radicals (1954) [12]. They noted that the effects of oxygen toxicity share many common features with those of ionizing radiation, and found that radiation and oxygen toxicity are synergistic. They further showed that many radioprotectants also were protective against oxygen poisoning and that they were, in fact, well-known antioxidants in lipid chemistry [12]. It was known that radiation involved homolytic bond cleavage and free radical formation, but the idea that such a process could occur in vivo was revolutionary, because free radicals are by their very nature highly reactive and transient. Gershman's work would have a profound influence on the formulation of the free radical theory of aging.

About 50 years ago, Denham Harman took the ideas developed by Gershman and co-workers one step further and argued, based on divergent strains of evidence, not only that oxygen free radicals (specifically HOO$^{\bullet}$ and OH$^{\bullet}$) are formed in vivo, but also that the cumulative damage caused by those free radicals is the ultimate cause of aging (1956) [16]. So was born the free radical theory of aging. The theory was extremely controversial, because many of its cornerstones were regarded as unproven at best: the idea that free radicals could exist in a biological environment had no experimental support. In subsequent years, electron paramagnetic resonance work did show the in vivo existence of carbon-centered free radicals [17], but the real spark of attention to the theory was prompted by the discovery by Joe McCord and Irwin Fridovich of superoxide dismutase [18]. The fact that nature has developed such highly abundant enzymes to eliminate specifically superoxide free radicals strongly supports the existence of such free radicals in vivo. At about the same time, it was demonstrated that hydrogen peroxide (H_2O_2) was generated by mitochondria as a by-product of aerobic respiration (i.e., oxidative phosphorylation) [19] which originates from the dismutation of superoxide. It was a clear demonstration of reactive oxygen species (ROS) production in a normal biological process. Oxidative phosphorylation, as noted, is a very efficient, oxygen-dependent way to produce metabolic energy, taking place in mitochondria. Because most oxygen (85–90%) consumed by animals is used by mitochondria, it is now generally thought that mitochondria are the major site of superoxide production during oxidative metabolism in the aerobic cell. Thus, oxygen is a double-edged sword for aerobic life: necessary for energy production, but inevitably "stealing" electrons, thus generating superoxide and other reactive oxygen species.

1.2 Development and Derivatives of Free Radical Theory

Over the years, the original free radical theory has been modified and refined. The term *free radical theory* is somewhat of a misnomer, because many oxidants such as H_2O_2 are not free radicals. Harman soon offered another insight, that mitochondrial production of ROS (a collective term for superoxide, H_2O_2, and other highly reactive partially reduced species of oxygen) determines aging [20], which was followed by a proposal of the more specific version of the free radical theory: the mitochondrial theory. Mitochondrial theory posits that mitochondria are not only the major producers of ROS, but also that they are the major targets, due to their proximity to the ROS production sites. Damage to mitochondria and mitochondrial DNA (which exists in the matrix) would result in accumulation of the mitochondria that produce even more ROS, causing a vicious cycle [21].

The more general version of the free radical theory is the oxidative stress (or oxidative damage) theory. The somewhat vague term *oxidative stress* has become common in recent years, referring to the balance between production of ROS (and their related products) and defense against ROS. Under conditions of oxidative stress, accumulation of oxidative damage would accrue. It is now clear that there are endogenous sources of ROS other than mitochondria. Thus, "oxidative stress theory"—though it may inevitably be loosely defined—is flexible and less restricted than the mitochondria-centered version of the theory.

Another major influence on Harman's development of the free radical theory was the rate-of-living theory of aging. Ancient philosophers believed that we were born with a limited quantity of "vital substances" and that our life span was determined by how quickly we used it up. Priestley, the first person to observe respiration, wondered "… as a candle burns out much faster in dephlogisticated (i.e., oxygen) than in common air, so we might, as may be said, live out too fast, and the animal powers be too soon exhausted in this pure kind of air." [22]. The rate-of-living theory was established with empirical evidence in the early 1900s by Max

Rubner [23], and subsequently by Raymond Pearl [24]; it is understood to be the oldest scientific theory of aging. Rubner, whose interest included the correlation between body size, surface area and basal metabolic rate, realized that life span correlates inversely with body size in mammals, and reasoned that this was due to basal metabolic rate [23, 25].

Pearl expanded the theory alongside studies in *Drosophila*, in which he examined the effects of elevated metabolic rate (culturing at a higher temperature) on life span [24]. The rate-of-living theory states that life span is a function of metabolic rate: the faster that rate, the shorter the life span. Although it is now recognized that this simple relationship between metabolic rate and life span does not always hold true [26–30], the free radical theory has been suggested to offer a molecular mechanism for the rate-of-living theory, because an organism's metabolic rate is essentially proportional to oxygen consumption.

1.2.1 Aging

In biological terms, aging is defined fundamentally as an age-related increase in susceptibility to diseases and death (see the definitions of Comfort (1960, p 8) [31], Maynard Smith (1962, p 115) [32], Frolkis (1982, p 4) [33] [Box 1.1]). Aging is deleterious, progressive, intrinsic, and universal [34]. Aging is a multifactorial and heterogeneous phenomenon. As such, characteristics of aging have the potential to differ vastly between species, and are nonuniform even between individuals of the same species. Although death is a clear endpoint, aging is much more difficult to describe.

There are many hypotheses to explain mechanisms of aging apart from the oxidative stress theory (Table 1.1) (see [35] for detailed examinations of different theories). Theories can be formulated at different levels of the biological hierarchy, including the molecular, cellular, and systemic levels, and they are not necessary mutually exclusive. There is often cross-talk between theories at different levels: one theory at the molecular level may underlie aspects of another theory at the systemic level, and so on. The oxidative stress theory fits well with other molecular theories. For example, oxidative stress might be involved in the "somatic mutation theory" because oxidative damage causes DNA mutagenesis, or in the "error catastrophe theory" because oxidative damage could compromise ribosome function, or in the "cross-linking/glycosylation theory" because oxidative damage can generate reactive aldehydes.

2   Oxidative Stress in Aging: Format of the Book

The oxidative stress theory remains one of the most popular explanations of aging at the molecular level. The purpose of this book is to give an authoritative account of the current status of the theory, 50 years after it was first proposed.

Box 1.1   Historic definitions of aging

Comfort A (1960) [31]   Aging is "an increased liability to die, or an increasing loss of vigour, with increasing chronological age, or with the passage of life cycle."

Maynard Smith J (1962) [32]   Aging processes are "those which render individuals more susceptible as they grow older to the various factors, intrinsic or extrinsic, which may cause death."

Frolkis VV (1982) [33]   "Aging is a naturally developing biological process which limits the adaptive possibilities of an organism, increases the likelihood of death, reduces the life span and promotes age pathology."

Table 1.1   Some theories of aging

Molecular and cellular level

- Wear and tear: Wearing out of cells and tissues by accumulation of normal insults from daily living.
- Waste accumulation: Accumulation of waste products in the cell which perturb normal cellular function.
- Error catastrophe: Errors in transcriptional and translational processes.
- Somatic mutation and DNA damage: Somatic mutations and damages to DNA alter genetic information.
- Telomeres and Hayflick limit: Possible number of cell division is limited (determined by the length of telomere) and cells become senesced.
- Cross-linking/glycosylation: Cross-linking of proteins with glucose alter their functions.
- Dysdifferentiation: Impaired regulations of gene activation and repression mechanisms.
- Free radical/oxidative stress: Accumulation of damage caused by free radicals that are produced as metabolic by-products.

Systemic level

- Metabolic (rate of living): Life span inversely correlates with metabolic rate.
- Neuroendocrine: Failure in neuroendocrine control of physiological homeostasis.
- Immune function: Decline in immune function that results in decreases in resistance to infectious diseases and increases in incidence of autoimmunity.

Section I. Introduction

In Chapter 2 we present the basics of oxidative biochemistry. It is intended to familiarize the reader with the biochemistry of reactive oxygen species and of the antioxidant system, since subsequent chapters will assume some knowledge of this area.

Section II. Role of Oxidative Stress in Aging

The fundamental idea of the oxidative theory is straightforward and simple. It has been tested in a wide range of "aging" model systems and in comparative studies. It is probably the most extensively tested theory on aging and has been the subject of a formidable body of literature.

Part II-A   Different Model Systems

As with many areas of biology, the use of model systems has proved invaluable to the oxidative stress theory of aging. Because of their short life span, model organisms make testing of specific interventions both logistically and temporally feasible. The ease of genetic manipulation of specific model organisms (e.g., mice, *Drosophila* species) has further increased their utility. The free radical theory of aging has received extensive experimental testing in model organisms, with sometimes contradictory results.

We have grouped the results by model organism, with each chapter (Chapters 3–8) being written by an expert evaluating the evidence for the free radical theory in a specific model system. This approach greatly simplifies evaluation of the evidence, considering the contradictory nature of the literature. Furthermore an investigator typically focuses his or her career on one model organism; our format enables evaluation of the evidence based on years of practical experience, great depth of knowledge of the literature, and critical unpublished observations. Naturally, the experimental results obtained in model systems can be complementary, but extrapolation and generalization of results between species must take phylogeny and physiological differences into account.

Part II-B   Comparative Approach

The free radical theory has been extensively evaluated by comparative physiology and biochemistry. The comparative method examines whether there are particular physiological or biochemical traits that correlate with life span across different species. Chapter 9 reviews the comparative biochemistry of oxidative damage, mitochondrial ROS production, and higher vertebrate animal life span.

Section III. Oxidative Stress in Human Aging and Diseases

A critical outgrowth of the free radical theory of aging is the free radical theory of disease, i.e., the realization that ROS may not only contribute to aging per se, but also to the pathogenesis of a variety of age-related and non-age-related diseases.

Human aging is characterized by increased susceptibility to age-related diseases. One important trend is the emergent role of oxidative stress in specific age-related (as well as non-age-related) diseases, rather than aging in general. Moreover, much more evidence is available on the pathogenic role of ROS in specific diseases than in aging generally. Thus, we devote an entire section to the role of oxidative stress in specific diseases. In addition, Section III includes a chapter (Chapter 15) of epidemiological reviews on effects of antioxidant supplementation on prevention of diseases; the evidence from the clinical trials is weak, and the possible reasons are discussed.

Section IV. Future

A relatively recent development in the long history of the free radical theory is the finding that ROS can act as signalling molecules, and not just as deleterious metabolic by-products. Although ROS have long been known to be purposefully produced by the immune system to kill invading pathogens, it is now recognized that an entire family of enzymes, called Nox, purposefully generate ROS at cell membranes, likely for cell signaling. Chapter 16 gives an overview of the current knowledge on ROS as signaling molecules.

We end the book with our evaluation of the current status of the free radical theory and where we believe it is headed in the future.

References

1. Kasting JF. Earth's early atmosphere. Science 1993;259(5097):920–6.
2. De Marais DJ. Evolution. When did photosynthesis emerge on Earth? Science 2000;289(5485):1703–5.
3. Hedges SB, Chen H, Kumar S, Wang DY, Thompson AS, Watanabe H. A genomic timescale for the origin of eukaryotes. BMC Evol Biol 2001;1:4.
4. Hedges SB, Blair JE, Venturi ML, Shoe JL. A molecular timescale of eukaryote evolution and the rise of complex multicellular life. BMC Evol Biol 2004;4:2.
5. Feng DF, Cho G, Doolittle RF. Determining divergence times with a protein clock: update and reevaluation. Proc Natl Acad Sci U S A 1997;94(24):13028–33.
6. Knoll AH. The early evolution of eukaryotes: a geological perspective. Science 1992;256(5057):622–7.
7. Falkowski PG. Evolution. Tracing oxygen's imprint on earth's metabolic evolution. Science 2006;311(5768):1724–5.
8. Raymond J, Segre D. The effect of oxygen on biochemical networks and the evolution of complex life. Science 2006;311(5768):1764–7.
9. Margulis L. Origin of eukaryotic cells. New Haven, Connecticut: Yale University Press; 1970.
10. Halliwell B. Antioxidant defence mechanisms: from the beginning to the end (of the beginning). Free Radic Res 1999;31(4):261–72.
11. Fridovich I. Oxygen is toxic! BioScience 1977;7:462–6.
12. Gerschman R, Gilbert DL, Nye SW, Dwyer P, Fenn WO. Oxygen poisoning and x-irradiation: a mechanism in common. Science 1954;119(3097):623–6.

13. Binger CAL, Faulkner JM, Moore RL. Oxygen poisoning in mammals. J Exp Med 1927;45:849–64.
14. Bert P. La pression barométrique: recherches de physiologie expérimentale. Paris, France: Masson; 1878.
15. Smith JL. The pathological effects due to increase of oxygen tension in the air breathed. J Physiol 1899;24(1):19–35.
16. Harman D. Aging: a theory based on free radical and radiation chemistry. J Gerontol 1956;11(3):298–300.
17. Commoner B, Townsend J, Pake GE. Free radicals in biological materials. Nature 1954;174(4432):689–91.
18. McCord JM, Fridovich I. Superoxide dismutase. An enzymic function for erythrocuprein (hemocuprein). J Biol Chem 1969;244(22):6049–55.
19. Boveris A, Oshino N, Chance B. The cellular production of hydrogen peroxide. Biochem J 1972;128(3):617–30.
20. Harman D. The biologic clock: the mitochondria? J Am Geriatr Soc 1972;20(4):145–7.
21. Miquel J, Economos AC, Fleming J, Johnson JE Jr. Mitochondrial role in cell aging. Exp Gerontol 1980;15(6):575–91.
22. Priestley J. Observations on different kinds of air. Philos Trans 1772;62:147–264.
23. Rubner M. Das Problem der Lebensdauer und seine Beziehungen sum Wachstum und Ernahrung. Munich: Oldenbourg; 1908.
24. Pearl R. The rate of living, being an account of some experimental studies of the biology of life duration. New York: Alfred A. Knopf; 1928.
25. Chambers WH. Max Rubner: June 2, 1854–April 27, 1932. J Nutr 1952;48(1):3–12.
26. Yan LJ, Sohal RS. Prevention of flight activity prolongs the life span of the housefly, *Musca domestica*, and attenuates the age-associated oxidative damage to specific mitochondrial proteins. Free Radic Biol Med 2000;29(11):1143–50.
27. Magwere T, Pamplona R, Miwa S, et al. Flight activity, mortality rates, and lipoxidative damage in *Drosophila*. J Gerontol A Biol Sci Med Sci 2006;61(2):136–45.
28. Holloszy JO, Smith EK. Longevity of cold-exposed rats: a reevaluation of the "rate-of-living theory." J Appl Physiol 1986;61(5):1656–60.
29. Holloszy JO, Schechtman KB. Interaction between exercise and food restriction: effects on longevity of male rats. J Appl Physiol 1991;70(4):1529–35.
30. Holloszy JO. Exercise increases average longevity of female rats despite increased food intake and no growth retardation. J Gerontol 1993;48(3):B97–B100.
31. Comfort A. Discussion session. I. Definition and universality of aging. In: Strehler BL, ed. The biology of aging. Washington DC: AIBS; 1960.
32. Maynard Smith J. Review lectures on senescence. I. The causes of ageing. Proc R Soc Lond Ser B 1962;157:115–27.
33. Frolkis VV. Aging and life-prolonging processes. Translated from the Russian by Nicholas Bobrov. Vienna: Springer-Verlag; 1982.
34. Strehler BL. Time, cells and aging. New York: Academic Press; 1982.
35. Arking R. Biology of aging: observations and principles. Sunderland, Massachusetts: Sinauer Associates, Inc; 1998.

2

The Basics of Oxidative Biochemistry

Satomi Miwa, Florian L. Muller, and Kenneth B. Beckman

1 Chemistry of Reactive Oxygen Species

Although a thorough knowledge of oxygen chemistry is necessary to understand the biochemistry of reactive oxygen species, that topic is beyond the scope of this book (see excellent works by Sawyer [1, 2] and the outstanding textbook of Halliwell and Gutteridge [3], Chapter 1). A large number of different reactive oxygen species (ROS) and reactive nitrogen species are now known to exist in biological systems, the most relevant to the present book being superoxide ($O_2^{\cdot-}$/ $HO_2^{\cdot}$), hydrogen peroxide (H_2O_2), hydroxyl radical ($OH^{\cdot}$), organic (lipid, alkyl, or short chain) hydroperoxides and hydroperoxide radicals (ROOH, $ROO^{\cdot}$), peroxynitrite and peroxynitrous acid ($ONOO^-$/ONOOH), carbonate radical ($CO_3^{\cdot-}$), and reactive aldehydes of varying lengths. Various oxo-metal complexes, (such as those formed from heme and H_2O_2), may also be relevant. We limit our discussion to the chemical properties relevant to the biological toxicity of the best known reactive species: superoxide, hydrogen peroxide, and the hydroxyl radical. Although we focus on these three species, this is not to say that the many other known ROS (and the downstream products thereof) are not relevant to oxidative stress and aging.

1.1 Superoxide

A one-electron reduction of O_2 yields superoxide ($HO_2^{\cdot}/O_2^{\cdot-}$; $pK_a = 4.88$), with a redox potential $O_2/O_2^{\cdot-}$ of $-160\,mV$ at pH 7.0 under physiological conditions [1, 2, 4, 5]. Thus, a low-potential, single electron carrier is the ideal catalyst of superoxide formation (in biological systems, iron-sulfur clusters, semiquinones, and cytochromes are low-potential, single electron carriers [6, 7]). Superoxide spontaneously dismutates to H_2O_2, with a pH-dependent second order rate of ~$10^5\,M^{-1}\,s^{-1}$ (at pH 7.0 [4]). After the discovery of superoxide dismutase, many questioned whether superoxide dismutase (SOD) could be physiologically relevant, because superoxide is not particularly reactive [8], and it decays spontaneously at such a

From: *Aging Medicine: Oxidative Stress in Aging: From Model Systems to Human Diseases*   11
Edited by: S. Miwa, K.B. Beckman, and F.L. Muller © Humana Press, Totowa, NJ

fast rate: this misconception (ultimately proven wrong by the drastic phenotypes brought about by SOD knockouts, in diverse organisms [9–17]) arises from the fact that superoxide dismutation by SOD is first order with regard to superoxide concentration, whereas spontaneous dismutation is second order. Indeed, it is now known that superoxide is extremely toxic (much more so than H_2O_2) and that intracellular concentrations in the picomolar range are lethal. The pH-dependent equilibrium constant for superoxide dismutation to H_2O_2 is $K\sim10^{18}$ at pH 7.0, so the reaction can be considered irreversible except in circumstances of very high levels of SOD and very low levels of superoxide [18]. Apart from dismutation, superoxide can act both as a reductant (e.g., superoxide reduces high-potential ferricytochromes to their ferrous states, with rate constants of 10^5 to 10^6 M^{-1} s^{-1} [4]) and as an oxidant (especially in its protonated $HO_2^{\bullet}$ form [4, 5]) via the oxidation of low-potential iron sulfur clusters, thereby liberating redox-active iron. The liberation of Fe^{2+} from iron-sulfur clusters is believed by some investigators to be the main pathway of biological superoxide toxicity [19, 20]: the availability of redox active iron is thought to be the rate limiting step in $OH^{\bullet}$ formation (see Subsection 1.3) by Fenton chemistry. Superoxide, as $HO_2^{\bullet}$ [21–23], can also initiate lipid peroxidation. Other oxidations of potential biological relevance include the reaction with catechols ($\sim10^4$ M^{-1} s^{-1}) and thiols (from 10 to 10^4 M^{-1} s^{-1}), the latter of which can proceed in a peroxidation-like chain reaction, consuming several thiols per superoxide molecule [24]. A third fate of superoxide is of interest: being a free radical, superoxide can react with other free radicals in annihilation reactions, which can proceed at diffusion limited (very fast) rates, because there is no activation energy barrier [25]. The best example of this type of reaction is that of superoxide with $NO^{\bullet}$, forming peroxynitrite/peroxynitrous acid [26]. Although beyond the scope of this review, peroxynitrite can undergo subsequent chemical rearrangements, yielding very strong oxidants ($NO_2^{\bullet}$, $OH^{\bullet}$, $CO_3^{\bullet-}$ [27]). The production of these strong oxidants, coupled with the great speed of peroxynitrite formation, has led some to argue that peroxynitrite formation is the main pathway of biological superoxide toxicity; however, this view remains controversial [28, 29]. Peroxynitrite formation is by no means the only biologically relevant radical annihilation reaction. It is now recognized that many enzymes use stable carbon-centered or delocalized free radicals as part of their normal catalytic mechanism [30]. Superoxide could theoretically react very rapidly with such radicals, causing irreversible enzymatic inactivation. One well-documented instance of such a case is the reaction of superoxide with the tyrosyl radical of ribonucleotide reductase, an enzyme absolutely essential for DNA synthesis, leading to tyrosine peroxide formation and irreversible inactivation of this enzyme [31]. There are still other pathways of superoxide toxicity. Superoxide can initiate the oxidation of short chain sugars [32, 33], forming toxic α, β-dicarbonyls. Superoxide also can oxidize low-potential heme proteins, such as hemoglobin and myoglobin, yielding the oxygen-carrying incompetent met-forms [34]. Tyrosine peroxide formation may be another pathway of superoxide toxicity [25], and we speculate that there are others yet to be discovered.

1.2 Hydrogen Peroxide

Dismutation and oxidation reactions of superoxide yield hydrogen peroxide. Hydrogen peroxide, although more oxidizing than superoxide, is biologically less toxic: picomolar intracellular levels of superoxide are lethal, whereas micromolar levels of H_2O_2 can be tolerated. H_2O_2 is a potent oxidizer (although not always a fast oxidizer), and is much more diffusible than superoxide, because it is less reactive and is membrane permeable: $O_2^{\cdot-}$ is generally considered membrane impermeable [35, 36], except in its $HO_2^\cdot$ form, which is in low abundance at physiological pH. H_2O_2 is usually a slow two-electron oxidizer [24], and is rather stable. However, in the presence of a metal catalyst or heme, it can act as a very rapid and indiscriminate oxidant; some biological molecules are direct oxidation targets of H_2O_2, specifically those with low-potential cysteines. H_2O_2-induced thiol oxidation may be damaging as in the case of the glycolytic enzyme glyceraldehyde-3-phosphate dehydrogenase (GAPDH) [37–39], but it also may be relevant in the redox regulation of certain enzymes, a well-documented case being the PTP1B phosphatase [40, 41].

1.3 Hydroxyl Radical

The full oxidizing strength of H_2O_2 can be harnessed if it is single-electron reduced to $OH^\cdot$, which is one of the most potent oxidizing agents known to chemistry, with redox potential $+2.40\,V$ [3]. This full strength can be achieved by Haber-Weiss chemistry, i.e., superoxide can react with H_2O_2, but the rate constants for this reaction are considered too low for biological significance [42]. Alternatively, H_2O_2 can react with reduced metal ions (most notably, Cu^+ and Fe^{+2} [43, 44]) to generate $OH^\cdot$ in a reaction termed *Fenton chemistry* [20, 42, 44, 45]. $OH^\cdot$ reacts at diffusion-limited rates with almost everything found in the cell (p 58 in [3]): as such, its toxicity is nonselective and its diffusion distance is very short. Because $OH^\cdot$ reacts rapidly and indiscriminately, there is little that antioxidants can do, unless present in prohibitively high amounts. Thus, unlike the cases of H_2O_2 and $O_2^{\cdot-}$, there is no enzyme that specifically detoxifies $HO^\cdot$, and it seems that biological systems tightly regulate the availability of Fenton chemistry-capable metal ions to minimize $OH^\cdot$ formation.

2 Antioxidant System

The eukaryotic antioxidant system has grown from the three classical enzymes—SOD, catalase, and glutathione peroxidase—to include a diverse and still expanding, enzymatic and nonenzymatic group of players. Furthermore, there are several enzymes that mitigate the effect of ROS by repairing oxidative damage, especially with regard to DNA oxidative damage (see Chapter 12). Although it is beyond the scope of this review to detail each of these enzymes, we provided a brief introduction to the $O_2^{\cdot-}$ and H_2O_2 detoxifying branches of the antioxidant network, because these enzymes are discussed extensively throughout this book, and most chapters assume some familiarity with them. At the same time, we alert readers that the understanding of the antioxidant network is still incomplete, with new members still being discovered.

2.1 SOD Accelerates Dismutation of Superoxide Radical

The enzyme SOD has special significance for the free radical theory of aging. Its discovery strongly suggested that superoxide (and by extension, oxygen free radicals) was produced in vivo. SOD is the among the best known antioxidant enzymes, and knockout studies in *Escherichia coli*, yeast, flies, and mice all testify to its importance (for review, see [46, 47]). The discovery by Fridovich and McCord that erythrocuperin catalyzes the dismutation of the superoxide anion led to an explosion of interest in oxygen free radicals from medical professionals and chemists alike. Indeed, SOD may be one the most well-studied enzymes. In mammalian systems, there are three different SOD isozymes encoded by three different genes [9, 48–51]. *Sod1* encodes copper-zinc SOD, which is found both in the cytoplasm and in the mitochondrial intermembrane space [52, 53]. *Sod2* encodes manganese superoxide dismutase, which is exclusively located in the mitochondrial matrix [50, 54]. *Sod3* is a copper-zinc SOD that is secreted and is extracellular in localization [49].

The general mechanism of superoxide dismutases can be summarized as follows, where M stands for metal (Cu^{2+} or Mn^{3+}), the superscripted minus sign ($-$) stands for negative electric charge, and the dot represents an unpaired electron. The rate constants do not vary in the pH range from 5.5 to 9.0.

$$O_2^{\cdot-} + \text{SOD-M} \rightarrow O_2 + \text{SOD-M}^- \quad (k = 10^9 \text{ M}^{-1} \text{ s}^{-1})$$
$$O_2^{\cdot-} + 2H^+ + \text{SOD-M}^- \rightarrow H_2O_2 + \text{SOD-M} \quad (k = 10^9 \text{ M}^{-1} \text{ s}^{-1})$$

The spontaneous decomposition of superoxide is pH dependent, and it can be written as follows (net reaction):

$$O_2^{\cdot-} + O_2^{\cdot-} + 2H^+ \rightarrow H_2O_2 + O_2 \quad (k = 10^5 \text{ M}^{-1} \text{ s}^{-1} \text{ at pH } 7.0)$$

Thus, SOD accelerates the destruction of superoxide by increasing the rate constant for spontaneous dismutation by >1,000-fold, and also by making the rate of superoxide decay a first order rather than second order process with respect to superoxide concentration. This means that SOD is more efficient at accelerating the decomposition of superoxide, compared with spontaneous dismutation, at low superoxide concentration. However, because an enzyme cannot change the position of equilibrium, but only the rate at which it is achieved, there is a lower limit of superoxide concentration below which the enzyme cannot reduce it.

In addition to catalyzing the (reversible) dismutation of $O_2^{\cdot-}$, SOD also can undergo a variety of radical-producing side reactions [55–58], most notably generating highly oxidizing $OH^{\cdot}$ and $CO_3^{\cdot}$ in an H_2O_2-dependent manner termed *peroxidase reaction* (best documented in CuZn-SOD [58]). In addition, it also can decompose peroxynitrite to the highly reactive $NO_2^{\cdot}$ radical [56]. SOD can catalyze the CO_2 and H_2O_2-dependent oxidation of a variety of molecules, including the dye 5-(and-6)-carboxy-2,7-dichlorodihydrofluorescein [55]. Finally, CuZn-SOD can catalyze the oxygen-dependent oxidation of thiols, in cysteine and glutathione [57]. These deleterious pro-oxidative side reactions may or may not be relevant in a non-pathological state, but they cannot be ignored when SOD is overexpressed at high levels in transgenic mice or flies.

2.2 Peroxiredoxins are Major Scavengers of Endogenously Produced H_2O_2

Although the superoxide-detoxifying system is compact with only three SOD enzymes, and the relative contribution of each enzyme is more or less understood, this is not the case for the cellular H_2O_2 detoxification system. There are multiple enzymes involved, including glutathione peroxidases, catalase, and peroxiredoxins. Catalase acts as a H_2O_2 "dismutase," turning two H_2O_2 molecules into O_2 and H_2O. Peroxidases perform two-electron reduction of H_2O_2, with its reducing equivalents ultimately originating from NAPDH.

It is traditionally held that glutathione peroxidase and catalase are the main scavengers of cellular H_2O_2. Based on genetic studies in yeast and mice, this view is no longer tenable. In the absence of catalase, oxidative stress and deleterious phenotypic consequences are minimal in yeast, mice, or even humans (patients with acatalasemia have no major pathology [13, 59, 60]). Nonetheless, lack of catalase does render cells susceptible to, and is the major kinetic sink for, exogenously added H_2O_2 [59, 60]. Although glutathione peroxidase 1 (Gpx1) is an abundant enzyme, and it also reacts rapidly with H_2O_2, genetic ablation does not result in oxidative stress or deleterious phenotypes in either yeast or mice [61–63]; it is worth mentioning that Gpx1 is missing in *Drosophila* and even in certain mammals, such as the naked mole rat [64]. Gpx1-lacking mice are highly sensitive to bolus exogenous ROS generators such as paraquat and diquat [61–63], although they exhibit little or no elevation in basal levels of oxidative damage, and do not show any compensatory antioxidant up-regulations (Han, Muller, Perez, Van Remmen, Richardson, unpublished data). No deleterious phenotypes are evident in yeast lacking Gpx1 or catalase, even in environmental situations where excess ROS production is expected to occur, such as hyperoxia or the postdiauxic phase of yeast growth [13, 65]. It does not seem to be a question of redundancy either, because yeast strains lacking all three Gpx homologs are viable and they do not show any obvious phenotypic defects [66]. The reason why knocking out Cat or

Gpx1 does not affect phenotype or increase endogenous oxidative stress is, in our assessment, because endogenously produced H_2O_2 is detoxified by another enzyme system: the peroxiredoxins.

Peroxiredoxin was first discovered in yeast (a.k.a. thiol-specific antioxidant, *tsa1* [67, 68]), and homologs are found in all kingdoms of life [68, 69]. In contrast to the lack of phenotype resulting from glutathione peroxidase or catalase yeast knockouts, knocking out *tsa1* has deleterious consequences: increased oxidative damage, thermosensitivity, decreased growth under aerobic conditions, and reduced viability during the stationary phase; most intriguingly however, the rate of mutagenesis and genomic instability is dramatically (10- to 20-fold) increased [67, 70–74]. In contrast to the lack of phenotype in mice lacking catalase or Gpx1, knocking out peroxiredoxin 1 in mice causes increased oxidative damage, increased cancer incidence and shortened life span [75]. Knockout of peroxiredoxin 1 or 2 results in hemolytic anemia [75, 76].

Although Gpx1, catalase, and peroxiredoxins all react with H_2O_2, the chemical basis of their relative importance is still not fully understood. Catalase is the most economical way of removing H_2O_2, because no reducing equivalents are consumed. As such, it is very efficient at removing large quantities of H_2O_2, but at low concentrations of H_2O_2, it exhibits a nonspecific peroxidase activity, i.e., it can catalyze the H_2O_2-dependent oxidation of a variety of molecules [6]. Gpx1 consumes a two-electron reducing equivalent- for every H_2O_2 it reduces to water. What is unique about peroxiredoxins is that they are inactivated by high levels of H_2O_2, because their active site thiol is oxidized ("overoxidized") to sulfinic acid [37, 68, 77]. As such, peroxiredoxins are ineffective at removing high levels of H_2O_2. Conversely, they seem to be the preferred target of low levels of H_2O_2, likely because of their high abundance, low redox potential, and rapid rate constant of reaction with H_2O_2 [37, 68, 77, 78].

To summarize this section, genetic evidence and recent biochemical work indicate that glutathione peroxidase, catalase, and peroxiredoxins play nonoverlapping roles in scavenging H_2O_2. High H_2O_2 levels are handled by glutathione peroxidase and catalase, whereas the lower levels of H_2O_2 produced by normal endogenous metabolism are scavenged by peroxiredoxins. This makes peroxiredoxins of great interest to the free radical theory of aging.

2.3   How Many Unknown Antioxidant Genes are out There?

Keep in mind that although the major players have probably been identified, new members of the antioxidant network are still being discovered. This is especially important when considering the effect of antioxidant knockout and overexpression on life span. Below, we highlight three relatively new antioxidant systems that we believe to be of great physiological significance, and that will receive increased attention in the near future.

2.3.1 Biliverdin/Bilirubin and Biliverdin Reductase

Biliverdin is a degradation product of heme that is reduced by biliverdin reductase to bilirubin. Ames's group reported that bilirubin could act as a peroxidation chain-breaking antioxidant in a defined chemical system [79, 80]. Snyder's group reported that biliverdin/bilirubin and biliverdin reductase form a network that catalytically detoxifies H_2O_2 [81], ablation of which results in apoptotic cell death. Although more work needs to be done (e.g., what are the phenotypical consequences of biliverdin reductase knockout?), it is probable that bilirubin will turn out to be a key player in the antioxidant network.

2.3.2 Apolipoprotein D (ApoD)

During an overexpression screen of genes for resistance to hyperoxia (100% O_2) in *Drosophila* species, Walker et al. [82] identified Glial Lazarillo, homolog of ApoD. Overexpression of ApoD thus conferred extended life span under hyperoxia. In further support of this finding, an independent group of investigators has reported that knocking out ApoD leads to a 20% reduction in median life span and increased oxidative damage [83]. ApoD belongs to the lipocalin family, which bind and transport small hydrophobic molecules [84]; although the exact biological function of ApoD is unknown, it evidently plays an important role in protection from oxidative stress, by yet unknown mechanisms.

2.3.3 Sulfiredoxin and Sestrin

Lower oxidation states of cysteine (disulfides) are readily reversible, while higher oxidation states, such as sulfinic acid, were once considered irreversible, biologically speaking. This view changed with the discovery of sulfiredoxin, an enzyme that can reduce sulfinic acid back to thiol, in an ATP-dependent manner [85]. Additional work suggests that it plays a role in resolving mixed disulfides bonds [86]. Initially discovered in yeast, sulfiredoxin is conserved in all eukaryotes, including mammals. In a perfect example of how multiple gene names can confuse the field, sulfiredoxin (*Srxn1*) was already known as a gene of unknown function, cloned by differential display of an in vitro model of tumorigenesis, and termed *neoplastic progression 3* (*Npn3*), although nothing about its actual function was reported [87]. As a result, in most mouse microarray studies, sulfiredoxin is termed neoplastic progression 3, and typically classified as "cancer related" or "other" rather than as "antioxidant" [88, 89].

Npn3/*Srxn1* is up-regulated by an exceptionally large fold-magnitude in microarray studies of oxidative stress. *Npn3*/*Srxn1* is induced up to 32-fold by D3T (liver [88]), 12-fold by $CdCl_2$ (liver [89]), 4- to 10-fold by paracetamol (liver [90], supplemental materials), and 3.3-fold by paraquat (heart [91], supplemental materials). A survey of the Gene Expression Omnibus database also indicates that a large induction of *Npn3/Srxn1* is observed in injury to the lung by hyperoxia (data set GDS247, ID 102780_at) or phosgene (GDS1244, 1451680_at). Our own microarray data indicate *Npn3/Srxn1* is also strongly up-regulated in the liver of CuZn-SOD knockout mice (Han, Muller, Perez, Van Remmen, Richardson, unpublished data). That *Npn3* and *Sxrn1* are synonyms for the same gene has not been pointed out in any of the 15 papers written on *Srxn1* since its discovery. Furthermore, this example highlights the problems associated with multiple names of genes and the need for uniform gene nomenclature in the postgenomic age.

Because it was discovered so recently, the function of sulfiredoxin is not yet fully known, and because no knockout of sulfiredoxin in mice is yet available, its true physiological importance remains to be established.

A similar catalytic activity to sulfiredoxin (reducing sulfinic acid back to sulfhydryl) was recently ascribed for the p53 target gene P26/sestrin1 and sestrin 2 [92]. This finding is intriguing, considering the increasing attention p53 is receiving as a modulator of ROS in vivo [93]. Although no knockout of sestrin 1 or 2 is yet available, knocking out p53 in mice results in increased oxidative damage, shortened life span, and cancer, which can be significantly attenuated by feeding of the antioxidant *N*-acetyl cysteine [93].

2.4   ROS Sources

Although it is now recognized that reactive oxygen species can be formed under many different conditions and in many different cellular compartments, the number of enzymes that have been documented to generate $O_2{}^{\cdot-}$ (either deleteriously or purposefully) is relatively small. Superoxide can be produced by NADPH oxidases [94] (first thought to be unique to phagocytes but now known to be located on the plasma membrane of most cell types [95]), xanthine oxidase [96], aldehyde oxidase [97], cytochromes P450 [98], and the mitochondrial electron transport chain [99, 100]. Many more enzyme systems (most oxygenases, e.g., monoamine oxidase) are known to generate H_2O_2. A particularly active source of H_2O_2 includes enzymes involved in the oxidation of fatty acids in the peroxisomes [6].

Despite the variety of ROS sources, most interest from a gerontological perspective has centered on the mitochondrial electron transport chain. Although it was recognized early on, through in vitro biochemical work, that the mitochondrial electron transport chain is an exceptionally strong source of superoxide, definitive proof (and its underlying biological importance) came from the finding that mice lacking mitochondrial matrix Mn-SOD (*Sod2*) die several days after birth [9, 10]. Mice knockouts for cytoplasmic (and intermembrane space) CuZn-SOD (*Sod1*) and extracellular CuZn-SOD (*Sod3*), although certainly not normal [51,101–108], do not exhibit as dramatic a phenotype as $Sod2^{-/-}$ mice, thus indicating that the mitochondrial matrix is indeed the most important site of superoxide toxicity. This also seems to be true in *Drosophila melanogaster*; although the *Sod1*[-/-] phenotype is more severe in flies than in mice, only *Sod2*[-/-] results in true postnatal lethality [16, 109, 110]. Removing xanthine oxidase worsens, rather than ameliorates, the phenotype of *Sod1* knockout flies [111]. Thus, for the present review, the mitochondrial electron transport chain remains the focus for oxidative stress and aging; however, no one would seriously argue that other nonmitochondrial sites of superoxide generation are not relevant to aging (e.g., the plasma membrane oxidoreductase [112]).

The mitochondrial electron transport chain is a series of one-electron shuttles of progressively stronger oxidants, which couple the energy released during oxidation to the pumping of protons across the inner mitochondrial membrane, and to the eventual generation of ATP [113]. There are several low-potential carriers in this system, which can potentially donate electrons to superoxide (for review, see [7]). Mitochondrial superoxide generation was first discovered as H_2O_2 released from mitochondria treated with the respiratory inhibitors antimycin A or rotenone [114, 115]. Not until Loschen et al. discovered that superoxide radicals are generated by submitochondrial particles, under the same conditions as H_2O_2 formation, was it realized that mitochondrial H_2O_2 originates as a dismutation product of superoxide [99]. For convenience, mitochondrial superoxide production is still largely assayed indirectly by measuring H_2O_2 [115, 116]. Most investigators now believe that complex I and complex III are the main sources of superoxide [7, 113] in mammalian mitochondria; it is also known that mutations can turn complex II into a superoxide generator [117, 118]. It has also been suggested that α-ketoglutarate dehydrogenase generates superoxide and H_2O_2 [119, 120]. Whereas it may very well produce H_2O_2, inhibition of the respiratory chain with KCN, which would be predicted (based on dependence on high NADH-to-NAD$^+$ ratio) to increase superoxide production dramatically, actually rescues rather than kills a Mn-SOD lacking yeast ([12] and see Chapter 5). In mitochondria from *D. melanogaster*, glycerol 3-phosphate dehydrogenase is a very potent source of superoxide [121]. Complex I releases superoxide exclusively toward the mitochondrial matrix, whereas complex III and glycerol 3-phosphate dehydrogenase release superoxide toward both the matrix and the cytoplasm [121–123]. Considerable debate surrounds the importance of these sites, in terms of how much superoxide mitochondria truly produce in vivo. In vitro, a very large rate of superoxide production is observed when mitochondria respire on succinate and undergo reverse-electron transfer through complex I [124–126]. The often-quoted figure that "1-2% of electrons going through the respiratory chain are diverted to superoxide" comes from these studies. Although it is taken as self-evident by some investigators [127], whether or not reverse-electron transfer occurs in vivo has not yet been established. Under normal forward-electron transfer, by using the rate of H_2O_2 release as an indicator of superoxide production, that number is closer to 0.1% of electrons [47, 124]. Because oxygen tension is correlated with the rate of superoxide production [127, 128], even the 0.1% figure may be an overestimate, because oxygen tension in vivo is ~3% compared with 21% used in the above experiments. Some investigators have thus claimed that no H_2O_2 is released (and by extension, no superoxide is produced) from "normal" mitochondria in the absence of respiratory inhibitors [129, 130].

This statement is difficult to take seriously, considering that mice and flies lacking mitochondrial *Sod2* die shortly after birth [9, 16] amid massive oxidative stress. Thus, although the rate of mitochondrial superoxide production is likely low in vivo (perhaps as low as 1 electron in 40,000 diverted toward superoxide formation), this is still high enough to be incompatible with life, in the absence of SOD.

One of the best ways to test the free radical theory of aging would be to modify the rate of mitochondrial superoxide production [7, 131]. However, the molecular details of this process are poorly understood. Studies with respiratory inhibitors (e.g., antimycin A, rotenone) have shown that almost anything that interrupts the "smooth" electron transfer through the respiratory chain results in a dramatic stimulation of ROS production (for review, see [7, 124, 128, 132, 133]). One can speculate that the availability of both a reduced reactive intermediate (e.g., semiquinone, iron-sulfur cluster) and O_2, and the ability of the latter to reach the former, should determine the rate of $O_2^{\cdot-}$ production. Oxygen tension, as mentioned, has already been demonstrated experimentally to modulate superoxide production [14, 127, 128, 134], and the absence of oxygen (anaerobiosis or anoxia) is the only condition known to prevent superoxide formation. Not surprisingly, heat, which denatures proteins and exposes reactive intermediates to O_2 also dramatically stimulates superoxide formation [135] (in model organisms, anaerobiosis dramatically increases theromotolerance [136, 137]). Thus, heat shock proteins may minimize superoxide production by maintaining the electron transfer complexes in the properly folded state. Finally, it is known that the higher the mitochondrial membrane potential ($\Delta\Psi$), the higher the rate of superoxide production [6, 128, 132, 138]. Superoxide production by succinate-driven reverse-electron transfer is essentially eliminated by a drop of >10 mV in $\Delta\Psi$ [124, 139], which is so small that it is essentially undetectable using routine $\Delta\Psi$ probes such as safranin O [139]. Even forward-electron transfer with glutamate/malate is inhibited by uncouplers (carbonyl cyanide p-trifluoromethoxyphenylhydrazone), but the drop in membrane potential required is substantially greater [140]. The dependence of superoxide production on $\Delta\Psi$ is generally explained by a higher $\Delta\Psi$ causing accumulation of electrons on reactive intermediates (such as semiquinones, flavosemiquinones, and low-potential iron-sulfur clusters), which are required for proton pumping [132]. Thus, "mild" uncoupling has been proposed as a method to reduce superoxide production therapeutically [132]. It seems that nature has espoused this strategy and that it has devised special proteins, dubbed uncoupling proteins, to maintain the membrane potential at safe low levels [141].

2.5 Measuring Oxidative Damage

Because ROS have short half-lives, and they are found at low concentrations in vivo, they are exceedingly difficult to observe. The alternative that biochemists have gravitated toward is measuring the end products of ROS, i.e., oxidative damage. Measuring oxidative damage is almost as old as the free radical theory itself.

It has a history of controversy, and many techniques used at one point in time have subsequently proved inappropriate, usually due to artifactual oxidation of the sample during preparation (see discussion on pp 388 and 407 in [3]). In this section, we briefly discuss the main current techniques for the measurement of end product oxidative damage. What makes a good marker of oxidative damage? A good marker of oxidative damage must be increased when oxidative stress is present (i.e., when induced by known treatments or agents that cause oxidative damage, e.g., paraquat, diquat, ionizing radiation, hyperoxia), and it must remain unchanged when oxidative stress is absent. The marker must measure a product that is endogenously present, not produced during the isolation procedure. This latter requirement is not trivial, and it affects several assays: even if a difference between a treatment and a control is observed, if the majority of the signal is artifactually produced during isolation, it will be very difficult to conclude that the *difference between the samples did not arise during the preparation procedure* (rather than having been endogenously present).

2.5.1 Lipid Peroxidation

Lipid peroxidation is a chain reaction in which carbon-centered radicals at the allylic position of polyunsaturated fatty acids (PUFA) react with molecular oxygen (at near diffusion-limited rates), thereby forming a peroxyl radical (ROO·). ROO· can then abstract the allylic H-atom from nearby PUFA (becoming a lipid peroxide ROO-H), which creates another carbon-centered radical, thereby repeating the process described above (for review, see p 291 in [3]). The initial proton abstraction event ("the chain initiator") is thought to be initiated mainly by superoxide (in its protonated form, $HO_2^·$ [21, 22]). The higher the number of double bonds (unsaturation) in a fatty acid, the greater its propensity to peroxidize. In subsequent reactions, lipid peroxides can undergo a variety of reactions, yielding a myriad of end products, e.g., reactive aldehydes (malondialdehyde), alkanes, isoprostanes, and isoketals. Vitamin E (α-tocopherol) plays a critical role in minimizing lipid peroxidation; in fact, dietary induction of vitamin E deficiency results in profound oxidative stress, and if unchecked leads to death [142].

The oldest assay to measure lipid peroxidation is the thiobarbituric acid-reactive substances assay (p 407 in [3]). This assay measures thiobarbituric acid-reactive substances: the reactive end product aldehydes formed during peroxidation of polyunsaturated fatty acids. It is now understood that, although suitable for in vitro chemical systems, this assay is inappropriate to determine lipid peroxidation in vivo, because >90% of the signal actually originates from artifactual oxidation during the harsh isolation procedure.

In the last decade, many new assays to measure lipid peroxides in vivo have been developed. The most popular marker for measuring lipid peroxidation is the gas chromatography-mass spectrometry F_2-isoprostane assay developed by Roberts and Morrow [143, 144]. F_2-isoprostanes are cyclooxygenase-independent oxidation products of arachidonic acid, which are produced in every tissue where this fatty acid is present [145, 146]. F_2-isoprostanes are eliminated via the bloodstream, enabling estimation of whole-organism lipid peroxidation by measuring plasma F_2-isoprostanes [147]. F_2-isoprostanes have been reported to be elevated in a variety of human pathologies [148]. Since isoprostanes are terminal end products, isolation artifacts are minimized (though not eliminated, because F_2-isoprostane levels increase if tissues are stored below −80 °C). It is our opinion that plasma F_2-isoprostanes are currently the most robust marker for measurements of oxidative damage. It has been reported that F_2-isoprostanes are dramatically increased in situations of oxidative stress (e.g., diquat, CCl_4), with the increases being much higher than that with previously used markers (e.g., 8-oxo-7,8-dihydro-2′-deoxyguanosine [8-oxo-dG]). In addition, Jackson and Morrow have extended their initial findings with isoprostanes, demonstrating that peroxidation products of docohexanoic acid, F4 neuroprostanes [149], are also useful markers of lipid peroxidation, especially in neuronal tissues.

2.5.2 DNA Oxidative Damage

Much work has gone into exploring the hypothesis that oxidative damage to DNA causes mutations and cancer. At least 100 different types of oxidative DNA lesions have been reported, including base modifications (e.g., 8-oxo-dG, thymidine glycol, and 8-hydroxycytosine), single- and double-strand breaks and interstrand cross-links [150, 151]. Measuring DNA oxidative damage has a tortuous history that is still not fully settled [152]. Although DNA oxidation yields many products, only a few have been rigorously quantified in vivo [153, 154]; of these, the most popular is 8-oxo-dG. The levels of 8-oxo-dG are measured by high-performance liquid chromatography (HPLC), typically using an electrochemical (EC) detector. The applicability of this assay to estimating the [low] endogenous levels of DNA oxidative damage in vivo has been questioned [155], because large artifactual increases occur during DNA extraction. For example, the values of 8-oxo-dG for the same tissue from the same species have ranged over 3 orders of magnitude [152]. This assay is typically carried out on whole fresh tissue or frozen cells. The sample is first homogenized, and digested with proteinase K at 56 °C; DNA is extracted with phenol and subsequently hydrolyzed into nucleotides (using nuclease P1), then converted to nucleosides by using alkaline phosphatase. The nucleosides are then injected into the HPLC, and 8-oxo-dG is detected electrochemically. Considering the harshness of these treatments, the low redox potential of guanosine, and the large excess of deoxyguanosine (dG) vs. 8-oxo-dG (even a 0.01% artifactual oxidation of dG would translate into a 10-fold artifactual increase of 8-oxo-dG [156]), it is easy to see how artifactual oxidations could inflate the measured amount of 8-oxo-dG. Several factors causing artifactual oxidation have been identified [157], including light from fluorescent lamps [158]. The use of phenol during the DNA extraction procedure also seems problematic [159]; this can be avoided by substituting NaI in the DNA extraction protocol [157, 159]. Other strategies to minimize oxidation involve the addition of desferrioxime (to prevent Fenton chemistry) or antioxidant enzymes such as catalase [157, 160]. Using these optimizations, the level of 8-oxo-dG has been measured to be ~0.5 per 10^6 dG in human lymphocytes and ~2 per 10^6 dG in rat liver DNA [157, 160]. Even if artifactual oxidation during DNA extraction were reduced to zero, this would not address the possibility of artifactual DNA oxidation during tissue homogenization: this is relevant because homolytic bond cleavage and free radical formation have been demonstrated under those conditions. Experiments with $H_2^{18}O$ under anaerobic conditions may resolve this issue, because any 8-oxo-dG produced artifactually would have been distinguishable by mass spectroscopy [160]. Significantly, even the lowest values obtained using the HPLC-EC method are still an order of magnitude higher than those obtained with the formamidopyri-midine-DNA glycosylase (Fgp-glycosylase) comet assay [152, 161]. The comet assay does not require tissue homogenization and DNA hydrolysis, but can only be used on cells, not whole tissues [162, 163]. It quantifies the number of strand breaks after treatment with Fpg-glycosylase, which converts 8-oxo-dG into single-strand breaks. The ~10-fold discrepancy between these two assays has not yet been resolved, and it cannot be attributed to differences in endogenous endonuclease activity, since the difference persists even in *Ogg1* knockout mice, which lack base excision repair [164, 165].

2.5.3 Protein Oxidation

Numerous oxidative modifications have been documented in proteins. Proteins vary in their susceptibilities to different types of oxidants, and in the sites and degree of oxidation. Such differences are generally influenced by types of accessible amino acid residues in the proteins. The best-known types of protein oxidation are the nitration of tyrosine, the sulfoxidation of methionine, and the carbonylation of most amine-containing amino acid residues [166–169].

Peroxynitrite ($ONOO^-$), which can be formed by the reaction of $NO^\bullet$ with super-oxide, $NO^\bullet + O_2^{\bullet-} \rightarrow ONOO^-$ ($k = 7 \times 10^9 M^{-1} s^{-1}$), can convert tyrosine to nitroty-rosine. The hydroxyl group of certain tyrosine residues is critical in some enzymes and cell signalling molecules (e.g., tyrosine phosphorylation). It was shown that for glutamine synthetase in *E. coli*, nitration of either one of two different tyrosine residues inhibited the enzymatic function [170]. There is also evidence indicating that nitrotyrosilation of Mn-SOD decreases its activity [171, 172].

Sulfur-containing amino acids (methionine and cysteine) can be oxidized by hydrogen peroxide, superoxide, peroxynitrite, and perhaps by molecular oxygen itself. However, these are the only oxidative modifications of proteins that can be repaired. ROS-mediated oxidation of methionine (Met) residues leads to methionine sulfoxide (MetO), consisting of a mixture of the *S*- and *R*-epimers of MetO. Methionine sulfoxide reductases (Msrs) catalyze the thioredoxin-dependent reduc-tion of MetO back to Met. The *S*-epimer of MetO is reduced back to Met by MsrA, and the *R*-epimer by MsrB. Importantly however, the oxidized form of thioredoxin produced during the reduction of MetO can be converted back to reduced form by the enzyme thioredoxin reductase, in an NADPH-dependent reaction. This cyclic oxidation and reduction of methionine has been proposed to play an antioxidant role [173]. Reversible oxidation of cysteine (thiol) residues in turn mediates the antioxidant function of glutathione, thioredoxin, metallothioneins, glutaredoxin, and peroxiredoxins. Oxidation of thiols (sulfhydryl) yields disulfides, and further oxidation yields sulfinic and sulfenic acids, and eventually sulfones [174]. Certain enzymes, such as GAPDH, have active site cysteines that can become irreversibly oxidized [37–39, 175]. A newly discovered antioxidant gene, termed sulfiredoxin, can reduce sulfinic acid back to sulfhydril (thiol) [85, 176]. The sulfones are still considered irreversible.

Carbonylation of proteins can occur by several different oxidative pathways, including metal-catalyzed oxidation of specific amino acid side chains (histidine, arginine, proline, lysine, and threonine) and adduction of carbonyl-containing oxidized lipids (e.g., 4-hydroxynonenal, malondialdehyde) and sugars [177]. Carbonyl content of samples has been popularly used as a global indicator of protein oxidation levels, perhaps due in part to the simplicity of the assay (carbonyl groups can be detected spectrophotometrically after their reactions with 2,4-dinitrophenylhydrazine), although this assay is also not free from controversy [178]. Specific proteins, such as adenine nucleotide translocase [179] and aconitase [180], seem be preferentially carbonylated with age in flies. In mice, carbonic anhydrase 3 and CuZn-SOD are preferentially carbonylated during aging [178, 181]. Oxidative modification of bacterial glutamine synthetase results in its inactivation and degradation [182, 183]. Protein carbonylation also can be quantified by gas chromatography-mass spectrometry [184, 185]. For example, glutamic and aminoadipic semialdehyde are the markers of direct protein oxidation by ROS, and they are found to be the major constituent of the total protein carbonyl value [184], whereas N^ε-(malondialdehyde)lysine and N^ε-(carboxymethyl)lysine (CML) arise from lipid peroxidation products [186]. N^ε-(carboxyethyl)lysine and CML also can be formed through glycoxidative damage from sugars [186, 187].

2.6 How Does Oxidative Damage Kill or Compromise the Function of the Cell?

In the preceeding sections, we outlined the chemistry of some of the best-described ROS, and the most easily measureable types of damage that ROS can inflict on macromolecules. Here, we summarize how these oxidative modifications can compromise cellular function or lead to cell death.

2.6.1 Lipid Peroxidation

Lipid peroxidation of membranes has the immediate effect of decreasing membrane fluidity and increasing ionic permeability [188]. High enough levels of lipid peroxidation can directly lead to loss of membrane barrier function, cell lysis, and cell death. Because ionic gradients play a critical role in several physiological processes, including energy metabolism (mitochondrial inner membrane) and neuronal conductance (plasma membrane), even moderately increased lipid peroxidation has the potential to be highly disruptive to overall tissue and organism function. Deficiency of the chain-breaking antioxidant, vitamin E, results in many deleterious phenotypes, including fertility loss, neurological phenotypes [189], myopathy [142], lipofuscin accumulation [190], and erythrocyte lysis; if severe enough, vitamin E deficiency results in death [142]. Absence of the lipid hydroperoxide-scavenging enzyme glutathione peroxidase 4 results in early embryonic lethality in mice [191], testifying to the great toxicity of lipid peroxidation.

2.6.2 DNA Oxidative Damage

Damage to DNA by reactive oxygen species falls into two main categories: damage to nucleotide bases and strand breaks (which, biologically, predominantly result in mutagenesis and genomic instability, respectively). Mutations, in turn, contribute to carcinogenesis, whereas genomic instability contributes both to carcinogenesis and, more critically, to cell death. There is an increasing body of evidence in model organisms (e.g., yeast and *E. coli*) that oxidative stress is a strong contributor to both mutagenesis and genomic instability. For example, genetic ablation of CuZn-SOD and peroxiredoxin 1 (*tsa1*) cause a 5- and 10-fold increase, respectively, in mutagenesis in yeast, as measured by the canavanine method [15, 72, 192–194]. This increased mutagenesis is entirely prevented by growth under anaerobic conditions [15, 73] (complete absence of oxygen implies no formation of superoxide and H_2O_2). In fact, spontaneous mutagenesis also is decreased ~3-fold by anaerobic conditions in wild-type yeast, indicating that oxidative stress is an important driver of mutagenesis even under antioxidant enzyme sufficient conditions [73]. Similar results also have been obtained in *E. coli* [195]. Oxidative stress brought about by antioxidant enzyme ablation also increases genomic instability in yeast. Knockout of peroxiredoxin 1 (*tsa1*) resulted in a 10-fold increase in gross chromosomal rearrangements [73, 74]. Again, anaerobic conditions dramatically reduce the gross chromosomal rearrangement rate, not only in the antioxidant knockouts, but also in wild-type control yeast [73].

Finally, although knockout of neither *sod1* nor *tsa1* is lethal by itself in yeast grown under optimal conditions (rich media), it was recently reported that combining *sod1*−/− or *tsa1*−/− with a wide assortment of knockouts of DNA repair genes resulted in synthetic lethality [196]. For example, a double knockout of either *sod1* or *tsa1* with *rad51* (involved in double-strand break recombination repair) was synthetically lethal [73, 196]. In the case of *tsa1*−/−*rad 51*−/− double knockout, the synthetic lethality could be rescued by growth under anaerobic conditions [73].

These data in single-celled model systems complement basic chemical work, and they establish that ROS are a significant cause of DNA damage in vivo, driving both mutagenesis and genomic instability. However, it is worth remembering that yeast and *E. coli* grow at 21% O_2 (normal atmospheric oxygen tension), whereas most mammalian cells are exposed to ~10 times less oxygen in situ; the relative importance of DNA oxidative damage may therefore be less in higher organisms.

2.6.3 Oxidative Damage to Proteins

Oxidative damage to proteins can be physiologically detrimental by several pathways. We briefly discuss two such pathways here: inactivation of enzymatic function and stimulation of protein aggregation.

There are several well-described examples of direct, selective inactivation of enzymatic function by different ROS: aconitase (superoxide [197], reversible), ribonucleotide reductase (superoxide [31], irreversible), GAPDH (H_2O_2, irreversible [37–39, 175]), carbonic anhydrase [181], and glutamine synthetase [182]. In certain cases, the ROS-dependent inactivation has clearly demonstrable phenotypical and metabolic consequences, e.g., the lysine and methionine auxotrophies of $sod1^{-/-}$ yeast can directly be ascribed to the superoxide-dependent inactivation of iron sulfur-containing enzymes [198–200].

It is increasingly appreciated that protein aggregation plays a critical role in many (most prominently neurodegenerative) diseases [201]. The notion that ROS can contribute to protein aggregation has been long known, especially considering the role of ROS in formation of lipofuscin, the age pigment composed of aggregated, oxidized proteins and lipids [202, 203]. Increasing evidence indicates that, at least in vitro, oxidative modification increases thermodynamic instability and facilitates protein aggregation, e.g., α-synuclein (Parkinson's disease) [204] and CuZn-SOD (amyotrophic lateral sclerosis). It has been suggested that this is a key pathway by which oxidative damage contributes to aging [205].

3 Conclusions

Chemical studies indicate that ROS can damage a number of different cellular macromolecules, and physiological studies in model organisms indicate that ROS can compromise cell function and viability in many ways. However, an efficient and complex network of enzymatic and nonenzymatic players does exist; how much oxidative damage "escapes" the effects of this network under normal conditions is still not fully resolved. Finally, it is important to recall that oxygen tension in animals is typically 10 times lower than that to which the above-mentioned model organisms are exposed.

References

1. Sawyer DT. Oxygen chemistry. New York: Oxford University Press; 1991.
2. Sawyer DT. Oxygen: inorganic chemistry. In: King RB, ed. Encyclopedia of inorganic chemistry. Chichester: John Wiley & Sons; 1994:2947–88.
3. Halliwell B, Gutteridge J. Free radicals in biology and medicine, 3rd edn. New York: Oxford University Press; 1999.

4. Bielski BHJ. Reactivity of $HO_2/O_2^{\cdot-}$ radicals in aqueous solution. J Phys Chem Ref Data 1985;14(4):1041–91.
5. Afanas'ev IB. Superoxide anion: chemistry and biological implications. Boca Raton, Florida: CRC Press; 1989.
6. Chance B, Sies H, Boveris A. Hydroperoxide metabolism in mammalian organs. Physiol Rev 1979;59(3):527–605.
7. Muller F. The nature and mechanism of superoxide production by the electron transport chain: its relevance to aging. J Am Aging Assoc 2000;23:227–53.
8. Sawyer DT, Valentine JS. How super is superoxide? Acc Chem Res 1981;14:393–400.
9. Li Y, Huang TT, Carlson EJ, et al. Dilated cardiomyopathy and neonatal lethality in mutant mice lacking manganese superoxide dismutase. Nat Genet 1995;11(4):376–81.
10. Lebovitz RM, Zhang H, Vogel H, et al. Neurodegeneration, myocardial injury, and perinatal death in mitochondrial superoxide dismutase-deficient mice. Proc Natl Acad Sci U S A 1996;93(18):9782–7.
11. van Loon AP, Pesold-Hurt B, Schatz G. A yeast mutant lacking mitochondrial manganese-superoxide dismutase is hypersensitive to oxygen. Proc Natl Acad Sci U S A 1986;83(11):3820–4.
12. Longo VD, Liou LL, Valentine JS, Gralla EB. Mitochondrial superoxide decreases yeast survival in stationary phase. Arch Biochem Biophys 1999;365(1):131–42.
13. Longo VD, Gralla EB, Valentine JS. Superoxide dismutase activity is essential for stationary phase survival in Saccharomyces cerevisiae. Mitochondrial production of toxic oxygen species in vivo. J Biol Chem 1996;271(21):12275–80.
14. Guidot DM, McCord JM, Wright RM, Repine JE. Absence of electron transport (Rho 0 state) restores growth of a manganese-superoxide dismutase-deficient Saccharomyces cerevisiae in hyperoxia. Evidence for electron transport as a major source of superoxide generation in vivo. J Biol Chem 1993;268(35):26699–703.
15. Gralla EB, Valentine JS. Null mutants of Saccharomyces cerevisiae Cu,Zn superoxide dismutase: characterization and spontaneous mutation rates. J Bacteriol 1991;173(18):5918–20.
16. Duttaroy A, Paul A, Kundu M, Belton A. A Sod2 null mutation confers severely reduced adult life span in Drosophila. Genetics 2003;165(4):2295–9.
17. Reveillaud I, Phillips J, Duyf B, Hilliker A, Kongpachith A, Fleming JE. Phenotypic rescue by a bovine transgene in a Cu/Zn superoxide dismutase-null mutant of Drosophila melanogaster. Mol Cell Biol 1994;14(2):1302–7.
18. Liochev SI, Fridovich I. Reversal of the superoxide dismutase reaction revisited. Free Radic Biol Med 2003;34(7):908–10.
19. Srinivasan C, Liba A, Imlay JA, Valentine JS, Gralla EB. Yeast lacking superoxide dismutase(s) show elevated levels of "free iron" as measured by whole cell electron paramagnetic resonance. J Biol Chem 2000;275(38):29187–92.
20. Liochev SI, Fridovich I. The Haber-Weiss cycle–70 years later: an alternative view. Redox Rep 2002;7(1):55–7; author reply 9–60.
21. Bielski BH, Arudi RL, Sutherland MW. A study of the reactivity of $HO_2/O_2^{\cdot-}$ with unsaturated fatty acids. J Biol Chem 1983;258(8):4759–61.
22. Antunes F, Salvador A, Marinho HS, Alves R, Pinto RE. Lipid peroxidation in mitochondrial inner membranes. I. An integrative kinetic model. Free Radic Biol Med 1996;21(7):917–43.
23. Aikens J, Dix TA. Perhydroxyl radical (HOO·) initiated lipid peroxidation. The role of fatty acid hydroperoxides. J Biol Chem 1991;266(23):15091–8.
24. Winterbourn CC, Metodiewa D. Reactivity of biologically important thiol compounds with superoxide and hydrogen peroxide. Free Radic Biol Med 1999;27(3–4):322–8.
25. Winterbourn CC, Kettle AJ. Radical-radical reactions of superoxide: a potential route to toxicity. Biochem Biophys Res Commun 2003;305(3):729–36.
26. Beckman JS, Beckman TW, Chen J, Marshall PA, Freeman BA. Apparent hydroxyl radical production by peroxynitrite: implications for endothelial injury from nitric oxide and superoxide. Proc Natl Acad Sci U S A 1990;87(4):1620–4.

27. Lymar SV, Hurst JK. Radical nature of peroxynitrite reactivity. Chem Res Toxicol 1998;11(7):714–5.
28. Liochev SI, Fridovich I. The relative importance of HO• and ONOO⁻ in mediating the toxicity of $O_2\cdot^-$. Free Radic Biol Med 1999;26(5–6):777–8.
29. Liochev SI, Fridovich I. Second order rate constants: a cautionary note. Free Radic Biol Med 2003;35(7):833.
30. Fontecave M. Ribonucleotide reductases and radical reactions. Cell Mol Life Sci 1998;54(7):684–95.
31. Gaudu P, Niviere V, Petillot Y, Kauppi B, Fontecave M. The irreversible inactivation of ribonucleotide reductase from Escherichia coli by superoxide radicals. FEBS Lett 1996;387(2–3):137–40.
32. Okado-Matsumoto A, Fridovich I. The role of alpha,beta-dicarbonyl compounds in the toxicity of short chain sugars. J Biol Chem 2000;275(45):34853–7.
33. Mashino T, Fridovich I. Superoxide radical initiates the autoxidation of dihydroxyacetone. Arch Biochem Biophys 1987;254(2):547–51.
34. Sutton HC, Roberts PB, Winterbourn CC. The rate of reaction of superoxide radical ion with oxyhaemoglobin and methaemoglobin. Biochem J 1976;155(3):503–10.
35. Gus'kova RA, Ivanov, II, Kol'tover VK, Akhobadze VV, Rubin AB. Permeability of bilayer lipid membranes for superoxide ($O_2\cdot^-$) radicals. Biochim Biophys Acta 1984;778(3):579–85.
36. Takahashi MA, Asada K. Superoxide anion permeability of phospholipid membranes and chloroplast thylakoids. Arch Biochem Biophys 1983;226(2):558–66.
37. Baty JW, Hampton MB, Winterbourn CC. Proteomic detection of hydrogen peroxide-sensitive thiol proteins in Jurkat cells. Biochem J 2005;389(Pt 3):785–95.
38. Janero DR, Hreniuk D, Sharif HM. Hydroperoxide-induced oxidative stress impairs heart muscle cell carbohydrate metabolism. Am J Physiol 1994;266:C179–C188.
39. Cochrane CG. Cellular injury by oxidants. Am J Med 1991;91(3C):23S–30S.
40. Mahadev K, Zilbering A, Zhu L, Goldstein BJ. Insulin-stimulated hydrogen peroxide reversibly inhibits protein-tyrosine phosphatase 1b in vivo and enhances the early insulin action cascade. J Biol Chem 2001;276(24):21938–42.
41. Lee SR, Kwon KS, Kim SR, Rhee SG. Reversible inactivation of protein-tyrosine phosphatase 1B in A431 cells stimulated with epidermal growth factor. J Biol Chem 1998;273(25):15366–72.
42. Koppenol WH. The Haber-Weiss cycle–70 years later. Redox Rep 2001;6(4):229–34.
43. Czapski G, Goldstein S. When do metal complexes protect the biological system from superoxide toxicity and when do they enhance it? Free Radic Res Commun 1986;1(3):157–61.
44. Liochev SL. The role of iron-sulfur clusters in in vivo hydroxyl radical production. Free Radic Res 1996;25(5):369–84.
45. Neyens E, Baeyens J. A review of classic Fenton's peroxidation as an advanced oxidation technique. J Hazard Mater 2003;98(1–3):33–50.
46. Muller FL, Lustgarten MS, Jang Y, Richardson A, Van Remmen H. Trends in oxidative aging theories. Free Radic Biol Med 2007;43(4):477–503.
47. Fridovich I. Mitochondria: are they the seat of senescence? Aging Cell 2004;3(1):13–6.
48. McCord JM, Fridovich I. Superoxide dismutase. An enzymic function for erythrocuprein (hemocuprein). J Biol Chem 1969;244(22):6049–55.
49. Carlsson LM, Jonsson J, Edlund T, Marklund SL. Mice lacking extracellular superoxide dismutase are more sensitive to hyperoxia. Proc Natl Acad Sci U S A 1995;92(14):6264–8.
50. Weisiger RA, Fridovich I. Mitochondrial superoxide simutase. Site of synthesis and intramitochondrial localization. J Biol Chem 1973;248(13):4793–6.
51. Reaume AG, Elliott JL, Hoffman EK, et al. Motor neurons in Cu/Zn superoxide dismutase-deficient mice develop normally but exhibit enhanced cell death after axonal injury. Nat Genet 1996;13(1):43–7.
52. Okado-Matsumoto A, Fridovich I. Subcellular distribution of superoxide dismutases (sod) in rat liver. Cu,Zn-SOD in mitochondria. J Biol Chem 2001;276(42):38388–93.
53. Sturtz LA, Diekert K, Jensen LT, Lill R, Culotta VC. A fraction of yeast Cu,Zn-superoxide dismutase and its metallochaperone, ccs, localize to the intermembrane space of mitochondria.

A physiological role for sod1 in guarding against mitochondrial oxidative damage. J Biol Chem 2001;276(41):38084–9.

54. Weisiger RA, Fridovich I. Superoxide dismutase. Organelle specificity. J Biol Chem 1973;248(10):3582–92.
55. Liochev SI, Fridovich I. Copper,zinc superoxide dismutase as a univalent NO(–) oxidoreductase and as a dichlorofluorescein peroxidase. J Biol Chem 2001;276(38):35253–7.
56. Beckman JS, Carson M, Smith CD, Koppenol WH. ALS, SOD and peroxynitrite. Nature 1993;364(6438):584.
57. Winterbourn CC, Peskin AV, Parsons-Mair HN. Thiol oxidase activity of copper, zinc superoxide dismutase. J Biol Chem 2002;277(3):1906–11.
58. Liochev SI, Fridovich I. Mutant Cu,Zn superoxide dismutases and familial amyotrophic lateral sclerosis: evaluation of oxidative hypotheses. Free Radic Biol Med 2003;34(11):1383–9.
59. Ogata M. Acatalasemia. Hum Genet 1991;86(4):331–40.
60. Ho YS, Xiong Y, Ma W, Spector A, Ho DS. Mice lacking catalase develop normally but show differential sensitivity to oxidant tissue injury. J Biol Chem 2004;279(31):32804–12.
61. Cheng WH, Ho YS, Ross DA, Valentine BA, Combs GF, Lei XG. Cellular glutathione peroxidase knockout mice express normal levels of selenium-dependent plasma and phospholipid hydroperoxide glutathione peroxidases in various tissues. J Nutr 1997;127(8):1445–50.
62. Fu Y, Cheng WH, Porres JM, Ross DA, Lei XG. Knockout of cellular glutathione peroxidase gene renders mice susceptible to diquat-induced oxidative stress. Free Radic Biol Med 1999;27(5–6):605–11.
63. Fu Y, Cheng WH, Ross DA, Lei X. Cellular glutathione peroxidase protects mice against lethal oxidative stress induced by various doses of diquat. Proc Soc Exp Biol Med 1999;222(2):164–9.
64. Andziak B, O'Connor TP, Buffenstein R. Antioxidants do not explain the disparate longevity between mice and the longest-living rodent, the naked mole-rat. Mech Ageing Dev 2005;126(11):1206–12.
65. Outten CE, Falk RL, Culotta VC. Cellular factors required for protection from hyperoxia toxicity in *Saccharomyces cerevisiae*. Biochem J 2005;388:93–101.
66. Inoue Y, Matsuda T, Sugiyama K, Izawa S, Kimura A. Genetic analysis of glutathione peroxidase in oxidative stress response of *Saccharomyces cerevisiae*. J Biol Chem 1999;274(38):27002–9.
67. Chae HZ, Kim IH, Kim K, Rhee SG. Cloning, sequencing, and mutation of thiol-specific antioxidant gene of *Saccharomyces cerevisiae*. J Biol Chem 1993;268(22):16815–21.
68. Rhee SG, Chae HZ, Kim K. Peroxiredoxins: a historical overview and speculative preview of novel mechanisms and emerging concepts in cell signaling. Free Radic Biol Med 2005;38(12):1543–52.
69. Wood ZA, Schroder E, Robin Harris J, Poole LB. Structure, mechanism and regulation of peroxiredoxins. Trends Biochem Sci 2003;28(1):32–40.
70. Lee SM, Park JW. Thermosensitive phenotype of yeast mutant lacking thioredoxin peroxidase. Arch Biochem Biophys 1998;359(1):99–106.
71. Lee JH, Park JW. Role of thioredoxin peroxidase in aging of stationary cultures of Saccharomyces cerevisiae. Free Radic Res 2004;38(3):225–31.
72. Wong CM, Siu KL, Jin DY. Peroxiredoxin-null yeast cells are hypersensitive to oxidative stress and are genomically unstable. J Biol Chem 2004;279(22):23207–13.
73. Ragu S, Faye G, Iraqui I, Masurel-Heneman A, Kolodner RD, Huang ME. Oxygen metabolism and reactive oxygen species cause chromosomal rearrangements and cell death. Proc Natl Acad Sci U S A 2007;104(23):9747–52.
74. Smith S, Hwang JY, Banerjee S, Majeed A, Gupta A, Myung K. Mutator genes for suppression of gross chromosomal rearrangements identified by a genome-wide screening in *Saccharomyces cerevisiae*. Proc Natl Acad Sci U S A 2004;101(24):9039–44.
75. Neumann CA, Krause DS, Carman CV, et al. Essential role for the peroxiredoxin Prdx1 in erythrocyte antioxidant defence and tumour suppression. Nature 2003;424(6948):561–5.

76. Lee TH, Kim SU, Yu SL, et al. Peroxiredoxin II is essential for sustaining life span of erythrocytes in mice. Blood 2003;101(12):5033–8.

77. Low FM, Hampton MB, Peskin AV, Winterbourn CC. Peroxiredoxin 2 functions as a non-catalytic scavenger of low level hydrogen peroxide in the erythrocyte. Blood 2006.

78. Peskin AV, Low FM, Paton LN, Maghzal GJ, Hampton MB, Winterbourn CC. The high reactivity of peroxiredoxin 2 with H_2O_2 is not reflected in its reaction with other oxidants and thiol reagents. J Biol Chem 2007;282(16):11885–92.

79. Stocker R, Glazer AN, Ames BN. Antioxidant activity of albumin-bound bilirubin. Proc Natl Acad Sci U S A 1987;84(16):5918–22.

80. Stocker R, Yamamoto Y, McDonagh AF, Glazer AN, Ames BN. Bilirubin is an antioxidant of possible physiological importance. Science 1987;235(4792):1043–6.

81. Baranano DE, Rao M, Ferris CD, Snyder SH. Biliverdin reductase: a major physiologic cytoprotectant. Proc Natl Acad Sci U S A 2002;99(25):16093–8.

82. Walker DW, Muffat J, Rundel C, Benzer S. Overexpression of a *Drosophila* homolog of apolipoprotein D leads to increased stress resistance and extended lifespan. Curr Biol 2006;16(7):674–9.

83. Sanchez D, Lopez-Arias B, Torroja L, et al. Loss of glial lazarillo, a homolog of apolipoprotein D, reduces lifespan and stress resistance in *Drosophila*. Curr Biol 2006;16(7):680–6.

84. Rassart E, Bedirian A, Do Carmo S, et al. Apolipoprotein D. Biochim Biophys Acta 2000;1482(1–2):185–98.

85. Biteau B, Labarre J, Toledano MB. ATP-dependent reduction of cysteine-sulphinic acid by *S. cerevisiae* sulphiredoxin. Nature 2003;425(6961):980–4.

86. Findlay VJ, Townsend DM, Morris TE, Fraser JP, He L, Tew KD. A novel role for human sulfiredoxin in the reversal of glutathionylation. Cancer Res 2006;66(13):6800–6.

87. Sun Y, Hegamyer G, Colburn NH. Molecular cloning of five messenger RNAs differentially expressed in preneoplastic or neoplastic JB6 mouse epidermal cells: one is homologous to human tissue inhibitor of metalloproteinases-3. Cancer Res 1994;54(5):1139–44.

88. Kwak MK, Wakabayashi N, Itoh K, Motohashi H, Yamamoto M, Kensler TW. Modulation of gene expression by cancer chemopreventive dithiolethiones through the Keap1-Nrf2 pathway. Identification of novel gene clusters for cell survival. J Biol Chem 2003;278(10):8135–45.

89. Wimmer U, Wang Y, Georgiev O, Schaffner W. Two major branches of anti-cadmium defense in the mouse: MTF-1/metallothioneins and glutathione. Nucleic Acids Res 2005;33(18):5715–27.

90. Welch KD, Reilly TP, Bourdi M, et al. Genomic identification of potential risk factors during acetaminophen-induced liver disease in susceptible and resistant strains of mice. Chem Res Toxicol 2006;19(2):223–33.

91. Edwards MG, Sarkar D, Klopp R, Morrow JD, Weindruch R, Prolla TA. Age-related impairment of the transcriptional responses to oxidative stress in the mouse heart. Physiol Genomics 2003;13(2):119–27.

92. Budanov AV, Sablina AA, Feinstein E, Koonin EV, Chumakov PM. Regeneration of peroxiredoxins by p53-regulated sestrins, homologs of bacterial AhpD. Science 2004;304(5670):596–600.

93. Sablina AA, Budanov AV, Ilyinskaya GV, Agapova LS, Kravchenko JE, Chumakov PM. The antioxidant function of the p53 tumor suppressor. Nat Med 2005;11(12):1306–13.

94. Babior BM, Kipnes RS, Curnutte JT. Biological defense mechanisms. The production by leukocytes of superoxide, a potential bactericidal agent. J Clin Invest 1973;52(3):741–4.

95. Lambeth JD. NOX enzymes and the biology of reactive oxygen. Nat Rev Immunol 2004;4(3):181–9.

96. Knowles PF, Gibson JF, Pick FM, Bray RC. Electron-spin-resonance evidence for enzymic reduction of oxygen to a free radical, the superoxide ion. Biochem J 1969;111(1):53–8.

97. Kundu TK, Hille R, Velayutham M, Zweier JL. Characterization of superoxide production from aldehyde oxidase: an important source of oxidants in biological tissues. Arch Biochem Biophys 2007;460(1):113–21.

98. Sligar SG, Lipscomb JD, Debrunner PG, Gunsalus IC. Superoxide anion production by the autoxidation of cytochrome P450cam. Biochem Biophys Res Commun 1974;61(1):290–6.

99. Loschen G, Azzi A, Richter C, Flohe L. Superoxide radicals as precursors of mitochondrial hydrogen peroxide. FEBS Lett 1974;42(1):68–72.

100. Boveris A, Cadenas E. Mitochondrial production of superoxide anions and its relationship to the antimycin insensitive respiration. FEBS Lett 1975;54(3):311–4.

101. Ohlemiller KK, McFadden SL, Ding DL, et al. Targeted deletion of the cytosolic Cu/Zn-superoxide dismutase gene (Sod1) increases susceptibility to noise-induced hearing loss. Audiol Neurootol 1999;4(5):237–46.

102. Matzuk MM, Dionne L, Guo Q, Kumar TR, Lebovitz RM. Ovarian function in superoxide dismutase 1 and 2 knockout mice. Endocrinology 1998;139(9):4008–11.

103. Ho YS, Gargano M, Cao J, Bronson RT, Heimler I, Hutz RJ. Reduced fertility in female mice lacking copper-zinc superoxide dismutase. J Biol Chem 1998;273(13):7765–9.

104. McFadden SL, Ding D, Burkard RF, et al. Cu/Zn SOD deficiency potentiates hearing loss and cochlear pathology in aged 129,CD-1 mice. J Comp Neurol 1999;413(1):101–12.

105. McFadden SL, Ding D, Reaume AG, Flood DG, Salvi RJ. Age-related cochlear hair cell loss is enhanced in mice lacking copper/zinc superoxide dismutase. Neurobiol Aging 1999;20(1):1–8.

106. Shefner JM, Reaume AG, Flood DG, et al. Mice lacking cytosolic copper/zinc superoxide dismutase display a distinctive motor axonopathy. Neurology 1999;53(6):1239–46.

107. Yoshida T, Maulik N, Engelman RM, Ho YS, Das DK. Targeted disruption of the mouse Sod 1 gene makes the hearts vulnerable to ischemic reperfusion injury. Circ Res 2000;86(3):264–9.

108. Levin ED, Brady TC, Hochrein EC, et al. Molecular manipulations of extracellular superoxide dismutase: functional importance for learning. Behav Genet 1998;28(5):381–90.

109. Phillips JP, Campbell SD, Michaud D, Charbonneau M, Hilliker AJ. Null mutation of copper/zinc superoxide dismutase in *Drosophila* confers hypersensitivity to paraquat and reduced longevity. Proc Natl Acad Sci U S A 1989;86(8):2761–5.

110. Kirby K, Hu J, Hilliker AJ, Phillips JP. RNA interference-mediated silencing of Sod2 in *Drosophila* leads to early adult-onset mortality and elevated endogenous oxidative stress. Proc Natl Acad Sci U S A 2002;99(25):16162–7.

111. Hilliker AJ, Duyf B, Evans D, Phillips JP. Urate-null rosy mutants of *Drosophila melanogaster* are hypersensitive to oxygen stress. Proc Natl Acad Sci U S A 1992;89(10):4343–7.

112. de Grey AD. The reductive hotspot hypothesis of mammalian aging: membrane metabolism magnifies mutant mitochondrial mischief. Eur J Biochem 2002;269(8):2003–9.

113. Nicholls DG, Ferguson SJ. Bioenergetics 3. London: Academic Press; 2002.

114. Jensen PK. Antimycin-insensitive oxidation of succinate and reduced nicotinamide-adenine dinucleotide in electron-transport particles. I. pH dependency and hydrogen peroxide formation. Biochim Biophys Acta 1966;122(2):157–66.

115. Loschen G, Flohe L, Chance B. Respiratory chain linked H_2O_2 production in pigeon heart mitochondria. FEBS Lett 1971;18(2):261–4.

116. Boveris A, Oshino N, Chance B. The cellular production of hydrogen peroxide. Biochem J 1972;128(3):617–30.

117. Ishii N, Fujii M, Hartman PS, et al. A mutation in succinate dehydrogenase cytochrome b causes oxidative stress and ageing in nematodes. Nature 1998;394(6694):694–7.

118. Guo J, Lemire BD. The ubiquinone-binding site of the *Saccharomyces cerevisiae* succinate-ubiquinone oxidoreductase is a source of superoxide. J Biol Chem 2003;278(48):47629–35.

119. Starkov AA, Fiskum G, Chinopoulos C, et al. Mitochondrial alpha-ketoglutarate dehydrogenase complex generates reactive oxygen species. J Neurosci 2004;24(36):7779–88.

120. Tretter L, Adam-Vizi V. alpha-Ketoglutarate dehydrogenase: a target and generator of oxidative stress. Philos Trans R Soc Lond B Biol Sci 2005;360(1464):2335–45.

121. Miwa S, St-Pierre J, Partridge L, Brand MD. Superoxide and hydrogen peroxide production by *Drosophila* mitochondria. Free Radic Biol Med 2003;35(8):938–48.

122. Han D, Williams E, Cadenas E. Mitochondrial respiratory chain-dependent generation of superoxide anion and its release into the intermembrane space. Biochem J 2001;353(Pt 2):411–6.

123. Muller FL, Liu Y, Van Remmen H. Complex III Releases superoxide to both sides of the inner mitochondrial membrane. J Biol Chem 2004;279(47):49064–73.

124. Hansford RG, Hogue BA, Mildaziene V. Dependence of H_2O_2 formation by rat heart mitochondria on substrate availability and donor age. J Bioenerg Biomembr 1997;29(1):89–95.

125. Cino M, Del Maestro RF. Generation of hydrogen peroxide by brain mitochondria: the effect of reoxygenation following postdecapitative ischemia. Arch Biochem Biophys 1989;269(2):623–38.

126. Hinkle PC, Butow RA, Racker E, Chance B. Partial resolution of the enzymes catalyzing oxidative phosphorylation. XV. Reverse electron transfer in the flavin-cytochrome beta region of the respiratory chain of beef heart submitochondrial particles. J Biol Chem 1967;242(22):5169–473.

127. Kudin AP, Bimpong-Buta NY, Vielhaber S, Elger CE, Kunz WS. Characterization of superoxide-producing sites in isolated brain mitochondria. J Biol Chem 2004;279(6):4127–35.

128. Boveris A, Chance B. The mitochondrial generation of hydrogen peroxide. General properties and effect of hyperbaric oxygen. Biochem J 1973;134(3):707–16.

129. Staniek K, Nohl H. Are mitochondria a permanent source of reactive oxygen species? Biochim Biophys Acta 2000;1460(2–3):268–75.

130. St-Pierre J, Buckingham JA, Roebuck SJ, Brand MD. Topology of superoxide production from different sites in the mitochondrial electron transport chain. J Biol Chem 2002;277(47):44784–90.

131. Brand MD. Uncoupling to survive? The role of mitochondrial inefficiency in ageing. Exp Gerontol 2000;35(6–7):811–20.

132. Skulachev VP. Role of uncoupled and non–coupled oxidations in maintenance of safely low levels of oxygen and its one–electron reductants. Q Rev Biophys 1996;29(2):169–202.

133. Kwong LK, Sohal RS. Substrate and site specificity of hydrogen peroxide generation in mouse mitochondria. Arch Biochem Biophys 1998;350(1):118–26.

134. Zhang L, Yu L, Yu CA. Generation of superoxide anion by succinate-cytochrome c reductase from bovine heart mitochondria. J Biol Chem 1998;273(51):33972–6.

135. Davidson JF, Schiestl RH. Mitochondrial respiratory electron carriers are involved in oxidative stress during heat stress in Saccharomyces cerevisiae. Mol Cell Biol 2001;21(24):8483–9.

136. Davidson JF, Whyte B, Bissinger PH, Schiestl RH. Oxidative stress is involved in heat-induced cell death in Saccharomyces cerevisiae. Proc Natl Acad Sci U S A 1996;93(10):5116–21.

137. Benov L, Fridovich I. Superoxide dismutase protects against aerobic heat shock in Escherichia coli. J Bacteriol 1995;177(11):3344–6.

138. Korshunov SS, Skulachev VP, Starkov AA. High protonic potential actuates a mechanism of production of reactive oxygen species in mitochondria. FEBS Lett 1997;416(1):15–8.

139. Votyakova TV, Reynolds IJ. DeltaPsi(m)-Dependent and -independent production of reactive oxygen species by rat brain mitochondria. J Neurochem 2001;79(2):266–77.

140. Gyulkhandanyan AV, Pennefather PS. Shift in the localization of sites of hydrogen peroxide production in brain mitochondria by mitochondrial stress. J Neurochem 2004;90(2):405–21.

141. Brand MD, Esteves TC. Physiological functions of the mitochondrial uncoupling proteins UCP2 and UCP3. Cell Metab 2005;2(2):85–93.

142. Hill KE, Motley AK, Li X, May JM, Burk RF. Combined selenium and vitamin E deficiency causes fatal myopathy in guinea pigs. J Nutr 2001;131(6):1798–802.

143. Morrow JD, Roberts LJ, 2nd. Mass spectrometric quantification of F_2-isoprostanes in biological fluids and tissues as measure of oxidant stress. Methods Enzymol 1999;300:3–12.

144. Morrow JD, Awad JA, Boss HJ, Blair IA, Roberts LJ, 2nd. Non-cyclooxygenase-derived prostanoids (F_2-isoprostanes) are formed in situ on phospholipids. Proc Natl Acad Sci U S A 1992;89(22):10721–5.

145. Montuschi P, Barnes PJ, Roberts LJ 2nd. Isoprostanes: markers and mediators of oxidative stress. FASEB J 2004;18(15):1791–800.

146. Morrow JD. Quantification of isoprostanes as indices of oxidant stress and the risk of atherosclerosis in humans. Arterioscler Thromb Vasc Biol 2004.

147. Awad JA, Morrow JD, Takahashi K, Roberts LJ, 2nd. Identification of non-cyclooxygenase-derived prostanoid (F_2-isoprostane) metabolites in human urine and plasma. J Biol Chem 1993;268(6):4161–9.

148. Fam SS, Morrow JD. The isoprostanes: unique products of arachidonic acid oxidation–a review. Curr Med Chem 2003;10(17):1723–40.
149. Roberts LJ 2nd, Montine TJ, Markesbery WR, et al. Formation of isoprostane-like compounds (neuroprostanes) in vivo from docosahexaenoic acid. J Biol Chem 1998;273(22):13605–12.
150. Cadet J, Delatour T, Douki T, et al. Hydroxyl radicals and DNA base damage. Mutat Res 1999;424(1–2):9–21.
151. Hoeijmakers JH. Genome maintenance mechanisms for preventing cancer. Nature 2001;411(6835):366–74.
152. Collins AR, Cadet J, Moller L, Poulsen HE, Vina J. Are we sure we know how to measure 8-oxo-7,8-dihydroguanine in DNA from human cells? Arch Biochem Biophys 2004;423(1):57–65.
153. Cadet J, Douki T, Frelon S, Sauvaigo S, Pouget JP, Ravanat JL. Assessment of oxidative base damage to isolated and cellular DNA by HPLC-MS/MS measurement. Free Radic Biol Med 2002;33(4):441–9.
154. Beckman KB, Ames BN. Endogenous oxidative damage of mtDNA. Mutat Res 1999;424(1–2):51–8.
155. Colón W, Wakem LP, Sherman F, Roder H. Identification of the predominant non-native histidine ligand in unfolded cytochrome c. Biochem 1997;36(41):12535–41.
156. Halliwell B. Why and how should we measure oxidative DNA damage in nutritional studies? How far have we come? Am J Clin Nutr 2000;72(5):1082–7.
157. Ravanat JL, Douki T, Duez P, et al. Cellular background level of 8-oxo-7,8-dihydro-2′-deoxy-guanosine: an isotope based method to evaluate artefactual oxidation of DNA during its extraction and subsequent work-up. Carcinogenesis 2002;23(11):1911–8.
158. Arif JM, Gupta RC. Artifactual formation of 8-oxo-2′-deoxyguanosine: role of fluorescent light and inhibitors. Oncol Rep 2003;10(6):2071–4.
159. Hamilton ML, Guo Z, Fuller CD, et al. A reliable assessment of 8-oxo-2-deoxyguanosine levels in nuclear and mitochondrial DNA using the sodium iodide method to isolate DNA. Nucleic Acids Res 2001;29(10):2117–26.
160. Hofer T, Moller L. Optimization of the workup procedure for the analysis of 8-oxo-7,8-dihydro-2′-deoxyguanosine with electrochemical detection. Chem Res Toxicol 2002;15(3):426–32.
161. Cadet J, D'Ham C, Douki T, Pouget JP, Ravanat JL, Sauvaigo S. Facts and artifacts in the measurement of oxidative base damage to DNA. Free Radic Res 1998;29(6):541–50.
162. Sauvaigo S, Petec-Calin C, Caillat S, Odin F, Cadet J. Comet assay coupled to repair enzymes for the detection of oxidative damage to DNA induced by low doses of gamma-radiation: use of YOYO-1, low-background slides, and optimized electrophoresis conditions. Anal Biochem 2002;303(1):107–9.
163. Pouget JP, Ravanat JL, Douki T, Richard MJ, Cadet J. Measurement of DNA base damage in cells exposed to low doses of gamma-radiation: comparison between the HPLC EC and comet assays. Int J Radiat Biol 1999;75(1):51–8.
164. Minowa O, Arai T, Hirano M, et al. Mmh/Ogg1 gene inactivation results in accumulation of 8-hydroxyguanine in mice. Proc Natl Acad Sci U S A 2000;97(8):4156–61.
165. Osterod M, Hollenbach S, Hengstler JG, Barnes DE, Lindahl T, Epe B. Age-related and tissue-specific accumulation of oxidative DNA base damage in 7,8-dihydro-8-oxoguanine-DNA glycosylase (Ogg1) deficient mice. Carcinogenesis 2001;22(9):1459–63.
166. Stadtman ER, Van Remmen H, Richardson A, Wehr NB, Levine RL. Methionine oxidation and aging. Biochim Biophys Acta 2005;1703(2):135–40.
167. Stadtman ER. Protein oxidation in aging and age-related diseases. Ann N Y Acad Sci 2001;928:22–38.
168. Requena JR, Levine RL, Stadtman ER. Recent advances in the analysis of oxidized proteins. Amino Acids 2003;25(3–4):221–6.
169. Schoneich C. Protein modification in aging: an update. Exp Gerontol 2006;41(9):807–12.
170. Berlett BS, Friguet B, Yim MB, Chock PB, Stadtman ER. Peroxynitrite-mediated nitration of tyrosine residues in Escherichia coli glutamine synthetase mimics adenylylation: relevance to signal transduction. Proc Natl Acad Sci U S A 1996;93(5):1776–80.

171. Quint P, Reutzel R, Mikulski R, McKenna R, Silverman DN. Crystal structure of nitrated human manganese superoxide dismutase: mechanism of inactivation. Free Radic Biol Med 2006;40(3):453–8.

172. Cruthirds DL, Novak L, Akhi KM, Sanders PW, Thompson JA, MacMillan-Crow LA. Mitochondrial targets of oxidative stress during renal ischemia/reperfusion. Arch Biochem Biophys 2003;412(1):27–33.

173. Levine RL, Berlett BS, Moskovitz J, Mosoni L, Stadtman ER. Methionine residues may protect proteins from critical oxidative damage. Mech Ageing Dev 1999;107(3):323–32.

174. Rhee SG, Jeong W, Chang TS, Woo HA. Sulfiredoxin, the cysteine sulfinic acid reductase specific to 2-Cys peroxiredoxin: its discovery, mechanism of action, and biological significance. Kidney Int Suppl 2007(106):S3–8.

175. Woo HA, Jeong W, Chang TS, et al. Reduction of cysteine sulfinic acid by sulfiredoxin is specific to 2-cys peroxiredoxins. J Biol Chem 2005;280(5):3125–8.

176. Chang TS, Jeong W, Woo HA, Lee SM, Park S, Rhee SG. Characterization of mammalian sulfiredoxin and its reactivation of hyperoxidized peroxiredoxin through reduction of cysteine sulfinic acid in the active site to cysteine. J Biol Chem 2004;279(49):50994–1001.

177. Stadtman ER, Berlett BS. Reactive oxygen-mediated protein oxidation in aging and disease. Drug Metab Rev 1998;30(2):225–43.

178. Chaudhuri AR, de Waal EM, Pierce A, Van Remmen H, Ward WF, Richardson A. Detection of protein carbonyls in aging liver tissue: a fluorescence-based proteomic approach. Mech Ageing Dev 2006;127(11):849–61.

179. Yan LJ, Sohal RS. Mitochondrial adenine nucleotide translocase is modified oxidatively during aging. Proc Natl Acad Sci U S A 1998;95(22):12896–901.

180. Yan LJ, Levine RL, Sohal RS. Oxidative damage during aging targets mitochondrial aconitase. Proc Natl Acad Sci U S A 1997;94(21):11168–72.

181. Cabiscol E, Levine RL. Carbonic anhydrase III. Oxidative modification in vivo and loss of phosphatase activity during aging. J Biol Chem 1995;270(24):14742–7.

182. Levine RL. Oxidative modification of glutamine synthetase. I. Inactivation is due to loss of one histidine residue. J Biol Chem 1983;258(19):11823–7.

183. Levine RL, Oliver CN, Fulks RM, Stadtman ER. Turnover of bacterial glutamine synthetase: oxidative inactivation precedes proteolysis. Proc Natl Acad Sci U S A 1981;78(4):2120–4.

184. Requena JR, Chao CC, Levine RL, Stadtman ER. Glutamic and aminoadipic semialdehydes are the main carbonyl products of metal-catalyzed oxidation of proteins. Proc Natl Acad Sci U S A 2001;98(1):69–74.

185. Pamplona R, Portero-Otin M, Requena J, Gredilla R, Barja G. Oxidative, glycoxidative and lipoxidative damage to rat heart mitochondrial proteins is lower after 4 months of caloric restriction than in age-matched controls. Mech Ageing Dev 2002;123(11):1437–46.

186. Fu MX, Requena JR, Jenkins AJ, Lyons TJ, Baynes JW, Thorpe SR. The advanced glycation end product, Nepsilon-(carboxymethyl)lysine, is a product of both lipid peroxidation and glycoxidation reactions. J Biol Chem 1996;271(17):9982–6.

187. Ahmed MU, Brinkmann Frye E, Degenhardt TP, Thorpe SR, Baynes JW. N-epsilon-(carboxyethyl)lysine, a product of the chemical modification of proteins by methylglyoxal, increases with age in human lens proteins. Biochem J 1997;324 (Pt 2):565–70.

188. Stark G. Functional consequences of oxidative membrane damage. J Membr Biol 2005;205(1):1–16.

189. Yokota T, Igarashi K, Uchihara T, et al. Delayed-onset ataxia in mice lacking alpha-tocopherol transfer protein: model for neuronal degeneration caused by chronic oxidative stress. Proc Natl Acad Sci U S A 2001;98(26):15185–90.

190. Katz ML, Stone WL, Dratz EA. Fluorescent pigment accumulation in retinal pigment epithelium of antioxidant-deficient rats. Invest Ophthalmol Vis Sci 1978;17(11):1049–58.

191. Yant LJ, Ran Q, Rao L, et al. The selenoprotein GPX4 is essential for mouse development and protects from radiation and oxidative damage insults. Free Radic Biol Med 2003;34(4):496–502.

192. Liu XF, Elashvili I, Gralla EB, Valentine JS, Lapinskas P, Culotta VC. Yeast lacking superoxide dismutase. Isolation of genetic suppressors. J Biol Chem 1992;267(26):18298–302.

193. Huang ME, Rio AG, Nicolas A, Kolodner RD. A genomewide screen in *Saccharomyces cerevisiae* for genes that suppress the accumulation of mutations. Proc Natl Acad Sci U S A 2003;100(20):11529–34.

194. Huang ME, Kolodner RD. A biological network in *Saccharomyces cerevisiae* prevents the deleterious effects of endogenous oxidative DNA damage. Mol Cell 2005;17(5):709–20.

195. Benov L, Fridovich I. The rate of adaptive mutagenesis in *Escherichia coli* is enhanced by oxygen (superoxide). Mutat Res 1996;357(1–2):231–6.

196. Pan X, Ye P, Yuan DS, Wang X, Bader JS, Boeke JD. A DNA integrity network in the yeast *Saccharomyces cerevisiae*. Cell 2006;124(5):1069–81.

197. Gardner PR, Fridovich I. Superoxide sensitivity of the *Escherichia coli* aconitase. J Biol Chem 1991;266(29):19328–33.

198. Chang EC, Kosman DJ. O_2-dependent methionine auxotrophy in Cu,Zn superoxide dismutase-deficient mutants of *Saccharomyces cerevisiae*. J Bacteriol 1990;172(4):1840–5.

199. Jensen LT, Sanchez RJ, Srinivasan C, Valentine JS, Culotta VC. Mutations in *Saccharomyces cerevisiae* iron-sulfur cluster assembly genes and oxidative stress relevant to Cu,Zn superoxide dismutase. J Biol Chem 2004;279(29):29938–43.

200. Wallace MA, Liou LL, Martins J, et al. Superoxide inhibits 4Fe-4S cluster enzymes involved in amino acid biosynthesis: cross-compartment protection by CuZn-superoxide dismutase. J Biol Chem 2004;279(31):32055–62.

201. Cleveland DW, Liu J. Oxidation versus aggregation–how do SOD1 mutants cause ALS? Nat Med 2000;6(12):1320–1.

202. Terman A, Brunk UT. Oxidative stress, accumulation of biological 'garbage', and aging. Antioxid Redox Signal 2006;8(1–2):197–204.

203. Terman A, Brunk UT. Lipofuscin: mechanisms of formation and increase with age. APMIS 1998;106(2):265–76.

204. Abou-Sleiman PM, Muqit MM, Wood NW. Expanding insights of mitochondrial dysfunction in Parkinson's disease. Nat Rev Neurosci 2006;7(3):207–19.

205. Squier TC. Oxidative stress and protein aggregation during biological aging. Exp Gerontol 2001;36(9):1539–50.

Section II
The Role of Oxidative Stress in Aging

Part II-A
Different Model Systems

3
Retrograde Response, Oxidative Stress, and Cellular Senescence

João F. Passos and Thomas von Zglinicki

Summary Mitochondria are major sources of reactive oxygen species (ROS) in cells, and they have been proposed to have an important role in cellular ageing. Recent evidence suggests that these organelles play a causal role in telomere-dependent senescence of human cells in culture. However, the relation between mitochondrial ROS production and replicative senescence might be more complicated than just simple cause and effect, because mitochondrial dysfunction has been shown to induce a variety of genes involved in nuclear-mitochondrial cross talk. Evidence suggests that this retrograde response might have an important, yet still relatively unexplored role in cell ageing, affecting mitochondrial function, ROS production, and, consequently, the life span of cells.

Keywords Senescence, telomeres, mitochondria, retrograde signalling, oxidative stress, fibroblasts.

1 Introduction

Growing evidence supports a role of oxidative stress in ageing. Because mitochondria are the main cellular site of production of reactive oxygen species (ROS), it was hypothesized >30 years ago that their dysfunction could be responsible for ageing, due to progressive accumulation of oxidative damage [1].

Historically, the role of mitochondria in the model of replicative senescence has been largely disregarded, possibly due to early studies that failed to observe significant differences in mitochondria between early and late passage cultured cells and a generalized notion that cellular division could dilute molecular damage. Moreover, the discovery that telomeres, the ends of chromosomes, shorten with cell division, suggested that senescence was the result of a counting mechanism, a deterministic biological clock, and as such, incompatible with ROS-generated random molecular damage.

Recently, the role of free radicals in the model of replicative senescence has generated substantial interest. Various studies have proposed that mitochondria

From: *Aging Medicine: Oxidative Stress in Aging: From Model Systems to Human Diseases*   39
Edited by: S. Miwa, K.B. Beckman, and F.L. Muller © Humana Press, Totowa, NJ

have a considerable impact on the replicative life span of cells. Moreover, it has been proposed that ROS (mainly from nonmitochondrial sources) also could contribute to the establishment and maintenance of the senescent phenotype, possibly acting as a tumor suppressor mechanism. Here, we analyze evidence for the role of ROS in the model of replicative senescence and the importance of mitochondria in this process.

2   Role of Mitochondrial ROS in Telomere-Dependent Senescence

Hayflick and Moorhead showed that embryo-derived fibroblasts can divide 50 ± 10 times before reaching replicative senescence when cultured in vitro [2]. The potential number of divisions became known as the "Hayflick limit." It was soon suggested that the shortening of telomeres counted the finite number of cell divisions and triggered replicative senescence in normal diploid cells [3, 4]. Later, this idea was confirmed experimentally [5] and supported by the observation that ectopic expression of the catalytic subunit of telomerase, an enzyme able to elongate telomeres, leads to immortalization of cells in culture [6]. The concept of a defined molecular mechanism that could count cell divisions, like beads on an abacus, led some enthusiasts to believe that this was evidence for an ageing clock. But soon, it was found that environmental stress, in particular, the level of ROS in a cell, also would affect the rate of telomere shortening [7, 8] and that individual cells could reach senescence much earlier than the intended "Hayflick limit" with surprisingly short telomeres [9]. Moreover, it was found that these early senescent cells possessed damaged mitochondria, producing high levels of ROS and that this affected telomeres directly [10].

Is there evidence supporting a causal role for mitochondria in replicative senescence? Several studies showed that cell senescence was associated with high levels of endogenous ROS [11, 12] and with accumulation of oxidation products, such as protein carbonyls and lipofuscin [13, 14]. It also was shown that senescent fibroblasts have impaired metabolism, with strong reduction of ATP and other nucleotide triphosphates [15].

Recently, we showed that senescent cells display mitochondrial dysfunction, characterized by lower mitochondrial membrane potential, mitochondrial DNA (mtDNA) damage, and increased superoxide production [10]. Clearly, this evidence is merely correlative and cannot indicate causality. However, several studies suggest a causal role for mitochondrial ROS production in replicative senescence.

First, improvement of mitochondrial function by diverse interventions has been shown to extend life span: Selective targeting of antioxidants directly to the mitochondria counteracted telomere shortening and increased life span in fibroblasts under mild oxidative stress [16]. Continuous treatment with nicotinamide, which changes mitochondrial function and reduces ROS generation, has been reported to extend life span by decelerating telomere shortening [17]. Also, pharmacologic mild chronic uncoupling of mitochondria that reduced the production of superoxide anion improved telomere maintenance and extended telomere-dependent life span [10]. A similar effect of mild uncoupling has been observed in yeast *Saccharomyces cerevisiae* [18].

Second, damage to the mitochondria has been shown to decrease the replicative life span of cells. Mitochondrial dysfunction generated by severe mitochondrial depolarization by using an uncoupling agent (carbonyl cyanide p-trifluoromethoxy-phenylhydrazone [FCCP]) led to an increased production of ROS, telomere attrition, telomere loss, and chromosome fusion in mouse embryos [19]. Also, mild oxidative stress by exposure to hyperoxia decreases the replicative life span of cells, and it accelerates telomere shortening [8, 20]. There is evidence supporting that oxidative stress generated by hyperoxia is dependent on use of oxygen by the mitochondria [21]. Moreover, it is known that mitochondrial function is severely affected with replicative senescence.

Thus, from these data, mitochondrial ROS do play, at least partially, a causal role in telomere-dependent replicative senescence. Nevertheless, it is still unclear what the trigger is for mitochondrial ROS generation. According to the mitochondrial theory of ageing, first proposed by Harman in 1972, damage to the mitochondrial DNA would be the main cause for free radical production [1]. However, new evidence suggests that this might not be as simple as initially formulated.

Recent work has shown that by creating homozygous knock-in mice that express a proofreading-deficient version of the nucleus-encoded catalytic subunit of mtDNA polymerase γ (PolgA), an extremely high level of mtDNA mutations and deletions could be attained. These knock-in mice showed a significant decrease in life span, apparently supporting Harman's theory [22]. However, the mechanism of this phenomenon remains unknown, because no increased markers of oxidative stress or defects in cellular proliferation were found in these animals [23, 24]. Importantly, mice that are heterozygous for PolgA function show no significant reduction in life span despite an mtDNA mutation burden 30 times higher than in old wild-type animals [25]. These studies suggest that mtDNA mutation load does not limit life span of wild-type mice and that mtDNA mutations, even at very high levels, do not necessarily lead to increased mitochondrial ROS generation. Mitochondrial dysfunction and generation of ROS might be consequences of other still unexplored factors occurring with cell ageing.

Other recent studies have begun to explore uncharted territories in the relation between ageing and mitochondrial biology, way beyond the simple ROS-leading to damage-leading to ROS cycle. Recently, research has been focused upon the impact of nuclear genes and their effect on mitochondrial function with respect to the ageing process. Caloric restriction, which is able to extend life span in various species, has been related to increased mitochondrial biogenesis [26]. Resveratrol, a sirtuin activator, has been shown to lead to increased mitochondrial biogenesis via activation of the nuclear transcription factor PGC-1α and to extension of life span [27]. It also has been shown that PGC-1α expression is required for induction of antioxidant enzymes, such as that glutathione peroxidase 1 and superoxide dismutase 2 [28].

Changes in mitochondrial function and biogenesis due to alterations in nuclear gene expression have been first described in budding yeast and named retrograde response. In Section 3, we review evidence relating retrograde response to replicative senescence.

3   Is Retrograde Signaling Part of a Senescence Signature? Does It Have a Causal Role?

Respiratory dysfunctional yeast cells compensate for mitochondrial dysfunction by up-regulation of a defined set of nuclear genes through activation of the transcription factors Rtg1p–Rtg3p. Microarray studies have shown that these cells increase transcription of genes coding for glycolytic enzymes, peroxisomal biogenesis genes, glyoxylate pathway genes, and genes involved in anaplerotic pathways [29, 30]. Also, these pathways are activated as a consequence of mitochondrial dysfunction in normal yeast replicative ageing [31].

In yeast respiratory deficient cells, the induction of the retrograde response compensates for a nonfunctional tricarboxylic acid (TCA) cycle. Because succinate cannot be oxidized to fumarate, the cell responds by inducing production of oxaloacetate and acetyl-CoA by anaplerotic pathways, i.e., alternative pathways to replenish TCA intermediates. In yeast, it has been suggested that the main purpose of these metabolic rearrangements is the maintenance of glutamate supplies.

Mitochondrial retrograde signaling in mammalian cells was initially described in C2C12 skeletal myoblasts [32] and in human lung carcinoma A549 cells [33]. The main trigger for this response in mammalian cells seems to be the decline of mitochondrial membrane potential. Experimentally, retrograde response was induced by either treatment of cells with ethidium bromide to deplete them of mtDNA or by using mitochondria-specific ionophores, such as carbonyl cyanide m-chlorophenylhydrazone. Because mitochondria also function as major calcium stores, a drop in mitochondrial membrane potential leads to calcium release from the mitochondria (and compromised calcium uptake), increasing cytosolic free Ca^{2+}. Calcium seems to be the link between mitochondrial function and nuclear gene expression, as several studies have shown.

In both C2C12 skeletal myoblasts and human lung carcinoma cells, mitochondrial dysfunction caused increased expression of a number of genes involved in calcium homeostasis, such as ryanodine receptor I or II (RyR1 or RyR2), calcineurin, calreticulin, and calsequestrin [32, 33]. Arnould et al. [34] have shown that increase of cytosolic free Ca^{2+} induced by mitochondrial dysfunction was responsible for activation of Ca^{2+}/calmodulin-dependent protein kinase type IV, and consequent activation of cAMP response element-binding protein by phosphorylation (Table 3.1).

There is uncertainty concerning the initial trigger of retrograde signaling. It is known that production of superoxide anion in the inner mitochondrial membrane can activate uncoupling proteins [35]. Moreover, cellular oxidative stress can

Table 3.1 List of genes proteins, or both involved in metabolism, Ca^{2+}-dependent signaling, mitochondrial function and assembly, and stress response and apoptosis, with changed expression after mitochondrial stress induced by depletion of mtDNA genome [Rho(0)], metabolic inhibitors, senescence, and hyperoxia (40% O_2) in different mammalian cell lines. Names of genes are written as found in respective reference

Function	Gene	Method	How mitochondrial stress was induced	Cell type	Reference
Metabolism	HK1	RT-PCR	Rho (0)	T143B, ARPE19, and GMO6225	[44]
	GLUT1				
	PCK2				
	GCK				
	LDHA				
	LDHB				
	FBP1				
	GAPD				
	ENO1				
	PKM2				
	PC				
	CS				
	CKB				
	ACAT2				
	ACACA				
	ACACB				
	PEX6				
	DIA1				
	Glut4	NB	Rho (0)	C2C12 rhabdomyoblasts and A549 human lung carcinoma cells	[42]
	PEP carboxy kinase				
	Hexokinase				
	ACSL5				
	PTE1	MA	Senescence/hyperoxia	MRC-5 fibroblasts	[10]
	GLS				
	HSD17B12				
	GALNT10				
	AK3				
	HK2				
	LOX				
	PDP2				
	RODH				
	OGDH				
	ALDH1A1				
	PDK4				
	FABP4				
	ATP8B1				
	GLS				
	GCLM				

(continued)

Table 3.1 (continued)

Function	Gene	Method	How mitochondrial stress was induced	Cell type	Reference
	HBG1				
	HSD17B2				
	CPT1A				
	ASK				
	SCD				
	GLUL				
	GRIA4				
	CYP1B1				
	FECH				
	TKT				
Ca-dependent signaling	RYR-1	RT-PCR	Rho (0)	T143B, ARPE19, and GMO6242	[44]
	ikB	WB	Rho (0) + mitochondrial metabolic inhibitors	Mouse C2C12 myocytes	[32]
	Cn A				
	RelA				
	NFATc				
	ATF2				
	RyR-2				
	Cathepsin-L	WB	Rho (0)	Human pulmonary carcinoma A549 cells	[33]
	TGFβ1				
	Calreticulin				
	Calsequestrin				
	Calcineurin				
	Egr-1				
	ATF2				
	NFAT				
	Phospho-p44 mapk				
	Phospho-p42 mapk				
	ERK1				
	ERK2	WB	FCCP	Rat pheochromacytoma PC12 cells	[49]
	CamkIV				
	CREB	WB	Rho (0)	143B human osteosarcoma cell line and MERRF cybrid cell line mutated for tRNA(Lys) (A8344G)	[34]
	RyR1				

(continued)

Table 3.1 (continued)

Calsequestrin	NB	Rho (0)	C2C12 rhabdomy-oblasts and A549 human lung carci-noma cells	[42]
Calreticulin				
Cathepsin				
TGFβ1				
Epiregulin				
ATP2C1				
MAP2K3	MA	Senescence	MRC-5 fibroblasts	[10]
PLCB4				
CABYR				
TGFB2				
PLCB4				
IGFBP3				
CABYR				
PRKCA				
BAMBI				
VAPNS2				
CAP2				
PRKCH				
CLECSF2				
CAV2				
CREB3L1				
TGFB2				
CALM2				
EREG				
DCG2				
GHR				
CEBPE				
GUCA1A				
IGFBP3				
SULF1				
TGFBP2				
PDE4DIP				
ADCY3				
TGIF2				
GPSM2				
CREB3L4				
PDE1A				
ARHGAP11A				
GUCY1B3				
ITPKB				
PDE1A				
LTBP4				
THBS2				
NFAT5				
MAP3K8				

(continued)

Table 3.1 (continued)

Function	Gene	Method	How mitochondrial stress was induced	Cell type	Reference
	S100A4				
	ADCY3				
	NFYB				
	IQGAP3				
	RICS				
	PCBP2				
	ARHGAP19				
	CREBBP				
	PDE1A				
	THBS1				
	MAP3K1				
	AURKB				
	RIN3				
	IGF1				
	TGFBR3				
	IGF1				
Mitochondrial function and assembly	MDH1	RT-PCR	Rho (0)	T143B, ARPE19, and GMO6225	[44]
	MDH2				
	SDHA				
	COXVB				
	ATP5B				
	ANT1				
	ANT2				
	CKMT1				
	CKMT2				
	CYTC				
	COXVB	WB	Rho (0) + mitochondrial metabolic inhibitors	Mouse C2C12 myocytes	[32]
	VDAC	NB	Rho (0)	C2C12 rhabdomyoblasts and A549 human lung carcinoma cells	[42]
	6 MtDNA transcript				
	Tim44				
	Tom40				
	CPS1				
	NRF1	MA	Senescence	MRC-5 fibroblasts	[10]
	UAP1L1				

(continued)

Table 3.1 (continued)

	MIPEP				
	MTSUS1				
	MAOA				
	SLC5A3				
	UCP2				
	PHB				
Stress response and Apoptosis	SCARB1	RT-PCR	Rho (0)	T143B, ARPE19, and GMO6225	[44]
	SCARB2				
	TXN				
	TXNR1				
	TXNR2				
	SOD2				
	CAT				
	GPX3				
	HSF1				
	HSPD1				
	HSPE1				
	Bcl-2				
	Bid	WB		Human pulmonary carcinoma A549 cells	[33]
	Bax				
	Bcl-XL				
	Bcl2				
	Bid	NB		C2C12 rhabdomy-oblasts and A549 human lung carci-noma cells	[42]
	Bax				
	P53				
	Bcl-2L1				
	PAWR	MA	Senescence	MRC-5 fibroblasts	[44]
	SOD2				
	Bcl-2L11				
	CCAR1				
	SOD2				
	Bcl-X	WB	Rho (0) + mitochon-drial metabolic inhibitors	C2C12 skeletal myoblasts	[43]
	Survivin				

WB, Western blot; NB, Northern blot; MA: microarrays; RT-PCR, reverse transcriptase-polymerase chain reaction.

transcriptionally induce expression of uncoupling protein (UCP)-2 and UCP-3 [10, 28], resulting in a decline in mitochondrial membrane potential [10]. Thus, it is possible that the main trigger of retrograde response is ROS generation, leading to expression and activation of UCPs, and decline of mitochondrial membrane potential and calcium release.

Transcriptional activation of the expression of nuclear encoded mitochondrial components as adaptation to perceived mitochondrial dysfunction seems to be a major function of retrograde signalling. Partial depletion of mtDNA and treatment with mitochondrial metabolic inhibitors has been shown to result in enhanced expression of various nuclear genes involved in control of mitochondrial inner membrane components [32], and uncoupling of mitochondria from HeLa cells has been reported to increase oxygen consumption and increase the protein levels of δ-aminolevulinic acid synthase, an early marker for mitochondrial biogenesis [36]. Nuclear respiratory NRF-1, a transcription factor that has been linked to the transcriptional control of many genes involved in mitochondrial function and biogenesis, has been shown to be up-regulated in HeLa cells depleted of mtDNA [37]. Also, both NRF-1 and mitochondrial transcription factor A, mtDNA-binding protein essential for maintenance, replication, and transcription of mtDNA, are up-regulated in response to mitochondrial lipopolysaccharide-induced damage [38].

Various lines of evidence suggest that retrograde signalling is also a feature of replicative senescence.

First, senescent cells display clear mitochondrial dysfunction, i.e., increased superoxide production despite low mitochondrial membrane potential, together with mtDNA damage [10].

Second, it has been described that when cells reach the end of their replicative life span, they suffer changes in calcium homeostasis. Senescent human skin and lung fibroblasts have been shown to have higher cytosolic calcium concentrations and lower capacity to store calcium than fibroblasts in early passage [10, 39]. This is consistent with evidence showing that impaired mitochondrial metabolism [15] and lower mitochondrial membrane potential due to expression of UCP-2 in senescent fibroblasts [10].

Third, increased glutamate production has been found in senescent fibroblasts together with significant changes in glucose metabolism, including increased activity of the glycolitic pathway [15]. These changes are highly reminiscent of what has been shown in yeast retrograde response.

Fourth, senescent fibroblasts also show increased mitochondrial mass and mtDNA copy number [10, 40], which could be interpreted as an adaptive response to mitochondrial dysfunction.

Finally, apoptosis is another feature linking retrograde signaling and senescence. Senescent fibroblasts have been shown to be resistant to apoptosis, and this has been related to expression of antiapoptotic protein Bcl-2 [41]. mtDNA-depleted cells have been shown to be significantly more resistant to etoposide-mediated apoptosis than control cells, consistent with an increase in Bcl-2 expression [42, 43].

In conclusion, it seems likely that retrograde response is activated in senescent cells. However, studies of retrograde senescence in mammalian cells have not yet yielded an actual retrograde response signature in terms of gene expression pattern.

In mammalian cells, gene expression in response to depletion of mitochondrial DNA or metabolic inhibitors is largely cell type specific, as several studies have demonstrated [42, 44–46] (Table 3.1). Furthermore, the processes by which mitochondrial dysfunction is induced also might modify gene expression patterns. In fact, a gene expression profiling study of senescent fibroblasts did identify increased expression of genes involved Ca^{2+} binding and Ca^{2+}-mediated signaling, glycolysis, and Krebs cycle enzymes; mitochondrial biogenesis and function; and stress response/apoptosis that thus might be regarded as candidate signature genes for retrograde response in fibroblast senescence [10], even if they show essentially no overlap with candidates found in other systems at the single gene level (Table 3.1).

One question remains, What is the functional consequence of retrograde response in replicative senescence? Does it extend the doubling potential of cells? In the budding yeast *S. cerevisiae*, early activation of retrograde response led to extension of replicative life span [47]. However, retrograde response also has been observed to occur progressively with age in yeast, accompanied by a decline in mitochondrial function. Therefore, it is possible that with age, the beneficial effects of retrograde response are not able to counteract the progressive deterioration of cells. There are other lines of evidence supporting this hypothesis in another model organism. Directed RNA interference inactivation of components of the mitochondrial electron transport chain early in development have been shown to increase longevity in *Caenorhabditis elegans* [48]. However, no effect is observed if mitochondrial dysfunction is induced in adult animals.

In mammalian systems, there are few available data concerning the effects of retrograde response on longevity. Mitochondrial dysfunction induced by a high dose of FCCP has been shown to decrease cell survival in mouse embryos [19]. Mitochondrial dysfunction and retrograde response accompany replicative senescence in human fibroblasts [10]. However, mild uncoupling by dinitrophenol (DNP) has been shown to extend replicative life span in human fibroblasts [10]. Although this seems superficially similar to the situation in yeast, it needs to be stressed that in this latter study, there is no evidence that mild uncoupling led to activation of retrograde signaling. In fact, long-term exposure to DNP resulted in a lower mitochondrial mass than controls when cells were approaching the end of their replicative life span. Increased mitochondrial mass has been described as a marker of induction of retrograde response in both yeast and mammalian cells [31, 32].

4 Conclusions

Even though recent data have questioned whether mtDNA mutations do have functional consequences in the ageing process, mitochondria do play an important role in ageing, which surpasses the role of their genome-encoded components. The ability to store calcium and its dependence on mitochondrial membrane potential, the role of uncoupling proteins, and mitochondrial biogenesis are starting to be recognized as important players in both cellular and organismal ageing. In cellular senescence, it is now evident that several processes involved in the response to mitochondrial dysfunction and differential expression of several genes may impact on replicative longevity, both in budding yeast and in human primary cells.

References

1. Harman D. The biologic clock: the mitochondria? J Am Geriatr Soc 1972;20(4):145–7.
2. Hayflick L, Moorhead PS. The serial cultivation of human diploid cell strains. Exp Cell Res 1961;25:585–621.
3. Olovnikov AM. Principle of marginotomy in template synthesis of polynucleotides. Dokl Akad Nauk SSSR 1971;201(6):1496–9.
4. Watson JD. Origin of concatemeric T7 DNA. Nat New Biol 1972;239(94):197–201.
5. Harley CB, Futcher AB, Greider CW. Telomeres shorten during ageing of human fibroblasts. Nature 1990;345(6274):458–60.
6. Bodnar AG, Ouellette M, Frolkis M, et al. Extension of life-span by introduction of telomerase into normal human cells. Science 1998;279(5349):349–52.
7. von Zglinicki T. Oxidative stress shortens telomeres. Trends Biochem Sci 2002;27(7):339–44.
8. von Zglinicki T, Saretzki G, Docke W, Lotze C. Mild hyperoxia shortens telomeres and inhibits proliferation of fibroblasts: a model for senescence? Exp Cell Res 1995;220(1):186–93.
9. Martin-Ruiz C, Saretzki G, Petrie J et al. Stochastic variation in telomere shortening rate causes heterogeneity of human fibroblast replicative life span. J Biol Chem 2004;279(17):17826–33.
10. Passos JF, Saretzki G, Ahmed S, et al. Mitochondrial dysfunction accounts for the stochastic heterogeneity in telomere-dependent senescence. PloS Biol 2007;5(5):e110.
11. Hutter E, Unterluggauer H, Uberall F, Schramek H, Jansen-Durr P. Replicative senescence of human fibroblasts: the role of Ras-dependent signaling and oxidative stress. Exp Gerontol 2002;37(10–11):1165–74.
12. Allen RG, Tresini M, Keogh BP, Doggett DL, Cristofalo VJ. Differences in electron transport potential, antioxidant defenses, and oxidant generation in young and senescent fetal lung fibroblasts (WI-38). J Cell Physiol 1999;180(1):114–22.
13. Sitte N, Merker K, von Zglinicki T, Grune T. Protein oxidation and degradation during proliferative senescence of human MRC-5 fibroblasts. Free Radic Biol Med 2000;28(5):701–8.
14. Sitte N, Merker K, Grune T, von Zglinicki T. Lipofuscin accumulation in proliferating fibroblasts in vitro: an indicator of oxidative stress. Exp Gerontol 2001;36(3):475–86.
15. Zwerschke W, Mazurek S, Stockl P, Hutter E, Eigenbrodt E, Jansen-Durr P. Metabolic analysis of senescent human fibroblasts reveals a role for AMP in cellular senescence. Biochem J 2003;376(Pt 2):403–11.
16. Saretzki G, Murphy MP, von Zglinicki T. MitoQ counteracts telomere shortening and elongates lifespan of fibroblasts under mild oxidative stress. Aging Cell 2003;2(2):141–3.
17. Kang HT, Lee HI, Hwang ES. Nicotinamide extends replicative lifespan of human cells. Aging Cell 2006;5(5):423–36.
18. Barros MH, Bandy B, Tahara EB, Kowaltowski AJ. Higher respiratory activity decreases mitochondrial reactive oxygen release and increases life span in *Saccharomyces cerevisiae*. J Biol Chem 2004;279(48):49883–8.
19. Liu L, Trimarchi JR, Smith PJ, Keefe DL. Mitochondrial dysfunction leads to telomere attrition and genomic instability. Aging Cell 2002;1(1):40–6.
20. Balin AK, Goodman DB, Rasmussen H, Cristofalo VJ. The effect of oxygen and vitamin E on the lifespan of human diploid cells in vitro. J Cell Biol 1977;74(1):58–67.
21. Li J, Gao X, Qian M, Eaton JW. Mitochondrial metabolism underlies hyperoxic cell damage. Free Radic Biol Med 2004;36(11):1460–70.

22. Trifunovic A, Wredenberg A, Falkenberg M, et al. Premature ageing in mice expressing defective mitochondrial DNA polymerase. Nature 2004;429(6990):417–23.
23. Kujoth GC, Hiona A, Pugh TD et al. Mitochondrial DNA mutations, oxidative stress, and apoptosis in mammalian aging. Science 2005;309(5733):481–4.
24. Trifunovic A, Hansson A, Wredenberg A et al. From the cover: somatic mtDNA mutations cause aging phenotypes without affecting reactive oxygen species production. Proc Natl Acad Sci U S A 2005;102(50):17993–8.
25. Vermulst M, Bielas JH, Kujoth GC, et al. Mitochondrial point mutations do not limit the natural lifespan of mice. Nat Genet 2007;39:540–3.
26. López-Lluch G, Hunt N, Jones B, et al. Calorie restriction induces mitochondrial biogenesis and bioenergetic efficiency. Proc Natl Acad Sci U S A 2006;103(6):1768–73.
27. Baur JA, Pearson KJ, Price NL, et al. Resveratrol improves health and survival of mice on a high-calorie diet. Nature 2006;444(7117):337–42.
28. St-Pierre J, Drori S, Uldry M, et al. Suppression of Reactive oxygen species and Neurodegeneration by the PGC-1 transcriptional coactivators. Cell 2006;127(2):397–408.
29. Epstein CB, Waddle JA, Hale W IV et al. Genome-wide responses to mitochondrial dysfunction. Mol Biol Cell 2001;12(2):297–308.
30. Traven A, Wong JMS, Xu D, Sopta M, Ingles CJ. Interorganellar Communication. Altered nuclear gene expression in a yeast mtDNA mutant. J Biol Chem 2001;276(6):4020–7.
31. Lai C-Y, Jaruga E, Borghouts C, Jazwinski SM. A mutation in the ATP2 gene abrogates the age asymmetry between mother and daughter cells of the yeast *Saccharomyces cerevisiae*. Genetics 2002;162(1):73–87.
32. Biswas G, Adebanjo OA, Freedman BD, et al. Retrograde Ca2+ signaling in C2C12 skeletal myocytes in response to mitochondrial genetic and metabolic stress: a novel mode of interorganelle crosstalk. EMBO J 1999;18(3):522–33.
33. Amuthan G, Biswas G, Ananadatheerthavarada HK, Vijayasarathy C, Shephard HM, Avadhani NG. Mitochondrial stress-induced calcium signaling, phenotypic changes and invasive behavior in human lung carcinoma A549 cells. Oncogene 2002;21(51):7839–49.
34. Arnould T, Vankoningsloo S, Renard P, et al. CREB activation induced by mitochondrial dysfunction is a new signaling pathway that impairs cell proliferation. EMBO J 2002;21:53–63.
35. Echtay KS, Roussel D, St-Pierre J et al. Superoxide activates mitochondrial uncoupling proteins. Nature 2002;415(6867):96–9.
36. Li B, Holloszy JO, Semenkovich CF. Respiratory uncoupling induces delta-aminolevulinate synthase expression through a nuclear respiratory factor-1-dependent mechanism in HeLa cells. J Biol Chem 1999;274(25):17534–40.
37. Miranda S, Foncea R, Guerrero J, Leighton F. Oxidative stress and upregulation of mitochondrial biogenesis genes in mitochondrial DNA-depleted HeLa cells. Biochemical and Biophysical Research Communications 1999;258(1):44–9.
38. Suliman HB, Carraway MS, Welty-Wolf KE, Whorton AR, Piantadosi CA. Lipopolysaccharide Stimulates mitochondrial biogenesis via activation of nuclear respiratory factor-1. J Biol Chem 2003;278(42):41510–8.
39. Papazafiri P, Kletsas D. Developmental and age-related alterations of calcium homeostasis in human fibroblasts. Exp Gerontol 2003;38(3):307–11.
40. Lee HC, Yin PH, Chi CW, Wei YH. Increase in mitochondrial mass in human fibroblasts under oxidative stress and during replicative cell senescence. J Biomed Sci 2002;9(6 Pt 1):517–26.
41. Wang E. Senescent human fibroblasts resist programmed cell death, and failure to suppress bcl 2 is involved. Cancer Res 1995;55(11):2284–92.
42. Biswas G, Guha M, Avadhani NG. Mitochondria-to-nucleus stress signaling in mammalian cells: nature of nuclear gene targets, transcription regulation, and induced resistance to apoptosis. Gene 2005;354:132–9.
43. Biswas G, Anandatheerthavarada HK, Avadhani NG. Mechanism of mitochondrial stress-induced resistance to apoptosis in mitochondrial DNA-depleted C2C12 myocytes. Cell Death Differ 2005;12(3):266–78.

44. Miceli MV, Jazwinski SM. Common and cell type-specific responses of human cells to mito-
    chondrial dysfunction. Exp Cell Res 2005;302(2):270–80.
45. Miceli MV, Jazwinski SM. Nuclear gene expression changes due to mitochondrial dysfunction
    in ARPE-19 cells: implications for age-related macular degeneration. Invest Ophthalmol Vis
    Sci 2005;46(5):1765–73.
46. Wang H, Morais R. Up-regulation of nuclear genes in response to inhibition of mitochondrial
    DNA expression in chicken cells. Biochim Biophys Acta 1997;1352(3):325–34.
47. Kirchman PA, Kim S, Lai C-Y, Jazwinski SM. Interorganelle Signaling is a determinant of
    longevity in *Saccharomyces cerevisiae*. Genetics 1999;152(1):179–90.
48. Dillin A, Hsu A-L, Arantes-Oliveira N, et al. Rates of behavior and aging specified by mito-
    chondrial function during development. Science 2002;298(5602):2398–401.
49. Luo Y, Bond JD, Ingram VM. Compromised mitochondrial function leads to increased
    cytosolic calcium and to activation of MAP kinases. Proc Natl Acad Sci U S A 1997;
    94:9705–9710.

4
Reactive Oxygen Species in Molecular Pathways Controlling Aging in the Filamentous Fungus *Podospora anserina*

Heinz D. Osiewacz and Christian Q. Scheckhuber

Summary To generate ATP via different types of respiration, the impact of reactive oxygen species (ROS) as by-products generated at the inner mitochondrial membrane is well analyzed and documented. Moreover, other pathways of ROS generation seem to be of relevance, but they are currently less explored. It now seems that ROS not only play a key role in the age-related damaging of biomolecules that are essential for proper cellular function but also in age-related signaling pathways. In particular, pathways related to apoptosis that finally bring life of senescent *P. anserina* cultures to an end are currently emerging to play a crucial role. Moreover, cellular defense and maintenance functions were demonstrated to effectively modulate life span. Among these functions, ROS scavenging, the tight control of cellular copper levels, retrograde responses, and mechanisms maintaining a population of functional mitochondria are of crucial importance. The elucidation of additional candidate pathways is a challenge for future research to reveal the mechanisms governing aging and life span control in a holistic view.

Keywords Reactive oxygen species (ROS), mitochondria, respiration, retrograde response, apoptosis, copper homeostasis.

1   Introduction: Senescence in *Podospora anserina*

Podospora anserina is a filamentous fungus extensively studied as an aging model at the organismic level [1–4]. In contrast to many other fungi that seem to be immortal, the life span of *P. anserina* is limited. A given strain is characterized by a specific time in which the filamentous cells of a culture, the hyphae, grow at their tips and form a culture termed a "mycelium," which represents a macroscopically visible individium. During aging, the morphology of the mycelium changes (Fig. 4.1), the growth rate slows down, and finally growth ceases and the colony dies at the tips [5, 6]. Mean life span of defined strains is also dependent on environmental factors. For example, when the wild-type strain "s", a strain most thoroughly investigated by different research groups, is grown on rich

From: *Aging Medicine: Oxidative Stress in Aging: From Model Systems to Human Diseases*   53
Edited by: S. Miwa, K.B. Beckman, and F.L. Muller © Humana Press, Totowa, NJ

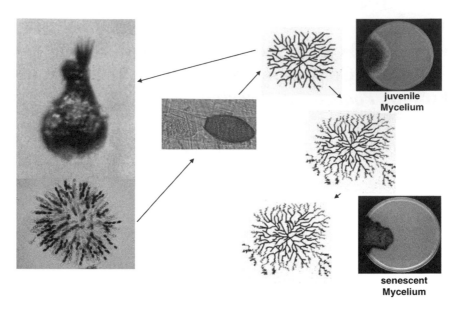

Fig. 4.1 Propagation and senescence of the filamentous ascomycete *P. anserina*. (*Left*) Fruiting bodies, termed perithecia, contain the products of sexual reproduction, the ascospores. After germination of an ascospore (*middle*) a mycelium is formed. Juvenile mycelia are characterized by aerial hyphae and regular growth. In the senescent stage, pigmentation increases and formation of aerial hyphae is strongly reduced. Mycelial tips swell and burst. Eventually, the mycelium stops growth completely and dies (*right*)

medium (e.g., cornmeal medium), the mean life span is 25 days, whereas on PASM, a specific synthetic medium, it is much shorter (14 days). Life spans of different wild-type isolates differ from each other, demonstrating that there is a genetic basis of aging, which is further demonstrated by a modification of the strain-specific life span of a given wild-type strain by mutations. Both mutations resulting in a life span reduction and an extension are known. In fact, when compared with animal models such *Caenorhabditis elegans* or *Drosophila melanogaster* in which life span extension is mainly rather moderate, some mutants of *P. anserina* were demonstrated to display a spectacular life span extension of several 100% to several 1,000% compared with the wild-type. Some mutants even seem to have acquired immortality (Table. 4.1). The characterization of such long-lived mutants proved to be very helpful to elucidate the molecular network governing aging and life span in this experimentally tractable aging model, and it provided data that can be translated, although not "one-to-one," to higher systems, including humans. Such analyses can efficiently be performed, because the fungus is easy to cultivate under laboratory conditions, and it is well accessible to genetic and molecular techniques. In the past several years, the availability of the complete

Table 4.1 Identified longevity genes and extrachromosomal mutants of the filamentous ascomycete *P. anserina*. "mat" denotes the mating type of homokaryotic isolates. BMM is a complex medium, whereas M2 and PASM are synthetic media. The mean life span of wild-type isolates on the corresponding type of medium was set to 100%

Longevity genes					
Nuclear gene	Condition	Pathway	Mean life span	Medium	Reference
PaAmid1	Knockout	Apoptosis	+59%	BMM	[49]
PaCox5	Knockout	PaCOX assembly	> +3,000%	M2	[30]
PaCox17	Knockout	PaCOX assembly	> +1,180%	BMM	[29]
PaCyc1	Mutation	Respiratory chain	> +2200%	M2	[50]
PaDnm1	Knockout	Mitochondrial division	+900%	BMM	[41]
PaeEF1A	Mutation	Translation	+218 % [mat +]	M2	[51]
PaGrisea	Mutation	Copper homeostasis	+60%	BMM	[52]
PaMca1	Knockout	Apoptosis	+148%	BMM	[49]
PaMca2	Knockout	Apoptosis	+78%	BMM	[49]
PaMth1	Overexpression	Unknown	+115%	PASM	[53]
PaTom70	Mutation	Protein import into mitochondria	> +2,700% [mat −]	M2	[54]
Extrachromosomal mutants					
Mutant					
AL2-1	Insertion of pAL2-1 into mtDNA	Respiratory chain	+1,360%	BMM	[55]
ex1	*PaCox1* deletion (complete)	Respiratory chain	Immortal	M2	[18]
ex2	*PaCox1* deletion (complete)	Respiratory chain	Immortal	M2	[31]
mex1	Deletion in *PaCox1* (first exon-intron transition)	Respiratory chain	Immortal	M2	[32]
mex5	Deletion in *PaCox1* (first exon-intron transition)	Respiratory chain	Immortal	M2	[32]
mex7	Deletion in *PaCox1* (first exon-intron transition)	Respiratory chain	Immortal	M2	[32]
mex16	Deletion in *PaCox1* (first exon-intron transition)	Respiratory chain	Immortal	M2	[33]
mid26	*PaCox1* deletion (pl-Intron)	Respiratory chain	+100%	M2	[56]
Wa32-LL	Impairment of *PaNd2* and *PaNd3* expression	Respiratory chain	> +750%	M2	[57]

genomic sequence and the establishment of effective techniques to construct transgenic and "knockout" strains in a reasonable time [7] provided important prerequisites for an efficient experimental use of the system.

Here, we summarize data related to several aspects of aging in *P. anserina* that are specifically linked to a role of reactive oxygen species (ROS). We provide a view that is emerging from recent investigations. There are several reviews [1-4, 8–11] that provide a summary of earlier investigations about this experimental aging model.

2   Mitochondrial DNA Instabilities

One of the first important molecular pathways affecting the life span of *P. anserina* was reported in the late 1970s and subsequently studied in more detail in the early 1980s. At that time, it was found that aging of all cultures isolated from the wild from different locations in Europe was characterized by instabilities of the standard mitochondrial DNA (mtDNA). One of the early hallmarks in aging cultures was found to be the accumulation of a circular DNA molecule, which, due to its physical characteristics was termed "plasmid-like" DNA (plDNA) [12] and due to the accumulation in senescent cultures as αsenDNA [13]. This autonomous DNA element was demonstrated to be derived from the first intron (pl-intron or intron α) of the gene coding for cytochrome *c* oxidase subunit I (COXI) [14]. In juvenile cultures, this piece of DNA is predominantly integrated into the *CoxI* gene. During aging, it becomes liberated and amplified [12, 15, 16]. It was shown that plDNA causes mtDNA instabilities due to its ability to reintegrate into the mtDNA. This leads to plDNA duplications, allowing efficient DNA recombination and rearrangement [17]. In senescent cultures, the standard mtDNA is almost quantitatively rearranged [16, 18]. As a consequence, the expression of mtDNA-encoded genes, needed for turnover of gene products is progressively impaired. Although for a long time it was believed that aging of *P. anserina* is caused by the age-related liberation and amplification of plDNA, later data demonstrated that this process, at least in laboratory strains, is not the case. This conclusion can be drawn from different mutants that were reported not to contain amplified plDNA in the senescent stage but still stop growth and die [19, 20]. However, because these mutants are strongly affected in vital functions (e.g., growth rate, fertility), it seems that they are unable to survive in nature, and aging in nature most likely is in fact caused by mtDNA instability. Regardless of this unsolved issue, the described mtDNA rearrangements are a hallmark of wild-type strain aging, and they do represent a useful biomarker of aging. Moreover, plDNA amplification clearly acts as a modulator accelerating the pace of aging because stabilization of the mtDNA is a predictable interference to extend life span in *P. anserina*. In nature, mtDNA instability leads to a strong pressure to efficiently reproduce within a very short time.

As mentioned, in the absence of pathways leading to mtDNA reorganization, life span of some *P. anserina* strains is still limited, indicating that other basic pathways with an impact on life span and aging must exist. In the past decades of research, a large body of data accumulated pointing to the impact of oxygenic energy transduction pathways and the generation of ROS.

3   Oxidative Stress

3.1   ROS Generation

The role of ROS in damaging biomolecules, and by doing so, impairing cellular function that leads to the aging of biological systems has been first put forward in the 1950s by D. Harman [21]. Since this time, numerous data have been collected that support the "free radical theory of aging" leading to a refined theory that stresses that the major source of oxidative stress are mitochondria [22]. However, even today the theory is not finally proven, and research is still aimed at challenging this theory.

An important source of ROS is the respiratory chain in the inner membrane of mitochondria, the powerhouse of every eukaryotic cell. During the oxygenic trans-duction of energy stored in biomolecules, such as lipids or carbon hydrates and the generation of ATP, electrons are transported from reduction equivalents originating from the dissimilation of reduced organic substrates (e.g., NADH + H^+) to oxygen as terminal electron acceptor, resulting in the production of water. This process takes place at the inner mitochondrial membrane, and it requires a group of proteins assembled into macromolecular complexes, termed complexes I–IV. During this process, protons are transferred across the inner membrane at three of the complexes, leading to the energization of the membrane. The resulting proton motive force finally leads to the generation of ATP at another supramolecular complex in the inner membrane, complex V or ATP synthase. During transport, electrons can escape the electron transport chain, and they are transferred as single electrons to oxygen, giving rise to the "free radical" superoxide anion. This process occurs at two sides of the respiratory chain, complex I and III [23–25].

The impact of the described processes on life span and aging has been exten-sively studied in *P. anserina* because this fungus, in contrast to yeast, which lacks complex I of the respiratory chain [26], respires via a standard respiratory chain, but, in contrast to higher systems, it is able to induce an alternative respiratory pathway [27–30] (Fig. 4.2). This situation is important when complexes of the standard respiratory chain are impaired, and it allows the selection of mutants defective in components of the standard respiratory chain. In other obligate aerobes with no alternative pathways of this kind (e.g., mammals), such mutations are lethal. In fact, in *P. anserina*, the analysis of a variety of different respiratory chain mutants revealed that such mutants display a long-lived phenotype (Table 4.1). In the past, several long-lived mutants have been generated and characterized, which, for different reasons, are affected in the respiratory chain. One group of mutants was found to be due to mutations in the mtDNA, and their phenotype consequently displays extrachromosomal inheritance. For example, a mutant termed ex1 is now growing for >20 years without any signs of senescence [31]. In this mutant, a large mtDNA region encoding part of the *CoxI* gene is deleted. Consequently, respiration via the cyanide-sensitive complex IV is not possible. In the mutant, an alternative cyanide-resistant and salicyl hydroxamic acid-sensitive respiratory pathway is induced. Via this pathway, less ROS but also less ATP are generated, apparently

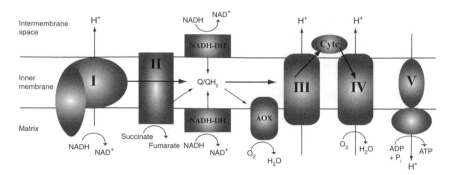

Fig. 4.2 Structure of the respiratory chain in *P. anserina* mitochondria. Electrons enter the respiratory chain via oxidation of NADH at NADH dehydrogenases (complex I, intermembrane space NADH-DH, matrix NADH-DH) or succinate at complex II, respectively. Ubiquinol (QH_2) can be oxidized by an alternative oxidase (AOX) or complex III. Finally, the electrons are transferred to molecular oxygen either via AOX or via cytochrome *c* (Cytc) and complex IV. Across the inner mitochondrial membrane, a proton motive force is created by pumping protons at complexes I, III, and IV. This gradient can be utilized by complex V to generate ATP

giving rise to lower molecular damage and immortality but also to reduced "fitness." The mutant is female sterile and grows much slower than the wild-type. Similar mutants with mutations in the mtDNA have been described [32, 33].

Another type of long-lived mutants with impaired mitochondrial electron transport is due to mutations in nuclear genes. These mutations are inherited according to the mendelian rules. As an example, mutant *cox5::ble* in which the nuclear gene for subunit V of the cytochrome *c* oxidase has been disrupted is characterized by an extreme life span extension [30]. This mutant, instead of 25 days, lives for >2 years. It does not display the wild-type–specific age-related reorganization of the mtDNA, respires via the alternative respiratory chain, and generates less ROS [30]. Again, also this mutant is characterized by impairments of fitness (e.g., reduced growth rate, reduced fertility).

Apart from mutations directly affecting the expression of genes encoding components of the respiratory chain, there are other mutants that, via more indirect pathways, affect the mitochondrial respiratory chain and that also lead to a switch in the type of respiration. An example of this type of mutant is long-lived mutant grisea, which has been selected in a screen of mutants selected after MMG mutagenesis [34]. The mutant is characterized by a hypopigmentation phenotype: Both ascospores and the mycelium are less pigmented than the wild-type [34–36]. Female fertility is reduced, and the mutant displays a long-life phenotype (60% increased mean life span) [36]. The mutation was found to be a single point mutation in the single intron of the *Grisea* gene, giving rise to a splice deficiency of the *Grisea* pre-mRNA [19, 35]. Consequently, the GRISEA protein is absent in the mutant strain. *Grisea* was found to code for a copper-regulated transcription factor controlling the expression of at least the two genes *PaCtr3* and *PaSod2* [37, 38]. Whereas *PaSod2* encodes a component of the ROS scavenging system (see below), *PaCtr3* codes for a high-affinity copper transporter involved in an efficient uptake of copper from the medium. In long-lived mutant grisea, *PaCtr3* is not expressed; therefore, copper uptake is limited to low copper-affinity uptake systems. In the grisea mutant, copper levels are too low to allow an efficient assembly of cytochrome *c* oxidase, which is dependent on binding of three copper ions. Consequently, the alternative oxidase is induced and electron transport via this pathway results in a reduced ATP generation and ROS generation [39]. Interestingly, although the mutant was originally reported to have a significantly lower growth rate than the wild-type, subcultivation over the years has led to a strain that, although respiring mainly via the alternative pathway, growth is almost as fast as the wild-type. It is reasonable to assume that a molecular pathway is induced in the mutant compensating the lower ATP generation via the alternative respiration. Such a pathway, however, remains to be demonstrated experimentally.

3.2 ROS Scavenging

Biological systems have developed multiple pathways to cope the negative effects of ROS. Among them, enzymatic systems converting superoxide anions to hydrogen peroxide and subsequently to water are substantially studied, and they are reported to affect life span and aging. Moreover, once manifested, damage and impairments may be repaired via specific repair mechanisms, or, if being too severe, compensating pathways may be induced.

In *P. anserina*, the impact of a ROS scavenging system on life span control has been demonstrated in several studies. In particular, the role of superoxide dismutases (SODs), leading to a disproportion of the superoxide anion, has been investigated. There are three SODs encoded in the genome of *P. anserina*. One is the copper/zinc-containing SOD (SOD1), which is active mainly in the cytoplasm but also to some extent in the mitochondrial intermembrane space [27, 37]. Another SOD seems to be a manganese-containing enzyme, which was suggested to be located in the mitochondrial matrix. The expression of this SOD is controlled by transcription factor GRISEA [37]. The role of a putative third SOD is currently investigated experimentally. Activity gel assays revealed that two SODs (most likely SOD1 and SOD2) are differentially active in juvenile and senescent strains. During aging of the wild-type, SOD1 activity increases but SOD2 activity decreases [27]. The increase in SOD1 activity seems to result from an increase in copper levels in the cytoplasm of senescent strains (see below). When present, this metal leads to a fast activation of the SOD1 apoprotein, which seems to be constitutively expressed in *P. anserina* [27].

4 Retrograde Response

A kind of a retrograde response, a mechanism compensating severe cellular impairments, is the induction of the alternative oxidase (AOX) in strains in which components of the respiratory chain are severely and irreversibly impaired.

This is the case in mutant strains, such as ex1 and cox5::ble in which structural genes coding for components of complex IV are mutated [30, 31]. It is also the case in other mutants in which the assembly of respiratory chain complexes is impaired for other reasons, such as the supply with cofactors [27-29]. Although the induction of the AOX has been demonstrated repeatedly in such mutants, basically nothing is known about signals and regulatory factors involved in this process. In plants, which also encode an alternative oxidase, induction of the corresponding gene was shown to result from ROS signaling. However, this possible pathway remains to be addressed in *P. anserina*. Because *P. anserina* also encodes different types of alternative NADH dehydrogenases [39], it will also be interesting to see whether and how these proteins can compensate impairment in the largest complex of the respiratory chain, complex I, which is known to be an important side of superoxide anion generation [23, 24]. Analysis of such a role cannot be performed in other aging models, because these models are not known to contain these kinds of alternative respiratory chain components, or, like yeast, they lack complex I [40].

5   Mitochondrial Dynamics

Recently, a role of mitochondrial morphology and of mitochondrial dynamics on life span and aging could be demonstrated in *P. anserina* [41]. Mitochondria are known to exist not exclusively as ellipsoid structures, as usually illustrated in textbooks. In contrast, mitochondria are highly dynamic [42, 43]. Individual mitochondria can fuse to form tubular structures and networks of branched filamentous entities [44, 45]. These entities, however, can divide by a process called fission. The extent of fission and fusion, because it is controlled by a complex molecular machinery of proteins, is responsible for the morphology of mitochondria in a given stage [45, 46]. Although the basis of the mechanisms of fission and fusion of mitochondria are currently extensively investigated, only little is known about the impact on aging. Interestingly, it was shown recently in *P. anserina* that mitochondrial morphology changes during aging from tubular networks in juvenile cultures to punctuate in senescent cultures [41]. *PaDnm1*, a gene encoding a putative dynamin-related GTPase involved in mitochondrial fission, was identified to be up-regulated in the long-lived mutant grisea. A *PaDnm1* knockout strain was found to have an 11-fold increased life span [41]. In the mutant, like in the wild-type, hydrogen peroxide is released in the senescent phase of the life cycle only. Most interestingly and in contrast to many other long-lived strains of *P. anserina* with a reduced fitness, the *PaDnm1* knockout strain is not impaired in growth and fertility. Thus, this strain is characterized by an increased "healthy" period of its life span and an example of healthy aging [41]. Interestingly, although wild-type cultures are found to be sensitive to etoposide, a frequently used inducer of apoptosis, this is not the case in the *PaDnm1* knockout strain [41]. This latter finding links mitochondrial dynamics and life span control with apoptotic pathways.

6 Age-Related Changes in Cytoplasmic Copper Levels

During the course of characterization of the copper metabolism in *P. anserina*, it was found that genes such as *PaCtr3* encoding a high-affinity copper transporter, or the metallothionein gene coding for a short cysteine-rich copper binding protein, are differentially expressed during aging of wild-type cultures [38, 47]. Moreover, activity of SOD1, which depends on cytoplasmic availability of copper, increases during aging [27]. These data suggest that during aging, cellular copper levels change in the different cellular compartments. It has been proposed previously that the changes in cytoplasmic copper levels result from release of copper from mitochondria, a view that gained additional support by the finding that the mitochondrial matrix represents a copper reservoir at least in yeast [48]. Because copper, via copper-sensing transcription factors such as GRISEA, affects gene expression, such a change may have yet unknown consequences (e.g., via the induction or repression of yet unknown genes). It is a challenge to further investigate this issue in *P. anserina* by examining the age-dependent expression of the complete set of copper-regulated genes and by analyzing whether the pathway emerging in *P. anserina* can be translated to other biological systems, including mammals.

7 ROS-Induced Apoptosis

From the investigations with the *PaDnm1* knockout strain, a first experimental set of data supporting the role of apoptosis in life span control in *P. anserina* became available recently [41]. Interestingly, within the genome several putative apoptosis factors are encoded [7, 49]. Both genes encoding components of caspase-dependent and caspase-independent pathways have been identified [7, 49]. In a recent study, the characterization of two cysteine/aspartate proteases with features of so-called metacaspases revealed the first important clues about a role of apoptosis as a final execution program in *P. anserina* [49]. Interestingly, in mutants in which either of the genes encoding metacaspase 1 or 2, respectively, were deleted, an increased life span was observed [49]. Metacaspase activity was measured in wild-type strains of *P. anserina* by using putative synthetic substrates. In addition, the activation of metacaspases via hydrogen peroxide in later stages of life became apparent [49]. Subsequent analysis of the apoptosis machinery in *P. anserina* is in progress and it will certainly provide new clues about the processes bringing life to an end in this aging model.

8 Conclusions and Outlook

In *P. anserina*, a body of experimental data has been generated in the past four decades of research demonstrating the impact of different molecular pathways on life span control and aging (Fig. 4.3). A critical role is played by the very basal

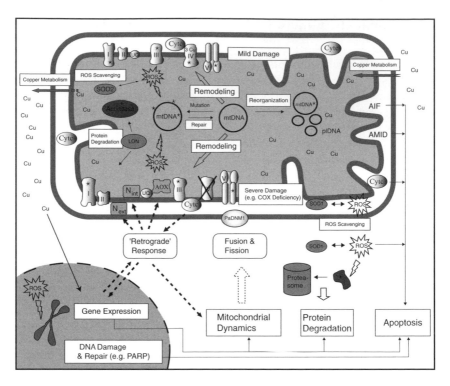

Fig. 4.3 Overview of factors that contribute to *P. anserina* aging. During aging, mtDNA is subject to reorganization and mutation, which leads to an impaired remodeling of the respiratory chain. Severe oxidative damage (*) leads to deficiency of crucial molecular components such as COX of the respiratory chain. ROS scavenging by SODs, degradation of oxidatively modified proteins by mitochondrial LON protease and cytosolic proteasomes as well as the copper-dependent induction of a retrograde response are among components that defend against the deleterious effects of oxidative stress. Mitochondrial dynamics (i.e., fusion and fission) also influences aging and is under control of proteins such as PaDNM1. Additionally, mitochondria seem to play a crucial role in apoptosis induction as a final execution program through the release of proapoptotic factors such as apoptosis-inducing factor (AIF) and AIF-like mitochondrion-associated inducer of death (AMID). The question how DNA damage and repair [e.g., by poly(ADP-ribose)polymerase (PARP)] impacts aging and apoptosis induction is currently under investigation

process related to energy transduction as they are carried out in mitochondria. *P. anserina*, comparable to higher systems such as humans, is strictly dependent on this function, and it has to deal with the consequences of ROS generated during this central process. From experimental details, important roles of various maintenance and compensation mechanisms can be envisioned. The details of their regulation and their interactions, however, remain to be deciphered in more detail. Also, other yet unexplored pathways (e.g., protein quality control systems, DNA repair) need to be thoroughly investigated. Once these data are available, it will be possible to generate a more holistic view about the aging process at the organismic level. Such a view will certainly be a good basis for translational research in other more complicated biological systems or in systems that, like humans, are not accessible to the kind of experimentation that is possible in *P. anserina*.

References

1. Esser K, Tudzynski P. Senescence in fungi. In: Thimann KV (ed) Senescence in plants. CRC Press, Boca Raton, Florida 1980:67–83.
2. Osiewacz HD, Scheckhuber CQ. Senescence in *Podospora anserina*. In: Osiewacz HD (ed) Molecular biology of fungal development. Marcel Dekker, New York 2002:87–108.
3. Lorin S, Dufour E, Sainsard-Chanet A. Mitochondrial metabolism and aging in the filamentous fungus *Podospora anserina*. Biochim Biophys Acta 2006; 1757(5–6):604–610.
4. Osiewacz HD, Scheckhuber CQ. Impact of ROS on ageing of two fungal model systems: *Saccharomyces cerevisiae* and *Podospora anserina*. Free Radic Res 2006; 40(12):1350 1358.
5. Rizet G. Sur l'impossibilité d'obtenir la multiplication végétative ininterrompue illimitée de l'ascomycète *Podospora anserina*. C R Acad Sci Paris 1953; 237:838–855.
6. Rizet G. Sur la longévité des souches de *Podospora anserina*. C R Acad Sci Paris 1953; 237:1106–1109.
7. Hamann A, Krause K, Werner A, Osiewacz HD. A two-step protocol for efficient deletion of genes in the filamentous ascomycete *Podospora anserina*. Curr Genet 2005; 48(4):270–275.
8. Osiewacz HD. Molecular analysis of aging processes in fungi. Mutat Res 1990; 237(1):1–8.
9. Osiewacz HD. Genetic regulation of aging. J Mol Med 1997; 75(10):715–727.
10. Osiewacz HD. Mitochondrial functions and aging. Gene 2002; 286(1):65–71.
11. Osiewacz HD. Aging in fungi: role of mitochondria in *Podospora anserina*. Mech Ageing Dev 2002; 123(7):755–764.
12. Stahl U, Lemke PA, Tudzynski P, Kück U, Esser K. Evidence for plasmid like DNA in a filamentous fungus, the ascomycete *Podospora anserina*. Mol Gen Genet 1978; 162(3):341–343.
13. Cummings DJ, Belcour L, Grandchamp C. Mitochondrial DNA from *Podospora anserina* and the occurrence of multimeric circular DNA in senescent cultures. In: Engberg J et al (eds) Specific eukaryotic genes. Munksgaard, Copenhagen, 1979:268–277.
14. Osiewacz HD, Esser K. The mitochondrial plasmid of *Podospora anserina*: a mobile intron of a mitochondrial gene. Curr Genet 1984; 8:299–305.
15. Stahl U, Kück U, Tudzynski P, Esser K. Characterization and cloning of plasmid like DNA of the ascomycete *Podospora anserina*. Mol Gen Genet 1980; 178(3):639–646.
16. Kück U, Stahl U, Esser K. Plasmid-like DNA is part of mitochondrial DNA in *Podospora anserina*. Curr Genet 1981; 3:151–156.
17. Sellem CH, Lecellier G, Belcour L. Transposition of a group II intron. Nature 1993; 366(6451):176–178.
18. Kück U, Osiewacz HD, Schmidt U et al. The onset of senescence is affected by DNA rearrangements of a discontinuous mitochondrial gene in *Podospora anserina*. Curr Genet 1985; 9(5):373–382.
19. Borghouts C, Kimpel E, Osiewacz HD. Mitochondrial DNA rearrangements of *Podospora anserina* are under the control of the nuclear gene grisea. Proc Natl Acad Sci USA 1997; 94(20):10768–10773.
20. Silar P, Koll F, Rossignol M. Cytosolic ribosomal mutations that abolish accumulation of circular intron in the mitochondria without preventing senescence of *Podospora anserina*. Genetics 1997; 145(3):697–705.
21. Harman D. A theory based on free radical and radiation chemistry. J Gerontol 1956; 11:298–300.
22. Harman D. Free radical theory of aging: history. EXS 1992; 62:1–10.

23. Liu Y, Fiskum G, Schubert D. Generation of reactive oxygen species by the mitochondrial electron transport chain. J Neurochem 2002; 80(5):780–787.
24. Genova ML, Ventura B, Giuliano G et al. The site of production of superoxide radical in mitochondrial complex I is not a bound ubisemiquinone but presumably iron-sulfur cluster N2. FEBS Lett 2001; 505(3):364–368.
25. Muller FL, Liu Y, Van RH. Complex III releases superoxide to both sides of the inner mitochondrial membrane. J Biol Chem 2004; 279(47):49064–49073.
26. Minagawa N, Yoshimoto A. The induction of cyanide-resistant respiration in *Hansenula anomala*. J Biochem (Tokyo) 1987; 101(5):1141–1146.
27. Borghouts C, Werner A, Elthon T, Osiewacz HD. Copper-modulated gene expression and senescence in the filamentous fungus *Podospora anserina*. Mol Cell Biol 2001; 21(2):390–399.
28. Lorin S, Dufour E, Boulay J, Begel O, Marsy S, Sainsard-Chanet A. Overexpression of the alternative oxidase restores senescence and fertility in a long-lived respiration-deficient mutant of *Podospora anserina*. Mol Microbiol 2001; 42(5):1259–1267.
29. Stumpferl SW, Stephan O, Osiewacz HD. Impact of a disruption of a pathway delivering copper to mitochondria on *Podospora anserina* metabolism and life span. Eukaryotic Cell 2004; 3(1):200–211.
30. Dufour E, Boulay J, Rincheval V, Sainsard-Chanet A. A causal link between respiration and senescence in *Podospora anserina*. Proc Natl Acad Sci USA 2000; 97(8):4138–4143.
31. Schulte E, Kück U, Esser K. Extrachromosomal mutants from *Podospora anserina*: permanent vegetative growth in spite of multiple recombination events in the mitochondrial genome. Mol Gen Genet 1988; 211:342–349.
32. Belcour L, Vierny-Jamet C. Variable DNA splicing sites of a mitochondrial intron: relationship to the senescence process in *Podospora*. EMBO J 1986; 5:609–614.
33. Sainsard-Chanet A, Begel O. Insertion of an LrDNA gene fragment and of filler DNA at a mitochondrial exon-intron junction in *Podospora*. Nucleic Acids Res 1990; 18(4):779–783.
34. Prillinger H, Esser K. The phenoloxidases of the ascomycete *Podospora anserina*. XIII. Action and interaction of genes controlling the formation of laccase. Mol Gen Genet 1977; 156(3):333–345.
35. Osiewacz HD, Nuber U. GRISEA, a putative copper-activated transcription factor from *Podospora anserina* involved in differentiation and senescence. Mol Gen Genet 1996; 252(1–2):115–124.
36. Borghouts C, Osiewacz HD. GRISEA, a copper-modulated transcription factor from *Podospora anserina* involved in senescence and morphogenesis, is an ortholog of MAC1 in *Saccharomyces cerevisiae*. Mol Gen Genet 1998; 260(5):492–502.
37. Borghouts C, Scheckhuber CQ, Werner A, Osiewacz HD. Respiration, copper availability and SOD activity in *P. anserina* strains with different lifespan. Biogerontology 2002; 3(3):143–153.
38. Borghouts C, Scheckhuber CQ, Stephan O, Osiewacz HD. Copper homeostasis and aging in the fungal model system *Podospora anserina*: differential expression of *PaCtr3* encoding a copper transporter. Int J Biochem Cell Biol 2002; 34:1355–1371.
39. Gredilla R, Grief J, Osiewacz HD. Mitochondrial free radical generation and lifespan control in the fungal aging model *Podospora anserina*. Exp Gerontol 2006; 41(4):439–447.
40. Friedrich T, Steinmüller K, Weiss H. The proton-pumping respiratory complex I of bacteria and mitochondria and its homologue in chloroplasts. FEBS Lett 1995; 367(2):107–111.
41. Scheckhuber CQ, Erjavec N, Tinazli A, Hamann A, Nyström T, Osiewacz HD. Reducing mitochondrial fission results in increased life span and fitness of two fungal ageing models. Nat Cell Biol 2007; 9(1):99–105.
42. Bereiter-Hahn J, Vöth M. Dynamics of mitochondria in living cells: shape changes, dislocations, fusion, and fission of mitochondria. Microsc Res Tech 1994; 27(3):198–219.
43. Yaffe MP. Dynamic mitochondria. Nat Cell Biol 1999; 1(6):E149–E150.
44. Shaw JM, Nunnari J. Mitochondrial dynamics and division in budding yeast. Trends Cell Biol 2002; 12(4):178–184.

45. Okamoto K, Shaw JM. Mitochondrial morphology and dynamics in yeast and multicellular eukaryotes. Annu Rev Genet 2005; 39:503–536.
46. Hoppins S, Lackner L, Nunnari J. The machines that divide and fuse mitochondria. Annu Rev Biochem 2007;76:751–780.
47. Averbeck NB, Borghouts C, Hamann A, Specke V, Osiewacz HD. Molecular control of copper homeostasis in filamentous fungi: increased expression of a metallothionein gene during aging of *Podospora anserina*. Mol Gen Genet 2001; 264(5):604–612.
48. Cobine PA, Ojeda LD, Rigby KM, Winge DR. Yeast contain a non-proteinaceous pool of copper in the mitochondrial matrix. J Biol Chem 2004;14447–14455.
49. Hamann A, Brust D, Osiewacz HD. Deletion of putative apoptosis factors leads to lifespan extension in the fungal ageing model *Podospora anserina*. Mol Microbiol 2007;65:948–958.
50. Sellem CH, Marsy S, Boivin A, Lemaire C, Sainsard-Chanet A. A mutation in the gene encoding cytochrome c_1 leads to a decreased ROS content and to a long-lived phenotype in the filamentous fungus *Podospora anserina*. Fungal Genet Biol 2007; 44(7):648–658.
51. Silar P, Picard M. Increased longevity of EF-1 alpha high-fidelity mutants in *Podospora anserina*. J Mol Biol 1994; 235(1):231–236.
52. Borghouts C, Kerschner S, Osiewacz HD. Copper-dependence of mitochondrial DNA rearrangements in *Podospora anserina*. Curr Genet 2000; 37(4):268–275.
53. Averbeck NB, Jensen ON, Mann M, Schägger H, Osiewacz HD. Identification and characterization of PaMTH1, a putative o-methyltransferase accumulating during senescence of *Podospora anserina* cultures. Curr Genet 2000; 37(3):200–208.
54. Jamet-Vierny C, Contamine V, Boulay J, Zickler D, Picard M. Mutations in genes encoding the mitochondrial outer membrane proteins Tom70 and Mdm10 of *Podospora anserina* modify the spectrum of mitochondrial DNA rearrangements associated with cellular death. Mol Cell Biol 1997; 17(11):6359–6366.
55. Hermanns J, Osiewacz HD. The linear mitochondrial plasmid pAL2-1 of a long-lived *Podospora anserina* mutant is an invertron encoding a DNA and RNA polymerase. Curr Genet 1992; 22(6):491–500.
56. Begel O, Boulay J, Albert B, Dufour E, Sainsard-Chanet A. Mitochondrial group II introns, cytochrome c oxidase, and senescence in *Podospora anserina*. Mol Cell Biol 1999; 19(6):4093–4100.
57. Maas MF, Sellem CH, Hoekstra RF, Debets AJ, Sainsard-Chanet A. Integration of a pAL2-1 homologous mitochondrial plasmid associated with life span extension in *Podospora anserina*. Fungal Genet Biol 2007; 44(7):659–671.

5
Oxidative Stress and Aging in the Budding Yeast *Saccharomyces cerevisiae*

Stavros Gonidakis and Valter D. Longo

Summary The production of reactive oxygen species is an unavoidable consequence of life in an aerobic environment. The accumulation of macromolecular damage caused by these highly reactive substances is thought to be one of the major contributors to the aging phenotype. Protection against this damage provided by antioxidant enzymes such as superoxide dismutases is critically important for the maintenance of an aging yeast population. The role of superoxide signaling in age-dependent apoptosis suggests that reactive oxygen species are involved in the regulation of multiple maintenance and repair systems during yeast aging. Enhanced protection against oxidative stress seems to be necessary but not sufficient for life span extension in model organisms ranging from yeast to mice.

Keywords Yeast, aging, oxidative stress, superoxide, apoptosis.

Although life can be sustained without molecular oxygen, most organisms have been evolving in its presence for the past 2 billion years. Therefore, natural selection has favored several tactics that allow aerobic organisms to consume oxygen while minimizing the damage caused by the toxic by-products of its metabolism. This chapter focuses on the involvement of reactive oxygen species (ROS) in the aging process of the budding yeast *Saccharomyces cerevisiae*.

1 The Model System

The molecular and cellular architecture of *Saccharomyces cerevisiae* shares remarkable similarities with that of higher eukaryotes. The powerful genomic and genetic tools available for its manipulation and its short life span under laboratory conditions make it an exceptionally well-suited model organism for biogerontological research. Unlike other model systems for aging research, where life span is only measured through the survival of an organisms or population over time, two different life span paradigms are being used in yeast. Replicative life span measures how

From: *Aging Medicine: Oxidative Stress in Aging: From Model Systems to Human Diseases*   67
Edited by: S. Miwa, K.B. Beckman, and F.L. Muller © Humana Press, Totowa, NJ

many "daughter" cells a single "mother" cell incubated onto solid medium can beget. Our lab introduced the survival of a nondividing population of yeast cells over time (chronological aging) as a method to study yeast aging more akin to the method used in higher organisms [1]. There have been instances where certain genetic manipulations produced opposite effects in the two life span assays [2]. However, the findings that the Ras and Sch9 pathways reduce survival in both paradigms [3] and that passages through stationary phase decrease replicative life span [4] suggest that some overlap exists between the mechanisms regulating aging in the those two different experimental regimes.

A typical chronological aging experiment is initiated by the inoculation of approximately 10^6 cells in synthetic complete medium containing 2% glucose. During the growth phase (approximately 10 h), energy is produced mainly from the fermentative catabolism of glucose. After exponential growth exhausts most of the extracellular glucose, the population switches from a fermentative to a largely respiratory metabolism by using the ethanol produced during log phase. Slow growth continues until approximately 48 h; the maximum cell density is measured 3 days after the inoculation. The aging population is maintained at 30 °C with shaking and its survival is subsequently followed by removing aliquots at regular time intervals and measuring the number of cells able to form colonies when plated on rich nutrient plates (colony-forming units [CFUs]). We have confirmed that the progressive decline in CFUs over time reflects a reduction in the viability of cells by observing an increase in protein concentration in the culture medium due to the death and lysis of aging cells [5] and also by observing a robust correlation between the number of culturable cells and the number of cells stained with FUN-1, which stains intravacuolar structures only in metabolically active organisms with intact plasma membranes (unpublished data).

The mean life span of wild-type strains grown in synthetic complete medium containing glucose (SDC) varies from 6-7 to 15-20 days, as opposed to yeast grown and incubated in the rich medium (YPD), which survive for longer periods in a low-metabolism stationary phase. Yeast grown in SDC maintain high metabolic rates during most of their life span. There also are data showing the large effect that the carbon source supporting growth can have on the subsequent chronological life span of a yeast population; populations grown on a respiratory carbon source have a longer life span compared with isogenic populations grown on dextrose, a fermentable carbon source [6]. *Saccharomyces cerevisiae* aging chronologically resemble postmitotic cells of multicellular organisms, because they have exited from the cell cycle and most of their energy is produced by mitochondrial respiration. The dependence of stationary phase cells on respiration for their energy production might explain the enhanced survival caused by growth on a respiratory carbon source, which would result in an aging population preadapted to a respiratory metabolism. It is also worth noting that the environment created during a chronological life span experiment in the laboratory resembles the "feast and famine" lifestyle that most microorganisms face in their natural ecological niches.

One of the major criticisms against the use of our experimental paradigm for the study of aging is that the death we observe is starvation- and not age-dependent.

Three observations argue against this possibility: First, stored glycogen and ethanol levels of the wild-type strain DBY746 adjusted for population density are not significantly different between days 3 and 7 of a typical viability experiment (unpublished data); therefore, yeast seem to be maintaining a high concentration of a major reserve carbon source during our aging experiments. Second, removal of all the nutrients present in expired medium and transfer of the population to sterile distilled water on day 3 results in a significant increase in the chronological life span of wild-type strains [7]. Last, adjustment of the pH of a yeast culture from 4.0 to 7.0 during the course of an aging experiment also results in a pronounced increase in the mean lifes span of the population [7].

2 ROS and Aging: A Lesson from Knockouts

Energy production from mitochondrial respiration with oxygen as the terminal electron acceptor results in the production of highly reactive, damaging substances collectively called ROS. Harman proposed that the time-dependent accumulation of macromolecular damage inflicted by these species is one of the major causes of aging in aerobic organisms [8]. Oxidative damage to cellular macromolecules has been observed during chronological and replicative yeast aging in several studies [9]. The importance of successful oxidative stress management for the maintenance of a healthy yeast life span has been made abundantly clear by the study of mutants lacking one or more antioxidant enzymes. Superoxide dismutases (SODs) hold a prominent position among those enzymes, and considerable effort has been devoted to the study of mutants lacking them in organisms as diverse as *Escherichia coli* and *Mus musculus*.

SODs catalyze the disproportionation of superoxide ($O_2^{\cdot-}$) to hydrogen peroxide and water in vivo. *Saccharomyces cerevisiae* expresses a cytosolic CuZn SOD coded by the *SOD1* gene and a Mn SOD localized in the mitochondrial matrix coded by *SOD2* [10]. SOD1 accounts for approximately 90% of the total SOD activity in cells growing on glucose [10]. The superoxide-dependent inactivation of the mitochondrial matrix-localized protein homoaconitase (Lys4) in the *sod1Δ* mutant [11] suggests that Sod1 also can provide protection against superoxide produced in the mitochondrial intermembrane space and diffusing to the mitochondrial matrix. The increase in mitochondrial protein carbonylation observed in *sod1Δ* provides further support for this notion [12]. Fridovich [13] suggested that the high levels of superoxide in the cytosol and intermembrane space of *sod1Δ* may react with endogenous nitric oxide, forming peroxynitrite that can diffuse into the mitochondrial matrix and inactivate Lys4 and other proteins.

Knockout strains lacking either of the two SODs exhibit reduced growth rates on respiratory carbon sources, amino acid auxotrophies, and shorter chronological life span. These multiple deficits reveal the importance of protection against oxidative stress under physiologically distinct conditions. *Escherichia coli* lacking both cytoplasmic SODs is unable to grow in glucose minimal media [14] and dies early in rich media [15]. The *E. coli* Fe SOD targeted to the mitochondria reverses the growth defect of yeast *sod2Δ* [16], demonstrating the functional conservation of the SODs during evolution. *Saccharomyces cerevisiae sod1* and *sod2* knockout (KO) strains have a reduced life span compared with wild type under both normoxic and hypoxic conditions. *Sod1Δ* has a shorter life span than *sod2Δ* in normoxic conditions, but the comparison is reversed in hypoxic conditions, with *sod1Δ* surviving better for the most part of the experiment [17]. These data illustrate the differential protective effect of cytoplasmic and mitochondrial SODs according to oxygen concentration. *Sod1Δ* also has a shorter replicative life span on all carbon sources tested [18].

It is generally accepted that the mitochondrial electron transport chain is the primary source of ROS in eukaryotic aerobes. The contribution of mitochondria-derived ROS in the accelerated demise of SOD-deficient yeast strains is shown by the following observations. First, the early survival of the *sod1Δ* strain is improved by deletion of the gene *COQ3*, which produces an enzyme necessary for the synthesis of ubiquinone [17]. The *coq3Δ* strain cannot make ubiquinone, and it is thus respiration-incompetent. Survival of the *coq3Δsod1Δ* strain can only be followed for the first few days of a typical life span experiment, because respiration is required for entry into stationary phase [17]. Experiments using respiratory inhibitors also support the role of SODs in the elimination of mitochondria-derived superoxide during yeast aging. Sodium cyanide and carbonyl cyanide *p*-trifluoromethoxyphenylhydrazone have been shown to reduce the production of superoxide in isolated mitochondria by blocking electron transport and uncoupling respiration from ATP synthesis, respectively. Addition of either of these substances significantly improves the survival of *sod2Δ* [19]. Conversely, chronological life span decreased in cells treated with antimycin A, a respiratory inhibitor that stimulates mitochondrial ROS generation [20].

Indices of respiratory competence (IRCs) provide a further clue for the role of mitochondrial SOD in the maintenance of aging yeast cells. IRCs are calculated by dividing the number of aging cells able to grow on a nonfermentable carbon source such as glycerol or lactate by the number of cells able to grow by fermenting glucose. This system exploits the ability of *S. cerevisiae* to grow and survive for several days without respiration to assess whether mitochondrial damage precedes damage in the cytosol. As shown in Fig. 5.1, whereas the IRC of the wild-type strain decreases to about 80% by day 7, that of *sod2Δ* decreases progressively to approximately 40% at the same time point [19]. This suggests that a large fraction of *sod2Δ* mutants lose the ability to respire in an age-dependent manner. Importantly, the respiratory incompetence of the Sod2-deficient cells does not seem to be caused by damage to the mitochondrial DNA, because all colonies grown on glucose medium grow well when replica-plated on medium containing a respiratory carbon source. Hence, loss of respiratory competence may be attributed to the accumulation of oxidatively damaged components necessary for efficient mitochondrial function. It is worth noting that the decrease in IRC preceded age-dependent death, suggesting that the loss of mitochondrial functionality is a cause and not a result of the aging process in yeast.

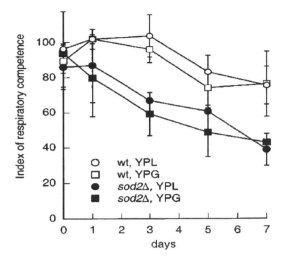

Fig. 5.1 Index of respiratory competence of wild type and *sod2Δ* during chronological aging calculated by dividing the number of cells able to grow on a respiratory carbon source (YPLactose or YPGlycerol) by the number of cells growing on a fermentable carbon source (YPDextrose). Figure taken from ref. 19, with permission from Elsevier

Among the mitochondrial defects caused by the lack of mitochondrial Sod, we observed an age-dependent reduction in the activities of the mitochondrial enzymes aconitase and succinate dehydrogenase. Interestingly, approximately 90% of the aconitase present in extracts from *sod2Δ* cells can be reactivated in vitro by the addition of Fe^{3+} and S^{2-}, which facilitate the reconstitution of the enzyme's Fe-S cluster. The measurement of the portion of the enzymatic activity that can be reactivated by Fe^{3+} and S^{2-} allows the estimation of the superoxide level in vivo [21]. Our results suggest that during yeast chronological aging, mitochondrial SOD prevents the superoxide-dependent release of iron from the 4Fe-4S cluster of aconitase and possibly other Fe-S cluster-containing enzymes.

Defective mitochondrial gene expression caused by a dominant-negative mutation in the *S. cerevisiae* mitochondrial RNA polymerase results in a number of defects shared with the *sod2Δ* strain, namely reduction of respiration, ROS elevation coupled with increased oxidative stress, and reduced chronological life span. Overexpression of either Sod greatly extends the life span of this mutant and enhances its ability to respire [22]. These results suggest the possibility that ROS produced by the mitochondria trigger a vicious circle of mutually amplifying macromolecular damage, misregulated gene expression, and decreased respiratory activity, which contributes to the chronological aging of a yeast population. The free iron released by the disassembly of 4Fe-4S clusters might be involved in the aforementioned loop, because a mutation specifically affecting the iron detoxification capacity of frataxin reduces chronological life span and precludes survival in the absence of Sod1 [23]. Mn SOD homozygous KO mice also showed a decrease in the activities of aconitase and succinate dehydrogenase, suffered from dilated cardiomyopathy, and died within 10 days of birth [24].

Along with defense against oxidation-induced damage, efficient functioning of the DNA repair machinery also is required for healthy yeast aging; the lack of proteins involved in the repair of oxidative DNA damage shortens chronological life span [25]. Apart from mitochondrial-related pathologies, Fig. 5.2 shows that yeast lacking mitochondrial SOD also show an increase in age-dependent damage to nuclear DNA [19]. This is quantified by measuring the fraction of the aging population that is able to grow in the presence of the toxic arginine analog canavanine (canavanine-resistant mutants arise specifically due to age-dependent mutations in the gene encoding the arginine permease, which are not known to confer any survival advantage during chronological aging) [26]. Because Mn SOD is localized to the mitochondrial matrix, the age-dependent increase in nuclear DNA damage in *sod2Δ* is hard to explain by using the simplistic view of ROS as agents of unavoidable, stochastic damage. One possibility is that a protonated form of $O_2^{\cdot-}$ or another membrane-permeable ROS is able to reach the nucleus. Alternatively, the escape of superoxide from mitochondria can be made possible by the opening of the permeability transition pore, which has been implicated in yeast apoptosis [27]. A third possible explanation entails the involvement of $O_2^{\cdot-}$ in the regulation of multiple maintenance and repair systems relevant to aging. For example, it is possible that the mitochondrial defects caused in the absence of Mn SOD are communicated to the nucleus via a form of retrograde signaling, causing the down-regulation of DNA repair systems and resulting in the observed increase in the age-dependent nuclear mutation frequency (this scenario is more extensively discussed in Section 3). Conversely, the increased DNA damage of *sod1Δ* mutants [7] can be explained by a combination of the altered signaling discussed above as and the enhanced damaging effect of superoxide in the absence of the cytosolic dismutase. Notably, the *sod1Δ* strain ranked among the top 33 mutators in a genome-wide screen for single gene knockouts with increased nuclear mutation frequency [28]. *Escherichia coli*

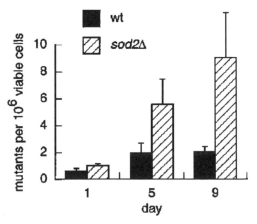

Fig. 5.2 Age-dependent nuclear mutation frequency of wild type and *sod2Δ* measured by the number of canavanine-resistant mutants. Figure taken from ref. 19, with permission from Elsevier

cells devoid of cytoplasmic SODs also have an increased mutation frequency that is reversed by anaerobiosis [29].

Catalases are responsible for the detoxification of hydrogen peroxide, one of the products of the dismutation of superoxide, with the other being water. *Saccharomyces cerevisiae* produces two catalases: one catalase located in the cytosol, catalase T, coded by *CTT1*; and catalase A, coded by *CTA1* and localized to the peroxisomes. The chronological life span of a *ctt1Δ* strain is unaffected [17], suggesting that the toxicity of hydrogen peroxide does not limit the chronological life span of yeast. However, both *ctt1Δ* and *cta1Δ* have a reduced replicative life span when grown on ethanol [30].

The data presented in this section have painted the following picture: mitochondrial and cytoplasmic SODs are important for the maintenance of an aging yeast population by protecting mitochondrial proteins, nuclear DNA, and many other targets against insults by toxic products of oxidative phosphorylation occurring in respiring mitochondria and also by ROS generated in the cytosol. Data from SOD knockout strains of *E. coli, S. cerevisiae*, and mice illustrate the crucial protective effect of these enzymes in evolutionarily distant organisms.

3   More Than Just Damage

As surmised in Section 2, ROS do not seem to simply cause damage during yeast aging, much like rust eventually erodes an exposed piece of unprotected metal. ROS have been shown to participate at multiple points in the signaling cascades of physiological and pathological apoptosis in metazoans. The major proapoptotic mammalian protein Bax induces apoptotic cell death in yeast, which was reversed by the expression of the antiapoptotic protein Bcl-2 [31]. Human Bcl-2 has been also shown to reverse the survival defects of SOD-deficient and to delay the death of wild-type yeast [32]. Two recent studies provided support for the intriguing hypothesis that the progressive decline of survivorship observed in aging yeast populations is mediated by a form of programmed cell death.

Herker et al. [33] showed that the viability decline observed during a yeast aging experiment was accompanied by a progressive accumulation of ROS, visualized using the stain dihydrorhodamine. They also found that overexpression of YAP1, a transcription factor activated in response to oxidative stress, enhanced the life span of wild-type yeast and decreased the age-dependent accumulation of ROS. The *yap1* deletion strain was also among those with the highest mutation frequencies in the genome-wide screen performed by Huang et al. [28].

The removal of damaged or redundant cells that would otherwise compromise the fitness of the organism provides an obvious explanation for the utility of apoptosis in multicellular organisms. However, it is much harder to attribute an adaptive value to the suicide of a unicellular organism. Our lab provided support for the hypothesis that the superoxide-dependent apoptosis observed during yeast chronological aging is a form of altruistic suicide ensuring the survival of a mutated clone at the expense of the self-sacrificing individuals [7].

We observed that chronologically aging yeast cells display typical molecular markers of apoptosis, such as phosphatidylserine externalization, cytosolic acidification, and chromatin condensation. Most wild-type yeast strains usually exhibit an increase in the number of CFUs after 90–99% of the population has died. This adaptive regrowth is made possible by the release of nutrients from the death and lysis of aged members of the population, and it is reminiscent of the well-studied growth advantage in stationary phase phenotype observed in *E. coli* and other bacteria [34]. We found that the occurrence of this regrowth phenotype was inversely correlated with protection against superoxide-mediated toxicity and that it was associated with increased release of nutrients in the medium and increased nuclear mutation frequency. Whereas regrowth of adapted mutant was observed in 48% of the wild-type populations, regrowth was observed in nearly all experiments with the short-lived *sod1Δ* strain and in none of those with long-lived mutants, including overexpressors of both SODs. Competition experiments between wild-type and Sod1-deficient cells mixed in a 1:1 ratio are also particularly revealing. Wild-type cells survive better during the first 9 days of the experiment, but they are eventually driven to extinction by the regrowing *sod1Δ* mutants. Conversely, the strain overexpressing the two SODs eventually succumbs in competition experiments with wild type. The competition and regrowth experiments demonstrate the existence of a trade-off at a population level between increased longevity and the ability to generate mutants able to survive and proliferate after most of the population has died. These results are also consistent with the existence of an altruistic aging program that ensures adaptation in the face of changing environmental conditions. Superoxide produced by respiring mitochondria probably acts as a signaling agent for this program, in a way similar to apoptosis in higher eukaryotes. Execution of the program entails an increase in nuclear mutation frequency, and it results in the release of nutrients in the extracellular medium that are used to support the proliferation of mutants that have acquired the mutation(s) that allow them to grow.

Along the lines of superoxide signaling in the regulation of age-dependent yeast apoptosis, a combination of observations invite speculation about the involvement of mitochondrial retrograde (RTG) signaling. Retrograde signaling refers to the molecular machinery that communicates the status of the mitochondria to the nucleus under normal and pathophysiological conditions, bringing about compensatory changes in nuclear gene expression [35]. The RTG pathway affects mitochondrial DNA maintenance through *ACO1*, the gene coding for aconitase. Aconitase was linked to mitochondrial DNA maintenance by two observations; first, it was one of the proteins found to be cross-linked to mitochondrial DNA; and second, the mitochondrial DNA loss phenotype in *aco1Δ* mutants is much more severe than that in any of the other mutants of the tricarboxylic acid (TCA) cycle [36]. Mammalian aconitase also has a function not directly related to its role as a TCA cycle enzyme; it can function as an iron regulatory protein (IRP) by binding to the untranslated region of target mRNAs that encode proteins involved in iron homeostasis. Because aconitase is one of the enzymes shown to be inactivated by superoxide during the aging of both wild-type and SOD-deficient strains, one could envisage a scenario whereby damaged aconitase serves as a signal (through released Fe^{3+} or the product of a reaction involving Fe^{3+}) of mitochondrial damage. This signal is transmitted to the nucleus via some form of retrograde signaling and causes nuclear gene expression changes that jointly contribute to the execution of apoptotic cell death. These changes might include a downregulation of multiple maintenance and repair systems that would render the cell more vulnerable to oxidation-induced damage to DNA and other cellular macromolecules. It is also possible that yeast mitochondrial aconitase acts directly as a regulator of mitochondrial gene expression in a manner dependent on the oxidation-sensitive status of its Fe-S cluster. The other possibility is that alterations in the Fe-S cluster of aconitase can modify gene expression or RNA/DNA stability in the mitochondria as shown for cytosolic IRPs. The observation of increased age-dependent nuclear DNA damage in *sod2Δ* is consistent with the scenaria presented above.

4 Age-Dependent Oxidative Damage Is Regulated

Since reduced protection against oxidation-induced damage decreases yeast life span, does enhanced protection afford a life span increase? The answer is a qualified yes. A large body of evidence accumulated over the past two decades has revealed the existence of conserved nutrient-sensing signaling pathways regulating longevity in organisms ranging from yeast to mice. The signaling pathways downstream of insulin/insulin-like growth factor are activated during periods of nutrient abundance, and they steer organisms toward a physiological state favoring growth and reproduction. Reduced signaling through these pathways, achieved by genetic interventions, results in extended life span in phylogenetically distant species [37]. Increased longevity correlates with enhanced resistance to thermal and oxidative stress, and it shares several common features with caloric restriction, a manipulation resulting in life span extension in all species thus far tested.

The selection of mutants resistant to heat shock and the superoxide generator paraquat from a transposon-mutagenized yeast library led to the discovery that Sch9 and Ras pathway components, homologs of the mammalian oncogenes Akt and Ras, regulate longevity and stress resistance in yeast [38]. The *sch9* strain survives three times as long as wild type. Sod2 mRNA is expressed at higher levels in *sch9* compared with wild type on days 5 and 6. Deletion of *SOD2* reverses the extended life span of *sch9* indicating that the enhanced expression of mitochondrial SOD is required for life span extension in this mutant [5]. Daf2 is the *Caenorhabditis elegans* ortholog of the mammalian insulin receptor and functions upstream of the yeast Sch9 homologs Akt1/2 [39]. Microarray analyses of gene expression in the long-lived DAF2 KO worm revealed the up-regulation of several genes conferring protection against thermal and oxidative stress, sod3, the *C. elegans* gene coding for the mitochondrial Mn SOD being one of them. However, deletion of SOD3 did not affect the life span of *daf2Δ* mutants [40].

Long-lived *Sch9* and *Ras2Δ* mutants are more resistant than wild-type to heat shock, hydrogen peroxide, and menadione (a superoxide producer) treatments. They also show a delay in the age- and superoxide-dependent inactivation of mitochondrial aconitase [38]. The extended life span of *Sch9* is also associated with a reduction in the age-dependent increase in nuclear DNA mutation frequency [7]. One could therefore argue that the enhanced life span of these mutants can be solely attributed to their apparent enhanced protection against the damaging effects of ROS. However, overexpression of both *SOD1* and *SOD2* by means of multicopy plasmids results in a life span extension of about 30% [5], which is far less pronounced than the 3-fold life span extension achieved by mutations in central nutrient-sensing pathways. Cytoplasmic SOD overexpression on its own also is reported to extend chronological life span; overexpression of CCS1, the chaperone responsible for Cu^{2+} loading of Sod1 in vivo, was required for this effect [41]. Another group reported a chronological life span extension caused by an increasing fraction of Sod1 localized to the mitochondrial intermembrane space due to overexpression of CCS1 [12]. Catalase overexpression does not have a life-prolonging effect [5], indicating that it is the toxicity of superoxide, and not hydrogen peroxide that limits the chronological life span of yeast. Mn SOD overexpression has been shown to decrease replicative life span, possibly due to an induced defect in mitochondria segregation from mother to daughter cells [42].

Again, results from *C. elegans* tend to mirror those obtained in yeast: Although *daf2* mutant worms live approximately 100% longer than wild-type, and they are resistant to thermal and oxidative stress, the role of SOD mimetics in the regulation of worm life span is controversial [43, 44]. Insulin-like growth factor-I receptor heterozygous knockout mice are resistant to the superoxide generator paraquat and live, on average, 25% longer than wild-type [45]. However, ubiquitous overexpression of CuZn SOD in mice also failed to produce a life span extension [46]. Apparently, inactivating mutations of genes involved in central nutrient-sensing pathways are much more effective than overexpression of antioxidant enzymes in extending life. These data suggest that enhanced protection against oxidative stress is necessary but not sufficient for life span extension in the model organisms thus far studied and point toward a more complex interaction between oxidative stress and aging than the theory put forward by Harman in 1956.

5   Conclusions

The relative contribution of oxidative stress in the aging process is one of the fundamental unresolved issues in molecular biogerontology. At this stage, many questions remain to be answered. Do long-lived yeast strains produce less superoxide as they age? Does ROS signaling mediate the age-dependent down-regulation of multiple protection and repair systems? Is mitochondrial retrograde signaling involved in the gene expression changes causing apoptosis during chronological aging? The application of improved analytical techniques in simple model systems such as yeast should reveal deeper mechanistic connections between oxidative stress, longevity, and its regulation. The increased understanding of conserved mechanisms regulating aging in simple organisms also will bring us closer to the lofty goal of a modest extension of human life span by the prevention of a number of age-related diseases.

References

1. Fabrizio P, Longo VD. The chronological life span of *Saccharomyces cerevisiae*. Aging Cell 2003;2(2):73–81.
2. Fabrizio P, Gattazzo C, Battistella L et al. Sir2 blocks extreme life-span extension. Cell 2005;123(4):655–67.
3. Fabrizio P, Pletcher SD, Minois N, Vaupel JW, Longo VD. Chronological aging-independent replicative life span regulation by Msn2/Msn4 and Sod2 in *Saccharomyces cerevisiae*. FEBS Lett 2004;557(1–3):136–42.
4. Ashrafi K, Sinclair D, Gordon JI, Guarente L. Passage through stationary phase advances replicative aging in *Saccharomyces cerevisiae*. Proc Natl Acad Sci U S A 1999;96:9100–5.
5. Fabrizio P, Liou LL, Moy VN et al. SOD2 Functions downstream of Sch9 to extend longevity in yeast. Genetics 2003;163(1):35–46.
6. Piper PW, Harris NL, MacLean M. Preadaptation to efficient respiratory maintenance is essential both for maximal longevity and the retention of replicative potential in chronologically ageing yeast. Mech Ageing Dev 2006;127(9):733–40.
7. Fabrizio P, Battistella L, Vardavas R et al. Superoxide is a mediator of an altruistic aging program in *Saccharomyces cerevisiae*. J Cell Biol 2004;166(7):1055–67.
8. Harman D. A theory based on free radical and radiation chemistry. J Gerontol 1956;11:298–300.
9. Reverter-Branchat G, Cabiscol E, Tamarit J, Ros J. Oxidative damage to specific proteins in replicative and chronological-aged *Saccharomyces cerevisiae*: common targets and prevention by calorie restriction. J Biol Chem 2004;279(30):31983–9.
10. Gralla EB, Kosman D. Molecular genetics of superoxide dismutases in yeasts and related fungi. Adv Genet 1992;30:251–319.
11. Wallace MA, Liou LL, Martins J et al. Superoxide inhibits 4Fe-4S cluster enzymes involved in amino acid biosynthesis. Cross-compartment protection by CuZn-superoxide dismutase. J Biol Chem 2004;279(31):32055–62.
12. Sturtz LA, Diekert K, Jensen LT, Lill R, Culotta VC. A fraction of yeast Cu,Zn-superoxide dismutase and its metallochaperone, CCS, localize to the intermembrane space of mitochondria. A physiological role for SOD1 in guarding against mitochondrial oxidative damage. J Biol Chem 2001;276(41):38084–9.
13. Liochev SI, Fridovich I. Cross-compartment protection by SOD1. Free Radic Biol Med 2005;38(1):146–7.
14. Carlioz A, Touati D. Isolation of superoxide dismutase mutants in Escherichia coli: is superoxide dismutase necessary for aerobic life? EMBO J 1986;5(3):623–30.
15. Dukan S, Nystrom T. Oxidative stress defense and deterioration of growth-arrested *Escherichia coli* cells. J Biol Chem 1999;274(37):26027–32.
16. Balzan R, Agius DR, Bannister WH. Cloned prokaryotic iron superoxide dismutase protects yeast cells against oxidative stress depending on mitochondrial location. Biochem Biophys Res Commun 1999;256(1):63–7.
17. Longo VD, Gralla EB, Valentine JS. Superoxide dismutase activity is essential for stationary phase survival in *Saccharomyces cerevisiae*. Mitochondrial production of toxic oxygen species in vivo. J Biol Chem 1996;271(21):12275–80.
18. Barker MG, Brimage LJ, Smart KA. Effect of Cu,Zn superoxide dismutase disruption mutation on replicative senescence in *Saccharomyces cerevisiae*. FEMS Microbiol Lett 1999; 177(2):199–204.
19. Longo VD, Liou LL, Valentine JS, Gralla EB. Mitochondrial superoxide decreases yeast survival in stationary phase. Arch Biochem Biophys 1999;365(1):131–42.
20. Barros MH, Bandy B, Tahara EB, Kowaltowski AJ. Higher respiratory activity decreases mitochondrial reactive oxygen release and increases life span in *Saccharomyces cerevisiae*. J Biol Chem 2004;279(48):49883–8.
21. Gardner PR, Fridovich I. Inactivation-reactivation of aconitase in Escherichia coli. A sensitive measure of superoxide radical. J Biol Chem 1992;267(13):8757–63.

22. Bonawitz ND, Rodeheffer MS, Shadel GS. Defective mitochondrial gene expression results in reactive oxygen species-mediated inhibition of respiration and reduction of yeast life span. Mol Cell Biol 2006;26(13):4818–29.
23. Gakh O, Park S, Liu G et al. Mitochondrial iron detoxification is a primary function of frataxin that limits oxidative damage and preserves cell longevity. Hum Mol Genet 2006;15(3):467–79.
24. Li Y, Huang TT, Carlson EJ et al. Dilated cardiomyopathy and neonatal lethality in mutant mice lacking manganese superoxide dismutase. Nat Genet 1995;11(4):376–81.
25. Maclean MJ, Aamodt R, Harris N et al. Base excision repair activities required for yeast to attain a full chronological life span. Aging Cell 2003;2(2):93–104.
26. Madia F, Gattazzo C, Fabrizio P, Longo VD. A simple model system for age-dependent DNA damage and cancer. Mech Ageing Dev 2007;128(1):45–9.
27. Manon S, Roucou X, Guerin M, Rigoulet M, Guerin B. Characterization of the yeast mitochondria unselective channel: a counterpart to the mammalian permeability transition pore? J Bioenerg Biomembr 1998;30(5):419–29.
28. Huang ME, Rio AG, Nicolas A, Kolodner RD. A genomewide screen in *Saccharomyces cerevisiae* for genes that suppress the accumulation of mutations. Proc Natl Acad Sci USA 2003;100(20):11529–34.
29. Farr SB, D'Ari R, Touati D. Oxygen-dependent mutagenesis in *Escherichia coli* lacking superoxide dismutase. Proc Natl Acad Sci U S A 1986;83(21):8268–72.
30. Van Zandycke SM, Sohier PJ, Smart KA. The impact of catalase expression on the replicative lifespan of *Saccharomyces cerevisiae*. Mech Ageing Dev 2002;123(4):365–73.
31. Ligr M, Madeo F, Frohlich E, Hilt W, Frohlich KU, Wolf DH. Mammalian Bax triggers apoptotic changes in yeast. FEBS Lett 1998;438(1–2):61–5.
32. Longo VD, Ellerby LM, Bredesen DE, Valentine JS, Gralla EB. Human Bcl-2 reverses survival defects in yeast lacking superoxide dismutase and delays death of wild-type yeast. J Cell Biol 1997;137(7):1581–8.
33. Herker E, Jungwirth H, Lehmann KA et al. Chronological aging leads to apoptosis in yeast. J Cell Biol 2004;164(4):501–7.
34. Zambrano MM, Siegele DA, Almiron M, Tormo A, Kolter R. Microbial competition: *Escherichia coli* mutants that take over stationary phase cultures. Science 1993;259(5102):1757–60.
35. Liu Z, Butow RA. Mitochondrial retrograde signaling. Annu Rev Genet 2006;40:159–85.
36. McCammon MT, Epstein CB, Przybyla-Zawislak B, McAlister-Henn L, Butow RA. Global transcription analysis of Krebs tricarboxylic acid cycle mutants reveals an alternating pattern of gene expression and effects on hypoxic and oxidative genes. Mol Biol Cell 2003;14(3):958–72.
37. Longo VD, Finch CE. Evolutionary medicine: from dwarf model systems to healthy centenarians. Science 2003;299:1342–6.
38. Fabrizio P, Pozza F, Pletcher SD, Gendron CM, Longo VD. Regulation of longevity and stress resistance by Sch9 in yeast. Science 2001;292(5515):288–90.
39. Kimura KD, Tissenbaum HA, Liu Y, Ruvkun G. *daf-2*, an insulin receptor-like gene that regulates longevity and diapause in *Caenorhabditis elegans*. Science 1997;277(5328):942–6.
40. Murphy CT, McCarroll SA, Bargmann CI, et al. Genes that act downstream of DAF-16 to influence the lifespan of *Caenorhabditis elegans*. Nature 2003;424(6946):277–83.
41. Harris N, Bachler M, Costa V, Mollapour M, Moradas-Ferreira P, Piper PW. Overexpressed Sod1p acts either to reduce or to increase the lifespans and stress resistance of yeast, depending on whether it is Cu(2+)-deficient or an active Cu,Zn-superoxide dismutase. Aging Cell 2005;4(1):41–52.
42. Harris N, Costa V, MacLean M, Mollapour M, Moradas-Ferreira P, Piper PW. Mnsod overexpression extends the yeast chronological (G(0)) life span but acts independently of Sir2p histone deacetylase to shorten the replicative life span of dividing cells. Free Radic Biol Med 2003;34(12):1599–606.
43. Keaney M, Matthijssens F, Sharpe M, Vanfleteren J, Gems D. Superoxide dismutase mimetics elevate superoxide dismutase activity in vivo but do not retard aging in the nematode *Caenorhabditis elegans*. Free Radic Biol Med 2004;37(2):239–50.

44. Melov S, Ravenscroft J, Malik S et al. Extension of life-span with superoxide dismutase/catalase mimetics. Science 2000;289(5484):1567–9.
45. Holzenberger M, Dupont J, Ducos B et al. IGF-1 receptor regulates lifespan and resistance to oxidative stress in mice. Nature 2003;421(6919):182–7.
46. Huang TT, Carlson EJ, Gillespie AM, Shi Y, Epstein CJ. Ubiquitous overexpression of CuZn superoxide dismutase does not extend life span in mice. J Gerontol A Biol Sci Med Sci 2000;55(1):B5–B9.

6
Oxidative Stress and Aging in the Nematode *Caenorhabditis elegans*

David Gems and Ryan Doonan

Summary The senescent decline that leads inevitably to death in most animal species is accompanied by a massive increase in molecular damage. Yet, the chain of events that initially causes this process, and the determinants of the rate at which it happens, remain poorly understood. For many years, much research on this topic has been guided by an interrelated set of theories that view oxidative damage as a potential primary cause of aging. These theories have framed the construction and interpretation of many studies in the nematode *Caenorhabditis elegans*. In this chapter, we critically survey these studies. Overall, these investigations have either disproved or, at least, failed to find clear evidence for many of the oxidative damage theories. In particular, they have failed to demonstrate any role of metabolic rate or mitochondrial superoxide ($O_2^{\cdot-}$) in aging. However, they have revealed a powerful influence of mitochondria on the rate of aging in *C. elegans*. This may or may not have something to do with mitochondrial $O_2^{\cdot-}$ production.

Keywords *Caenorhabditis elegans*, aging, oxidative stress, molecular damage, metabolism, mitochondria, antioxidant.

1 Introduction

Theories relating oxidative damage to aging, which have been reviewed previously [1, 2] and in Chapter 1 of this book, have motivated a large number of studies using *C. elegans*. These theories have been linked conceptually to form a theoretical framework (Fig. 6.1A). Briefly, aging is the result of molecular damage. This results in particular from reactions with reactive oxygen species (ROS), such as $O_2^{\cdot-}$ (superoxide) and its derivatives, which is produced mainly as a by-product of the activity of the mitochondrial electron transport chain (ETC). The rate of oxidative metabolism is a determinant of aging, because it affects the rate of production of ROS.

Although this is sometimes represented as a unified theory, it contains a number of distinct and testable propositions that, individually, may or may not be

From: *Aging Medicine: Oxidative Stress in Aging: From Model Systems to Human Diseases*   81
Edited by: S. Miwa, K.B. Beckman, and F.L. Muller © Humana Press, Totowa, NJ

A. Oxidative Damage Theory of Aging

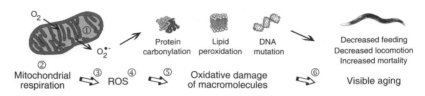

B. Support For Theory in *C. elegans*

Hypothesis	Likelihood	Evidence
① Mitochondrial function influences aging	Definitely	Many mitochondrial defects increase lifespan
② Metabolic rate determines aging rate	Unclear	Conflicting; support is correlative only
③ Mitochondrial superoxide causes aging	Unlikely	SOD mimetics do not increase lifespan
④ ROS-mediated damage is a cause of aging	Unclear	Weak; best evidence: *gst-10* studies
⑤ ROS are a cause of molecular damage	Likely	None
⑥ Molecular damage causes aging	Likely	Consistent, but correlative

Fig. 6.1 Oxidative stress and aging in the nematode *C. elegans*. (A) Oxidative damage theory of aging, as it pertains to *C. elegans*. This theory proposes that mitochondrial ROS (specifically $O_2^{\cdot-}$) are directly responsible for oxidative damage to cellular macromolecules. This damage ultimately manifests itself as deterioration of tissues, resulting in observable changes in behavior, morphology, and mortality rate associated with aging. Protein carbonylation, lipid peroxidation, and DNA mutation can all be assayed biochemically in *C. elegans*. In contrast, direct measurements of ROS are difficult, especially in vivo. Numbers pertain to hypotheses given in B. (B) Support for the oxidative damage theory of aging based on studies using *C. elegans* as a model system. Likelihood: overall assessment based on information given in this chapter. Evidence: brief comment on the nature of the evidence (i.e., not a full justification of likelihood). Note that much evidence is merely correlative, limiting the strength of proof or disproof for most hypotheses

true (Fig. 6.1B). For example, aging may or may not be caused by molecular damage, and this damage may or may not be caused by ROS. The main source of ROS may or may not be mitochondrial oxidative phosphorylation, and, more broadly, mitochondria may or may not be critical determinants of aging rate. Metabolic rate may or may not affect aging, and any links between metabolic rate and aging may or may not reflect effects on $O_2^{\cdot-}$ production. The use of *C. elegans* as an experimental model introduces another dimension to each of these questions, namely, Does the role of any of these factors show evolutionary conservation? As a hypothetical example, $O_2^{\cdot-}$ might be a major cause of the damage that underlies aging in mammals, but not in *C. elegans*.

In the following discussion, we critically assess each of these questions in turn by examining relevant experimental studies using *C. elegans*. There is a rich and complex scientific literature in this field, particularly due to the work of Siegfried Hekimi (McGill University, Canada), Naoaki Ishii (Tokai University, Japan), and Jacques Vanfleteren (Ghent University, Belgium), and their collaborators. Overall, these studies imply that some of the propositions above are true, some are half-truths, and some are false, at least as far as *C. elegans* is concerned.

2   Why Test Theories of Aging in *C. elegans*?

2.1   *C. elegans as a Model for Studies of Aging*

Caenorhabditis elegans is a free-living nematode of little economic importance, found in soil rich in organic matter. Experimentally, it has the advantage of being a complex animal, with a nervous system, reproductive system, and alimentary canal, yet one that is so small (adults are ~1.2 mm in length) that it may be handled like a microorganism, with the convenience and low cost that this implies. For studying aging, it has two particular advantages: its life span is very short (usually 2–3 weeks; Fig. 6.2), and there are no inbreeding effects on life span, which have complicated studies of the genetics of aging in *Drosophila* and the mouse. There are also the obvious advantages of an established genetic model system: availability of a well-annotated genome sequence, well-characterised mutations in large numbers of genes, and powerful molecular genetic methodologies. The latter include RNA-mediated interference to knock down gene expression, construction of transgenic animals, and use of fluorescent proteins to visualize gene expression within the transparent body of the nematode. The existence of a well-coordinated community of *C. elegans* researchers has led to the creation of central research resources. For example, information on *C. elegans* is collated into a central Web facility, WormBase (http://www.wormbase.org), and cooperatively written books on paper [3, 4], and, more recently, freely available online (http://www.wormbook.org). Experimental resources include strain distribution via the Caenorhabditis Genetics Center (http://biosci.umn.edu/CGC/CGChomepage.htm), the Fire Lab plasmid vector kit for preparation of transgenic lines (currently distributed commercially by Addgene), and a library of RNA interference (RNAi) feeding clones that includes most of the genes in the *C. elegans* genome [5].

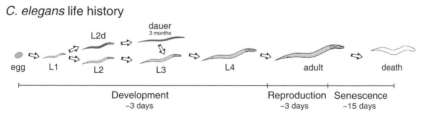

Fig. 6.2 *C. elegans* life history. This is broadly divisible into embryonic and larval development, reproduction, and senescence. Larval development has four stages of growth (e.g., L1, larval stage 1). Total life span is a mere 3 weeks at 20 °C (*C. elegans* life span is dependent on ambient temperature). Typically, life span measurements represent adult life span only, the mean being approximately 18 days. In contrast, at the L2 stage, larvae can enter an alternative, dormant state known as dauer. Dauer larvae can survive for at least 3 months without food, essentially a fourfold increase in longevity. After exposure to a food source, dauer larvae resume development and subsequently reproduce and senesce as normal

2.2   Approaches to Testing Oxidation-Related
       Theories of Aging

The role of oxidative stress in aging in C. elegans has been investigated in several different ways. First, correlations between aging and various aspects of oxidative metabolism have been examined. Such studies typically either examine age changes in wild-type nematodes, or differences between wild-type nematodes and mutants with altered rates of aging. Attempts also have been made to test theories of aging more directly by manipulating individual aspects of the relevant biology (e.g., antioxidant defense) and looking at effects on aging. The majority of studies have been of the less informative first type.

One of the strengths of C. elegans as a model for studying aging is the ease with which classical genetic approaches may be applied. Many genes have been identified where loss of function due to mutation or RNAi leads to altered life span. A problem with studies of short-lived strains is that a reduction in life span can result either from accelerated aging (progeria) or from pathologies unrelated to normal aging, and it can be difficult to distinguish the two. However, methods have been developed to identify likely instances of progeria [6, 7], and some studies of short-lived mutants have been informative. For example, the gene mev-1 encodes a subunit of complex II in the electron transport chain (ETC) [8]. Mutation of mev-1 results in hypersensitivity to oxidative stress, elevated production of mitochondrial ROS, and shortened life-span [9, 10]; for a recent review on mev-1, see Ishii et al. [11]. Most studies have focused on mutations that increase life span, such as those affecting the insulin/insulin-like growth factor (IGF)-1 signaling (IIS) pathway, which can more than double the adult life span [12]. For example, long-lived IIS mutants show resistance to oxidative stress and increased levels of the antioxidant enzymes superoxide dismutase (SOD) and catalase [13–15].

Many correlative studies of this type suggest a link between oxidative damage and aging, from which it is sometimes tempting to conclude: there is no smoke without fire, i.e., surely the oxidative damage theory must be true? However, it is not safe to conclude this. As yet, there is no direct evidence demonstrating, for example, control of normal aging in C. elegans by superoxide or SOD, or hydrogen peroxide (H_2O_2), or catalase. In fact, relatively few studies have been conducted that directly test oxidative damage theories of aging in C. elegans. Many more studies of this sort have been conducted in other models. For example, numerous studies of the effects on aging of overexpression of SOD and catalase have been conducted in Drosophila [16–18]. Ultimately, it is likely that only by means of such direct testing will theories of aging be verified or falsified in C. elegans. In the overview that follows, the evidence for and against each individual oxidation-related theory (Fig. 6.1B) is examined.

3 Is Aging in *C. elegans* Caused by Molecular Damage?

3.1 Age Increases in Damage to Protein, DNA, and Lipid

As in other organisms, levels of molecular damage increase with age in *C. elegans* (Fig. 6.1A). Age increases in levels of oxidized (carbonylated) proteins are seen in whole *C. elegans* extracts [19, 20]. One protein showing large age increases in carbonylation is the yolk protein vitellogenin 6 [21]. Vitellogenins accumulate to high levels during aging in *C. elegans* [6, 22]. Properties of some *C. elegans* vitellogenins suggest that they may form part of lipoprotein particles akin to mammalian apoB-dependent low-density lipoprotein (LDL) particles [23]. Oxidation of LDLs by ROS contributes to atherosclerosis in mammals. Together, this suggests distant molecular parallels between protein aging in *C. elegans* and mammalian atherosclerosis.

Recently, it was discovered that levels of carbonylated proteins increase with age in mitochondria but not the cytoplasm [20] (F. Matthijssens and J. R. Vanfleteren, personal communication). This intriguing result suggests various possibilities: that damage to mitochondria is critical in *C. elegans* aging, that molecular damage occurs more readily within mitochondria (perhaps due to $O_2^{\cdot-}$), and that damaged proteins are repaired or replaced more efficiently in the cytosol than in mitochondria.

Levels of DNA damage also increase with age in *C. elegans*. There are age increases in numbers of single-strand DNA breaks [24] and of deletions in mitochondrial DNA [25]. Age changes in other forms of molecular damage such as lipid peroxidation and glycation remain largely unexplored in *C. elegans*.

3.2 Age Increases in Blue Fluorescence

Accumulation of fluorescent material occurs during aging in a wide range of animal species, including humans. Such age pigment, or lipofuscin, is a complex agglomeration of damaged lipids, proteins, and carbohydrates [26]. Lipofuscin is thought to represent the residuum of damaged molecular matter that the cell is unable to dispose of, which typically accumulates in lysosomes and may contribute to aging [27, 28].

In *C. elegans*, age increases in fluorescent material also have been observed, either by spectrophotometric assays of nematode extracts [29–32] or whole animals [7] (Fig. 6.3A), or visual examination of animals by using epifluorescence microscopy [6, 33] (Fig. 6.3B–E). The latter approach reveals punctate blue fluorescence, particularly in the intestine (gut granules) (Fig. 6.3C). These puncta are probably secondary lysosomes [34].

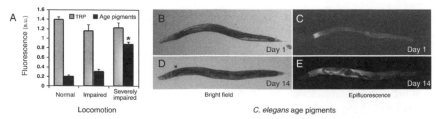

Fig. 6.3 Blue fluorescence as a biomarker of aging in *C. elegans*. (**A**) Age pigment fluorescence reflects physiological, rather than chronological age. As worms age, normal spontaneous locomotion progressively deteriorates to movement only when touched. Before death, only the head moves feebly when touched. Senescent adults of identical age were sorted as class A (normal), class B (impaired), and class C (severely impaired) based on locomotion, and blue fluorescence level was measured for each class. Note that all animals are of the same chronological age, suggesting that age pigment levels correlate with impending death rather than age. TRP, tryptophan fluorescence. Adapted from Gerstbrein et al. [7], with permission of Blackwell Publishing. (**B–E**) Before death, blue fluorescent material seems to be redistributed from the intestine to the pseudocelom. (**C**) Fluorescent gut granules in the intestine of a young (day 1) adult. (**E**) Prior to death (day 14), fluorescence increases dramatically with a rapid, redistribution of fluorescent material into the pseudocelomic space, apparently accompanying an organism-wide breakdown in tissue and organ integrity. Overall, these results suggest that in *C. elegans*, the age increase in blue fluorescence does not reflect the slow age increase in molecular damage, but rather is an indicator of impending death in individual nematodes

The age increase in blue fluorescence could reflect the broader age accumulation of molecular damage and might, in principle, contribute to aging. However, neither of these possibilities stands up well to scrutiny. Although fluorescent gut granules are highly visible even in late larvae and young adults, during early and mid-adulthood the population mean increases in blue fluorescence are modest [7] (A. Taylor and D. Gems, unpublished). More significantly, if aging nematodes are graded on the basis of impaired locomotion (which reflects declining life expectancy), only the most impaired show increased blue fluorescence (Fig. 6.3A; [7]) (D. Gems, unpublished). Moreover, this increase is not due to increased gut granule fluorescence, but rather to a sudden, large increase in fluorescent material in the pseudocelom of the worm in the days preceding its death (compare Fig. 6.3C and E) (A. Taylor and D. Gems, unpublished). In addition, *Escherichia coli*-fed *C. elegans* maintained in liquid culture, conditions that result in a normal life span, showed only marginal increases in blue fluorescence with age, yet they showed a normal life span [7]. That animals age normally in the absence of substantial increases in blue fluorescence suggests that it contributes little to aging. Another study in liquid culture saw a substantial age increase in blue fluorescence [32], perhaps due to higher food levels used (J. R. Vanfleteren, personal communication).

Together, these findings cast some doubt on the view that age increases in blue fluorescence reflect overall age increases in molecular damage that cause aging. The age increase in blue fluorescence in *C. elegans* may instead be a culture condition-dependent effect reflecting terminal pathology in nematodes as they approach death.

3.3 Molecular Damage in Mutants with Altered Life Span

Studies of whole *C. elegans* homogenates show that *daf-2* and *age-1* mutants (long-lived) accumulate protein carbonyls at a lower rate than wild type, whereas *mev-1* and *daf-16* mutants (short-lived) accumulate them more quickly [19, 35, 36]. Accumulation of protein carbonyls was also slower than wild type in isolated mitochondria from *daf-2(e1370)* animals [37]. *mev-1(kn1)* mutants also show increased levels of DNA damage (8-oxo-7,8-dihydro-2′-deoxyguanosine) and elevated nuclear mutation rate [38]. Thus, there is a clear general correlation between rate of damage accumulation and aging. Mutations which affect life span also affect age increases in blue fluorescence. In long-lived *daf-2* mutants, the age increase is slowed down, whereas in short-lived *daf-16* mutants it is accelerated [6, 7].

3.4 Conclusions

Overall, these findings are consistent with the view that accumulation of molecular damage causes aging. Yet, it remains unclear whether the accumulation of damage is really the cause of aging (i.e., of increased morbidity and mortality rate) or merely a noncausal correlate either of a different sort of damage that is causal, or some other age-associated change. If these or other sorts of damage *are* causal, it is not clear whether they are the primary cause of aging, or downstream, knock-on effects of some unknown primary cause.

4 Do Reactive Oxygen Species Cause Aging in *C. elegans*?

4.1 Alterations of Prooxidant Levels

If aging is caused by ROS, then manipulating ROS levels should affect aging rate. The effects of ambient oxygen concentration on life span and mortality rate have been tested in wild-type and *mev-1* mutant populations (Fig. 6.4A) [39]. In wild-type, these parameters were unaltered in 2, 8, and 40% oxygen relative to 21%. This is a striking result, because it implies either that levels of $O_2^{\cdot-}$ production are unaltered over this range, or that ROS are not a determinant of aging. However, very large changes in O_2 concentration can affect aging in wild type. In 1% O_2, mean life span was increased by 15% (Fig. 6.4A), and the Gompertz component of mortality was decreased [39]. Whether this effect is mediated by changes in $O_2^{\cdot-}$ production, metabolic rate, or some other factor is unclear. In 60% O_2, wild type mean life span was slightly reduced (by 14%) (Fig. 6.4A), perhaps due to increased oxidative damage. In contrast to wild type, in *mev-1* populations there is a direct relationship between oxygen concentration and life span (Fig. 6.4A); the mutation rate in *mev-1* mutants is also hypersensitive to effects of elevated oxygen [38].

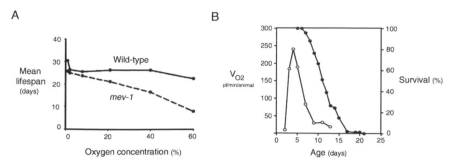

Fig. 6.4 Oxygen and aging in *C. elegans*. (**A**) Effect of ambient oxygen concentration on mean life span. Wild-type and *mev-1* mutants were cultured under various levels of ambient oxygen relative to atmospheric oxygen concentration (21%). Although increasing ambient oxygen presumably leads to increased oxidative stress, wild-type life span is relatively insensitive to ambient oxygen levels. This suggests that ROS are not a determinant of the rate of normal aging. In contrast, *mev-1* mutants are acutely sensitive to ambient oxygen concentration, consistent with the finding that *mev-1* mutants have defective electron transport and elevated ROS production. Adapted from Honda et al. [39], with permission of the Gerontological Society of America. (**B**) Rate of oxygen consumption (V_{O2}) (open circles) and survival (closed circles) of aging wild-type animals at 25 °C. Note that oxygen consumption drops dramatically after day 5 of adulthood. This suggests that age increases in oxidative damage are unlikely to be the result of age increases in mitochondrial superoxide production. Adapted from Suda et al. [101], with permission of Elsevier Publishing

Overall, these results suggest that, under normoxic conditions, $O_2^{\cdot-}$ levels determine aging in *mev-1* but not wild type. The possibilities that $O_2^{\cdot-}$ does not cause normal aging, whereas elevated $O_2^{\cdot-}$ levels can accelerate aging, are not necessarily contradictory. Aging in both cases may involve molecular damage, but with damage resulting from different causes. Indeed, other observations suggest mechanistic differences between aging in *mev-1* and wild type. In otherwise wild-type *C. elegans*, prevention of apoptosis (programmed cell death) by mutation of *ced-3* does not extend life span [6]. Thus, apoptosis does not contribute to normal aging. By contrast, mutation of *ced-3* increases life span of *mev-1* populations, apparently by preventing $O_2^{\cdot-}$-induced apoptosis [40]. However, this extension is the result of suppression of early mortality, and late-life survival was unchanged. *mev-1* mutants also have elevated lactic acid levels, suggesting that lactic acidosis might contribute to their mortality [10].

Direct effects of ROS on *C. elegans* are most often examined by administration of redox cycling compounds such as juglone, or, more commonly, paraquat (methyl viologen) [14, 41–44], which generate $O_2^{\cdot-}$ in vivo. $O_2^{\cdot-}$ production by redox cyclers can be measured as an increase in cyanide-independent O_2 consumption. Although 1 mM paraquat does not detectably increase cyanide-independent O_2 consumption by *C. elegans* [45], 2 mM paraquat does increase it, and this concentration is just sufficient to decrease adult life span [42]. In vivo, redox cyclers receive electrons from NADH or NADPH via the action of diaphorase enzymes, and this activity has been detected in *C. elegans* [45].

In conclusion, although several treatments predicted to increase intracellular ROS have been shown to reduce life span, it remains unclear whether such effects reflect accelerated aging, or whether effects of ROS limit normal aging.

4.2 Does Elevated ROS Accelerate Age Changes in Molecular Damage?

If ROS causes normal aging, one would expect that experimental elevation of ROS would accelerate age changes in molecular damage seen in normal aging (Fig. 6.1). This prediction has been little explored, although one report described increased blue fluorescence under hyperoxia [33].

Isolated mitochondria from *mev-1* animals show elevated levels of $O_2^{\cdot-}$ production [10]. Thus, $O_2^{\cdot-}$ production might be elevated in vivo and it might account for the shortened life span of *mev-1* under normoxia (Fig. 6.4A). The increased levels of protein oxidation in *mev-1* animals supports this hypothesis [19, 36]. *mev-1* also has been reported to elevate levels of blue fluorescence [33]. However, a recent study saw no such effect either in *mev-1* or *gas-1* animals [7].

4.3 Antioxidant Defense and Aging

Organismal defenses against oxidative damage include chemical and enzymatic antioxidants. If oxidative damage causes aging, then one might expect a correlation between antioxidant defense and longevity. Moreover, experimental enhancement of antioxidant defense should retard aging. Many studies have tested both of these expectations; yet in each case, establishing a causal role of oxidative damage in aging is difficult. For example, a correlation between level of an antioxidant agent and longevity could be coincidental. If experimentally induced elevation in levels of an antioxidant agent increases life span, the possibility remains that this occurs by some other mechanism than protection against molecular damage. Moreover, if increases in life span are not seen, it remains possible that multiple antioxidant defense mechanisms act in concert to protect against aging or that antioxidant mechanisms act in concert with other prolongevity mechanisms.

A range of genes and processes contribute to protection against oxidative damage [46, 47], any one of which may limit the rate of age accumulation of molecular damage, and its impact on homeostasis and survival. In the first line of defense are enzymes that detoxify primary prooxidant molecules. For example, SODs convert $O_2^{\cdot-}$ into H_2O_2 [48], and this is converted into water and O_2 by catalases and glutathione peroxidases (GPX). Numerous proteins affect ROS production levels, such as metal trafficking proteins. Free metal ions such as Fe^{3+} stimulate production of very damaging forms of ROS such as OH^-, and metallothioneins and ferritins counteract this production. The forms of molecular damage that can occur are extremely diverse, as are the enzymes that detoxify, repair, or remove damaged moieties. For example, peroxidised lipids are targets for numerous glutathione lipid hydroperoxidases and glutathione *S*-transferases (GSTs). In proteins, oxidation of just the amino acid methionine can be repaired by methionine sulfoxide reductase. Effects of oxidative damage to protein on protein function can, to some extent, be restored by the action of molecular chaperones. Finally, oxidized proteins can be removed by cellular turnover processes such as proteasome-dependent protein degradation and autophagy. Any of these enzymes and processes could, in principle, contribute to longevity assurance.

4.4 SOD and Catalase

The biology of SOD and catalase in *C. elegans* is unusual in several respects. For example, *C. elegans* has more isoforms of these enzymes than higher animals. Instead of one cytosolic Cu/Zn SOD there are two, encoded by *sod-1* and *sod-5* [13, 49, 50], and instead of one mitochondrial Mn SOD there are also two, encoded by *sod-2* and *sod-3* [51–53]. A combination of SOD activity assays in *sod* mutants, and studies of levels of mRNA and reporter expression imply that *sod-1* and *sod-2* are the major isoforms expressed during reproductive development, whereas *sod-3* and *sod-5* are dauer up-regulated isoforms [50, 54, 55] (J. J. McElwee, R. Doonan, and D. Gems, unpublished). Why there should be dauer-specific isoforms is unclear. SOD-2 and SOD-3 Mn SODs have similar specific activities [52], and either SOD-1 or SOD-5 Cu/Zn SOD can rescue the paraquat sensitivity of SOD-deficient yeast [50], which suggests that reproductive and dauer isoforms are not functionally different.

The SOD-1 and SOD-5 Cu/Zn SODs are unusual in other respects. To mature, Cu/Zn SODs must incorporate copper, and in all other eukaryotes, whether animals, fungi, or plants, this requires the copper chaperone of SOD (CCS) protein. Uniquely, *C. elegans* does not possess a CCS, and Cu/Zn SOD maturation does not require it, but instead depends on an unidentified glutathione-dependent pathway [50]. Studies of SOD-1 and SOD-5 expressed in yeast also hint that, in contrast to other eukaryotes, *C. elegans* might not have Cu/Zn SOD in the mitochondrial intermembrane space, although the evidence here is not conclusive [50].

The Cu/Zn SOD encoded by *sod-4* is similar to mammalian extracellular Cu/Zn SODs [56]. However, it is also different in that there are two predicted isoforms, products of alternative splicing of mRNA. SOD4-1 resembles a typical secreted Cu/Zn SOD, but SOD4-2 has an additional C-terminal sequence resembling a transmembrane domain. This suggests that this unique SOD is secreted from the cell, but then it remains tethered at the cell surface [56].

The *C. elegans* genome contains a tandem array of three genes encoding catalases (*ctl-1, ctl-2,* and *ctl-3*; [57]). By contrast, other metazoans have only a single catalase, whereas *Saccharomyces cerevisiae* have a peroxisomal and a cytosolic catalase. CTL-2 is a peroxisomal catalase, and it is responsible for ~80% of total catalase activity. It also has a lower pH optimum for activity and higher peroxidase activity than mammalian peroxisomal catalases [57–59]. Much of the *ctl-1* and *ctl-3* gene sequences are 100% identical. Studies of a CTL-1::green fluorescent protein (GFP) fusion protein imply that CTL-1 is a cytosolic catalase [58]. Although this paper was retracted (see below), it was for reasons unrelated to the CTL-1::GFP finding. One possibility is that CTL-1 acts as a cytosolic H_2O_2 scavenger because *C. elegans* lacks an H_2O_2-scavenging glutathione peroxidase [14] (J. R. Vanfleteren, personal communication). A promoter fusion test implies that *ctl-3* is expressed in pharyngeal muscle and neurons [57]. More work is needed to confirm and define the cellular localization of CTL-1 and CTL-3. In summary, compared, for example, with humans, *C. elegans* has a more elaborate armoury of SODs (six) and catalases (three) to detoxify ROS; yet, its life span is a mere few weeks.

Long-lived *daf-2* and *age-1* mutants show age increases in SOD and catalase activity levels, and in resistance to oxidative stress (e.g., paraquat and H_2O_2), increases that are not seen in the wild type [13–15, 54]. Northern blot analysis reveals a large increase in *sod-3* mRNA in *daf-2* mutants [54, 60], and microarray studies reveal additional, smaller increases in *sod-1* and *sod-5* mRNA [61–63]. *sod-3* levels are elevated throughout the life course in *daf-2* mutants, even in the developing embryo [54]. Microarray studies also show increases in expression of at least one catalase gene in *daf-2* mutants, but because of the high degree of similarity between *ctl* gene sequences, one cannot say which gene(s). This also complicates interpretation of RNAi studies [63]. Levels of SOD and catalase also are elevated in *C. elegans* subjected to dietary restriction, and, in contrast to insulin/IGF-1 signaling mutants, this increase does not depend on *daf-16* [64]. In dauer larvae (Fig. 6.2), levels of SOD activity are four- to fivefold higher than in young adults, and levels of *sod-3* mRNA are elevated [13, 54, 65]. Catalase levels also seem to be elevated in dauer larvae [66], although here there is conflicting evidence [13].

It seems likely that the elevated levels of antioxidant enzymes contributes to oxidative stress resistance, at least to some extent, but what about longevity? The effects on aging of manipulations of SOD and catalase levels have been investigated in *C. elegans*, although not as systematically as in *Drosophila*. RNAi knockdown of expression of *sod-3* has been reported to very weakly suppress *daf-2* longevity [63], but, surprisingly, RNAi of *sod-5* had the opposite effect [61]. More surprisingly, deletion of *sod-2* and *sod-3*, alone or in combination, has no effect on adult life span (J. J. McElwee and D. Gems, unpublished).

Whereas deletion of *ctl-1* (cytosolic catalase) has no effect on life span, deletion of *ctl-2* (peroxisomal catalase) shortens life span [57]. The authors interpreted this as progeria, although more evidence would be required to establish this with certainty. *ctl-2* mutants show abnormalities in peroxisomal morphology. Surprisingly, protein oxidation (protein carbonyl levels) increases more rapidly with age in wild-type than in *ctl-1* or *ctl-2* animals [57]. Mutation of *ctl-1* was also at one time thought to suppress the longevity of *daf-2* mutants [58], but the study concerned was subsequently retracted [67].

The effects of overexpression of *sod* genes has not been studied in any detail. Overexpression of a *sod-3::gfp* fusion protein did not affect life span, but, as the authors stressed, SOD activity level was not examined in this strain [68]. In one study, it was observed that loss of heat shock factor 1 (HSF-1) suppressed *daf-2* mutant longevity without suppressing the elevation in *sod-3* expression [69]. This suggests, at least, that elevated *sod-3* expression does not increase life span in animals deficient in HSF-1.

Administration of the SOD mimetic salen manganese compounds EUK-8 and EUK-134 to *C. elegans* results in significant increases in SOD activity levels (e.g., a fivefold increase in mitochondrial SOD activity) and resistance to paraquat [42, 70]. Although one study reported that these compounds also increased life span in *C. elegans* [71], other workers were unable to reproduce this effect, either in *C. elegans* [42, 72] or in *Drosophila* [73] or house flies [74]. Levels of EUK-8 that were optimal for protection against paraquat had no effect on life span. Higher levels of EUK-8 actually shortened life span [42, 72]. These results imply that $O_2^{\cdot-}$ does not contribute to normal aging in *C. elegans* (Fig. 6.1B).

4.5   Other Antioxidant Defenses

If studies of SOD and catalase in *C. elegans* aging are somewhat fragmented, the role of other antioxidant defense processes is a surface that has barely been scratched. In one study glutathione peroxidase (GPX) activity was not detected in *C. elegans* by using either *tert*-butyl hydroperoxide [14] or H_2O_2 (J. R. Vanfleteren, personal communication). However, the *C. elegans* genome contains a number of GPX-like proteins (e.g., C11E4.2, F26E4.12, F55A3.5, R03G5.5, R05H10.5, T09A12.2, and Y94H6A.4); possibly some or all of these are lipid hydroperoxidases.

There are hints of a possible role of metal trafficking proteins in longevity assurance. Exogenous iron shortens life span in *C. elegans* [75], and *daf-2* and *age-1* mutants are resistant to heavy metals (e.g., cadmium and copper), and they show elevated expression of *mtl-1*, which encodes a metallothionein [76]. RNAi of *mtl-1* slightly reduces *daf-2* mutant longevity [63]. *ftn-1*, which encodes a ferritin heavy chain, is also strongly up-regulated in *daf-2* mutants (Table 6.1) [62].

Global changes in gene expression in *daf-2* and *age-1* mutants have been studied using whole genome DNA microarrays [61–63, 77]. By one estimate, 2,348 genes are up- or down-regulated in *daf-2* animals relative to normal-lived *daf-16; daf-2* controls: in other words, some 12% of genes in the *C. elegans* genome [62]. This finding weakens the conclusions of studies showing correlations of expression of individual genes (e.g., *sod-3*) with IIS mutant longevity. In fact, the number of genes that are regulated by IIS is so high that it is possible to find evidence supporting most theories of aging [78].

One way to lessen the problem of bias in data interpretation is to screen for gene classes overrepresented among differentially expressed genes. One study combined this approach with a comparison of array data from *daf-2* mutants (compared with *daf-16; daf-2*) and dauer larvae (compared with recovered dauer larvae) [55, 62, 78]. The rationale here was that it is likely that *daf-2* mutants are long-lived by dint of

Table 6.1 Changes in mRNA levels in *daf-2* versus *daf-16; daf-2* of selected genes linked to antioxidant defense

Gene	Protein	Log-2 FC	*p*
sod-1 (C15F1.7)	Cu/Zn superoxide dismutase	**0.80**	**0**
sod-2 (F10D11.1)	Mn superoxide dismutase	0.37	0.096
sod-3 (C08A9.1)	Mn superoxide dismutase	**4.15**	**0**
sod-4 (F55H2.1)	Cu/Zn superoxide dismutase	0.44	0.16
sod-5 (ZK430.3)	Cu/Zn superoxide dismutase	**1.66**	**0.0015**
gst-4 (K08F4.7)	Glutathione *S*-transferase	**1.78**	**0.00013**
gst-10 (Y45G12C.2)	Glutathione *S*-transferase	*0.76*	*0.032[a]*
C46F11.2	Glutathione reductase	*−0.51*	*0.024*
mtl-1 (K11G9.6)	Metallothionein	**3.45**	**0**
mtl-2 (T08G5.10)	Metallothionein	**−0.80**	**0.00097**
cuc-1 (ZK652.11)	Copper chaperone	*−0.75*	*0.044*
F40G9.2	Copper chaperone	−0.29	0.46
ftn-1 (C54F6.14)	Ferritin heavy chain	**5.19**	**0**
ftn-2 (D1037.3)	Ferritin heavy chain	*−0.45*	*0.030*
F20D6.11	Ferredoxin reductase	**0.87**	**0.0098**
aco-1 (ZK455.1)	Aconitase/iron-regulatory protein	−0.40	0.24
pcs-1 (F54D5.1)	Phytochelatin synthase	−0.103	0.704
trxr-1 (C06G3.7)	Thioredoxin reductase	0.097	0.63
trxr-2 (ZK637.10)	Thioredoxin reductase	0.055	0.82
F43E2.5	Methionine sulfoxide reductase (MsrA)	−0.49	0.091

Data derived from mRNA profile analysis by using whole genome microarrays to compare *daf-2* (long-lived, DAF-16 on) with *daf-16; daf-2* (not long lived, DAF-16 off) strains [62]. Values where $p < 0.01$ are shown in bold; values where $0.001 < p < 0.05$ are in italics. Note that of the 8/20 genes where $p < 0.01$, all but one are up-regulated in *daf-2*. *ctl* (catalase) genes are not shown because similarity between the genes makes microarray data uninterpretable. However, one or more *ctl* gene shows up regulation in *daf-2* animals.
[a]Up-regulation of *gst-10* in *daf-2* mutants has also been demonstrated in an independent study [84].

misexpressing dauer longevity assurance processes. The few gene classes identified as associated with longevity included three involved in the phase 1, phase 2 biotransformation system (i.e., the xenobiotic metabolism or drug detoxification system). In addition, GSTs were strongly overrepresented among genes up-regulated in *daf-2* mutants, but not dauers. The biotransformation is a complex system of enzymes involved in multiple processes, particularly detoxification and clearance of a wide spectrum of endobiotic and xenobiotic toxins [79].

The comparison of transcript profiles from *daf-2* mutants and dauers implies that the biotransformation system is activated in these long-lived milieus and suggests the possibility that these detoxification processes contribute to longevity [78]. Recently, a comparison of transcript profiles from long-lived IIS mutant *C. elegans, Drosophila*, and mouse discovered evolutionary conservation in the up-regulation of three classes of biotransformation enzymes (particularly GSTs) and longevity [80].

GSTs are a highly diverse, rapidly evolving enzyme class (there are 51 putative GST-encoding genes in *C. elegans*). Among other things, GSTs use glutathione conjugation to detoxify endobiotic and xenobiotic toxins, including the products of oxidative damage [81]. A screen for genes up-regulated upon exposure to the $O_2^{\cdot-}$ generator paraquat identified *gst-4* [82]. Overexpression of *gst-4* resulted in increased resistant to paraquat, but not increased life span [83]. However, RNAi of *gst-4* slightly reduces *daf-2* mutant longevity [63], and microarray data implies that *gst-4* expression is increased in *daf-2* mutants (Table 6.1).

The *gst-10* gene is also up-regulated in *daf-2* mutants [84] (Table 6.1). GST-10 protein detoxifies 4-Hydroxy-2-nonenal (HNE), an abundant lipid peroxidation product resulting from oxidative stress [85]. RNAi of *gst-10* increased sensitivity to HNE toxicity, and it reduced life span in both wild-type and *daf-2* mutant populations. The effect of *gst-10* RNAi on *daf-2* mutant life span has been confirmed by us (D. Weinkove and D. Gems, unpublished). RNAi of *gst-5, gst-6, gst-8*, or *gst-24* also increased sensitivity to HNE toxicity, but of these genes only RNAi of *gst-5* reduced life span [86]. Overexpression in *C. elegans* of either *gst-10* or murine *mGsta4* (which also detoxifies HNE) lead to increased levels of HNE-conjugating activity, increased resistance to oxidative stress (e.g., paraquat and H_2O_2) and lowered levels of HNE-protein adducts. Interestingly, overexpression of *gst-10* or *mGsta4* increased median life span, by 22 and 13%, respectively [43]. Arguably, this is the most robust proof to date of a role in *C. elegans* longevity assurance of an enzyme involved in protection against oxidative damage.

Another enzyme contributing to oxidative stress resistance is mitochondrial nicotinamide nucleotide transhydrogenase (NNT). This catalyses the reduction of $NADP^+$ by NADH, providing NADPH for reduction of glutathione within mitochondria. This is important in animal mitochondria, which lack catalase. H_2O_2 generated by SOD is usually detoxified instead by mitochondrial GPX. Reduced glutathione in mitochondria is also a substrate for phospholipid hydroperoxidases. In *C. elegans, nnt-1* is widely expressed (e.g., in intestinal, hypodermal, and neuronal cells). Deletion of *nnt-1* leads to a greatly lowered glutathione (GSH)/glutathione disulfide ratio (58 *vs* 12 in wild type *vs* mutant) [44]. The large magnitude of this effect implies that cytosolic as well as mitochondrial GSH pools are affected. This results in increased sensitivity to paraquat but, oddly, not H_2O_2, and there is no effect on life span [44].

4.6   Noncatalytic Antioxidants

There is a long history of studies of the effects on aging of noncatalytic antioxidants (principally vitamin E), often generating inconclusive findings. Vitamin E studies have used its constituents α-tocopherol and tocotrienols, and the α-tocopherol derivative α-tocopherolquinone (α-TQ). An early study found that α-tocopherol and α-TQ both increase life span of *C. briggsae* (a sister species to *C. elegans*) by 31% [87]. Similarly, vitamin E increased life span in *C. elegans* [88]. However, in both studies nematodes were cultured in an axenic medium (i.e., without *E. coli*), which is nutritionally suboptimal; moreover, the effects of vitamin E on life span were exerted during development, not adulthood. Thus, these findings may reflect a nutritional effect on growth in axenic medium. In another study, vitamin E increased *C. elegans* life span by around 20%, but it also reduced fecundity and delayed the timing of reproduction [89]. Here, the authors concluded that effects on aging could reflect slight toxicity, which slowed development, growth, and aging. Yet, another study compared the effects of α-tocopherol and tocotrienols on levels of protein oxidation, resistance to oxidative damage (exerted by ultraviolet B irradiation), and longevity. Although α-tocopherol had no effect, tocotrienols had a protective effect against damage and stress, and they caused a slight increase in mean but not maximum life span [90]. A more recent report described a single trial where vitamin E increased life span in wild-type (+11%) but not *mev-1* animals [91]. Overall, and taking into account the tendency to publish only results showing positive effects, these studies provide little persuasive evidence that vitamin E supplementation protects against aging.

4.7 Conclusions

Aging in *C. elegans* is accompanied by an accumulation of molecular damage, but why this accumulation occurs is unclear. It is also unclear to what extent this damage is caused by ROS, or $O_2^{\cdot-}$ in particular. If molecular damage causes aging, it is unclear how important damage caused by $O_2^{\cdot-}$ is. Arguably, the strongest evidence that $O_2^{\cdot-}$ does contribute to *C. elegans* aging is that overexpression of HNE-conjugating GSTs can increase longevity, because $O_2^{\cdot-}$ contributes to HNE formation. The fact that SOD mimetics do not increase life span seems to contradict this; an alternative interpretation is that there exists a proportion of $O_2^{\cdot-}$ in cells whose level, by some unknown mechanism, is unaffected by increases in levels of SOD activity.

5 Do Mitochondria Play a Role in *C. elegans* Aging?

5.1 Does Superoxide Production by Mitochondria Contribute to Aging?

Studies of the source of oxidative damage in the cell have often focused on $O_2^{\cdot-}$ produced as a by-product of the reduction of O_2 by the mitochondrial ETC. Isolated mitochondria or submitochondrial particles can generate substantial amounts of $O_2^{\cdot-}$. For example, $O_2^{\cdot-}$ production by isolated rat liver mitochondria respiring in state 4 accounts for around 1–2% of oxygen consumed [92]. However, levels of mitochondrial $O_2^{\cdot-}$ production in vivo are much lower, in the 0.1–0.3% range [93, 94], and the relevance of mitochondrial $O_2^{\cdot-}$ to aging remains unclear [95, 96]. In addition, the relative importance of other sources of ROS as contributors to molecular damage and aging is unknown; ROS, including $O_2^{•-}$ and H_2O_2, also are produced in other ways, for example $O_2^{•-}$ by membrane-associated NADPH oxidase, cytochrome P450 oxidases, and xanthine oxidase. The notion that mitochondrial $O_2^{•-}$ causes aging remains very much a hypothesis.

5.2 Mitochondria, Superoxide, and Aging in C. elegans

C. elegans mitochondria are similar in many respects to those of higher animals. For example, their mitochondrial DNA is similar in terms of gene content and overall size [97, 98]. However, there are some significant differences (see below) so, as always with C. elegans, once should generalize cautiously. Little is known about levels of mitochondrial $O_2^{•-}$ production in vivo in C. elegans and whether it contributes to aging. However, isolated C. elegans mitochondria do produce $O_2^{•-}$ [10].

In mammalian cells, levels of mitochondrial $O_2^{•-}$ production increase with age, e.g., a 25% increase with age in isolated rat heart mitochondria [96]. One study has reported that there is no age increase in mitochondrial $O_2^{•-}$ in C. elegans [20]. A second recent study has even reported a decline with age in mitochondrial ROS production (measured as H_2O_2) [37]. Consistent with this, complex I activity drops by 60% between day 4 and day 12 [20].

As worms age, their rate of oxygen consumption drops dramatically (Fig. 6.4B) [20, 99–101]. For example, a recent study measured a drop in oxygen consumption from ~200 pl/min/worm in early adulthood to ~25 pl/min/worm by 9 days of age [101]. Taken together, these results suggest that in vivo levels of mitochondrial $O_2^{•-}$ production decrease substantially with age in C. elegans. Thus, the age increase in oxidative damage in C. elegans seems unlikely to be due to increased $O_2^{•-}$ production later in life.

5.3 Mitochondrial ETC Defects Can Increase
or Reduce Life Span

Based on the oxidative damage theory, one might predict that mutations affecting components of the ETC could, in principle, either decrease or increase life span. Abnormalities in electron flow might increase production of $O_2^{•-}$ and reduce life span; alternatively, overall reduction in electron flow might lower $O_2^{•-}$ and increase life span. In C. elegans, both effects of disruption of ETC genes on life span have been seen, but any role of $O_2^{•-}$ in this remains undemonstrated (for review, see [102]).

In several large scale RNAi screens for genes with effects on life span, genes encoding mitochondrial proteins predominated [103–106]. In particular, RNAi of many mitochondrial and nuclear genes encoding proteins of ETC complexes I–V caused substantial increases in life span. RNAi affecting other mitochondrial proteins, such as mitochondrial carriers, also increased life span. The combination of mitochondrial defects and increased life span is sometimes referred to as the Mit phenotype [107]. In most cases, Mit animals also show delayed development, reductions in body size, fertility, activity level, and feeding rate [103], and abnormalities in mitochondrial morphology [104]. For some genes, Mit animals have normal body size and feeding rates, but increased life span [105], implying that life extension is not causally connected to reduced body size or feeding rate. The loss of a single protein component of the large ETC protein complexes may cause accumulation of unfolded proteins in the mitochondria, and in many cases, Mit animals accumulate the mitochondrial chaperone HSP6 [106, 108]. Mit mutants also have been identified with mutations that either affect ETC genes directly [109, 110] or in the case of *lrs-2*, indirectly. *lrs-2* encodes a unique mitochondrial leucyl-tRNA synthetase, which is required for the expression of the 12 mitochondrially encoded polypeptides. The *lrs-2* mutation is predicted to block expression of all twelve of these polypeptides; maternally rescued mutants form small, sterile, long-lived adults [104]. Severe loss of ETC function often causes larval arrest and lethality [104, 110].

By what mechanisms might life span be extended in Mit animals? One interpretation is that it is due to reduced metabolic rate and perhaps also to lowered production of $O_2^{\cdot}$. Several observations are consistent with the first view: in Mit animals, there is usually a reduction in O_2 consumption rate [104], and ATP levels can be reduced to as little as 20% of wild type [103]. This might suggest that ATP levels limit the rate of processes that promote aging. However, a challenging study by Dillin et al. [103] suggests that something more complex is occurring. To test the timing of effects of Mit defects on aging, expression of ETC genes was selectively knocked down during larval development or in adulthood. Knockdown in larvae alone increased adult life span [103]. This is perhaps not surprising because mitochondrial number may be programmed during development: in the transition from L4 to adulthood alone there is a sixfold increase in the number of mitochondria [111].

More surprisingly, adult-specific knockdown of ETC gene expression reduced ATP levels but did not increase life span. The authors postulated that there exists in *C. elegans* a system that registers the rate of respiration during development and adjusts the rate subsequent of aging accordingly [103]. The reason this is quite surprising is that the timing of action of insulin/IGF-1 signaling and dietary restriction are exactly the opposite: during adulthood and not development. This finding warrants further investigation: for example, how does life-long, larva-specific and adult-specific RNAi of ETC genes compare in terms of effects on mitochondrial $O_2^{\cdot-}$ production, O_2 consumption, mitochondrial number and morphology, and HSP-6 expression?

Life extension in Mit animals is unlikely to involve the insulin/IGF-1 pathway, because knockdown of ETC genes increases life span both in *daf-16* and *daf-2* mutant animals [103–106]. Life span is also increased by RNAi of several genes encoding glycolytic enzymes such as phosphoglycerate mutase (F57B10.3) [104] and glucose-6-phosphate isomerase (Y87G2A.8) [105], suggesting that glycolysis somehow reduces life span. This seems to involve different mechanisms relative to Mit animals, because animals develop normally, body size is not reduced, mitochondria show normal morphology, and the extension in life span requires *daf-16* [104, 105].

Coenzyme Q (CoQ, or ubiquinone) plays a major role in the ETC. Production of $O_2^{\bullet-}$ by the mitochondrial ETC seems to be largely the result of transfer of electrons from ubisemiquinone to oxygen [2, 92]. Deficiency in CoQ can apparently also increase life span. For example, *clk-1* encodes a mitochondrial protein necessary for the final step in CoQ biosynthesis [112–114, for review, see 115]. Mutation of *clk-1* causes accumulation of the precursor of nematode CoQ, demethoxy-ubiquinone-9 [116], and increased life span [117]. CoQ varies between species in the number of isoprene units in its side chain. *E. coli* have an eight unit side chain (CoQ_8), *C. elegans* have CoQ_9, and mammals CoQ_{10}. Likewise, if *C. elegans* are fed on *E. coli* lacking CoQ_8, this increases their life span, too [118]. The above-mentioned findings might suggest that lowering CoQ levels reduces flux through the ETC, thereby lowering ROS production and increasing life span (but see below).

In a few cases, mutations affecting ETC proteins lead to a shortening of life span. *mev-1(kn1)* is a point mutation in the gene for succinate dehydrogenase cytochrome *b* in complex II, and it causes hypersensitivity to oxidative stress and shortened life span [9]. The mutation compromises electron transfer from succinate to ubiquinone and results in increased electron leakage to oxygen. In wild-type mitochondria, $O_2^{\bullet-}$ production results from electron leak at complex I and particularly III [2]. *mev-1(kn1)* disrupts complex II and results in $O_2^{\bullet-}$ production from complex II [10]. *gas-1* encodes a subunit of complex I and, like *mev-1*, mutation of *gas-1* results in hypersensitivity to oxidative stress and reduced life span under normoxia [119]. Unlike *mev-1, gas-1* does not increase nuclear mutation rate.

The short life span and sensitivity to prooxidants of *mev-1* animals was first attributed to the fact that SOD levels are half that of wild type [9]. Consistent with this, deletion of *sod-1*, the major Cu/Zn SOD in *C. elegans*, shortens life span (J. J. McElwee and D. Gems, unpublished). However, although it was reported that *mev-1* life span can be extended by administration of chemical mimetics of SOD [71], a further study was unable to replicate this finding (F. Matthijssens and J. R. Vanfleteren, personal communication).

mev-1 and *gas-1* are atypical among genetic interventions affecting mitochondria and life span in that they shorten rather than increase life span. Here, it is worth bearing in mind that *mev-1(kn1)* is a reduction-of-function allele, and not a null; RNAi of *mev-1* results in a high level of embryonic lethality [120]. *mev-1(kn1)* reduces activity of the ETC by 80%, but it does not affect succinate dehydrogenase activity [8]. The MEV-1 subunit of complex II contains a binding site for CoQ. Potentially, reduced affinity of CoQ to MEV-1 protein leads to increased mobility of CoQ and electron leak to oxygen. By contrast, in most cases knockdown of expression of genes encoding ETC proteins may simply reduce electron flow and $O_2^{\bullet-}$ formation.

5.4 Is Superoxide Production Important for Mitochondrial Effects on Aging?

Mutational studies have demonstrated that mitochondria can influence aging in *C. elegans*. One interpretation is that this reflects altered electron flux through the ETC and altered $O_2^{\cdot-}$ levels. If this were true, one would expect an accompanying alteration in metabolic rate. Moreover, one would not expect an increase in somatic maintenance mechanisms (e.g., antioxidant defense).

Is it the case that in Mit animals there is reduced electron flux and reduced $O_2^{\cdot-}$ production leading to retarded aging? This question has been extensively investigated in studies of *clk-1* and CoQ. Several findings suggest that the above-mentioned view is an oversimplification. First, if the longevity of *clk-1* mutants were due to an effect of lowered CoQ levels on electron flux, then this strain should have a reduced metabolic rate. In fact, neither metabolic rate nor ATP levels are lower in *clk-1* animals [31, 114, 121], although RNAi of other mitochondrial genes does lower ATP levels [103]. Moreover, succinate-cytochrome *c* reductase activity is almost normal in *clk-1* mutants, implying that DMCoQ$_9$ (perhaps supplemented with bacterially derived CoQ$_8$) functions as well as CoQ$_9$ [114, 116]. It has been suggested that DMCoQ$_9$ produces less $O_2^{\cdot-}$ than CoQ$_9$ [116], but this has not been tested directly.

What really complicates interpretation of the role of *clk-1* and CoQ in aging is that the reduced (quinol) form of CoQ can act as a lipid-soluble antioxidant that protects against lipid peroxidation [122–124], which is why it is marketed as a human dietary supplement. Consistent with this, CoQ$_{10}$ supplementation increases life span in *C. elegans* in both wild-type and *mev-1* animals [91] and also reduces $O_2^{\cdot-}$ production in isolated mitochondria.

Thus, these results seem to conflict with the observation that feeding *C. elegans* with *E. coli* lacking CoQ$_8$ increases their life span [118]. How may these findings be reconciled? One possibility is that different forms of CoQ have different effects on life span. The increases in life span seen by Ishii et al. [91] resulted from supplementation with CoQ$_{10}$, which may somehow promote longevity more than CoQ$_9$; possibly, CoQ$_8$ increases superoxide production more than CoQ$_9$. To explore this, *E. coli* strains were engineered that produce CoQ$_7$, CoQ$_8$, CoQ$_9$, or CoQ$_{10}$. *E. coli* producing CoQ$_9$, or CoQ$_{10}$ partially suppressed the reduced fertility of a weak *clk-1* mutant, but, surprisingly, effects of these *E. coli* strains on life span were not reported [125].

Another possibility is that DMCoQ$_9$ generates less $O_2^{\cdot-}$ than CoQ$_9$ [124]. RNAi of *sod-1* (cytosolic Cu/Zn SOD) and mutation of *sod-4* (putative extracellular Cu/Zn SOD) can partially suppress some *clk-1* mutant phenotypes. This, it has been suggested, may reflect reduced $O_2^{\cdot-}$ production by DMCoQ$_9$, which interferes with signaling pathways in which $O_2^{\cdot-}$ acts as a secondary messenger [23, 115]. However, it seems unlikely that the presence of DMCoQ$_9$ causes increased life span, because mutation of *rte-2* suppresses *clk-1* longevity without reducing levels of DMCoQ$_9$ [126].

If Mit longevity reflects a reduction in metabolism and ROS production, one would not expect any associated increase in stress resistance. However, this prediction is not well supported. For many genes encoding mitochondrial genes, long-lived animals subjected to RNAi proved to be resistant to H_2O_2 and heat stress, although not paraquat [127]. Moreover, mutation of *isp-1* in complex III elevates *sod-3* expression and increases paraquat resistance [109], and *clk-1* mutant animals show resistance to ultraviolet light [128] and increased catalase levels (but reduced SOD activity levels) [31]. This could imply that disruption of mitochondria stimulates stress resistance pathways, perhaps due to increased $O_2^{\cdot-}$ production [107]. Consistent with the latter idea, treatment with the drug antimycin A, which blocks complex III, increases $O_2^{\cdot-}$ production from isolated *C. elegans* mitochondria [10], and, interestingly, it seems to increase life span [103], although the effect is not large. Mitochondrial ROS production (measured as H_2O_2) is also elevated by mutation of *daf-2* [37]. There is also evidence that the SKN-1-dependent antioxidant system is activated in *clk-1* mutants [107]. Thus, there is mounting evidence that increased mitochondrial $O_2^{\cdot-}$ production can contribute to longevity.

5.5   *Uncoupling Proteins and Aging in C. elegans*

Production of $O_2^{\cdot-}$ by mitochondria is predicted to be highest when the ETC is fully reduced, in state 4. Uncoupling proteins (UCPs) or chemical protonophores such as dinitrophenol can uncouple electron transport from ATP synthesis, which increases heat production and lowers $O_2^{\cdot-}$ production [129]. A prediction of the oxidative damage theory is that increased uncoupling should reduce ROS production, thereby increasing life span. This has been investigated a little in *C. elegans*. The worm genome contains a single gene encoding a protein with sequence homology to mammalian UCPs, *ucp-4*, which is strongly expressed in muscle [130]. Absence of *ucp-4* function resulted in increased levels of ATP and cold sensitivity, consistent with function as an uncoupling protein. However, only a very slight increase in mitochondrial membrane potential was seen, and life span was not affected. It also has been suggested that mitochondria from *daf-2* mutants have a higher level of coupling, given the lower calorimetric/respirometric ratio and the higher levels of ATP and ROS production [32, 37, 131]. However, it is worth noting that this lowering of the calorimetric/respirometric ratio is not suppressed by mutation of *daf-16*, which does suppress *daf-2* longevity [32, 132]. Thus, one can at least say that this metabolic shift is not enough in itself to increase life span. Further studies seem warranted to establish the effects of mitochondrial uncoupling on aging in *C. elegans*.

5.6   *Conclusions*

Various genome-wide screens for genes with effects on aging have all pointed to the importance of mitochondria in aging. Disruption of mitochondrial function usually increases life span in *C. elegans*, but the mechanisms involved are unknown.

One possibility is that $O_2^{\cdot-}$ production in Mit animals is reduced, but this remains largely unexplored. In principle, reduced ATP production might seem a strong candidate mechanism, potentially linking the Mit phenotype, dietary restriction and rate-of-living effects; for example, ATP feeds growth, including protein synthesis, which promotes aging [133–135]. However, the importance of ATP levels in aging is not experimentally supported (see [103]). One alternative is that disruption of mitochondrial function activates somatic maintenance processes [107].

6 Is Metabolic Rate a Determinant of Aging in *C. elegans*?

6.1 Metabolic Rate and Superoxide Production

Central to many discussions of the rate-of-living theory is the idea that increased metabolic rate will lead to increased production of $O_2^{\cdot-}$, and, consequently, an increased rate of accumulation of molecular damage and of aging [136, 137]. This view has informed many studies of metabolic rate and aging in *C. elegans*, as elsewhere. However, this view is not necessarily correct. At lower rates of metabolism, the inner mitochondrial membrane potential increases, which can increase $O_2^{\cdot-}$ production. As metabolic rate increases, membrane potential and $O_2^{\cdot-}$ production drop [129]. Thus, all else being equal, one might expect life span to increase with increasing metabolic rate.

6.2 Effects of Temperature on Life Span

C. elegans do show striking rate-of-living effects insofar as life span is shorter at higher temperatures. For example, median life span of wild-type hermaphrodites is 24 days at 15 °C compared with 16 days at 22.5 °C [138]. This implies that processes whose rate determines the rate of aging occurs faster at higher temperatures; but the identity of these critical processes remain undetermined. To date, studies of rate-of-living effects have focussed on energy metabolism and production of $O_2^{\cdot-}$ by mitochondria. In *C. elegans*, metabolic rate does increase with increasing temperature [139], but a causal role of metabolic rate in determining aging rate has not been demonstrated.

6.3 Metabolic Rate in Long-Lived Nematodes

As a test of the rate-of-living theory, metabolic rate has been measured in long-lived nematodes, including *age-1* and *daf-2* mutants, *clk-1* mutants, and nematodes subjected to various forms of dietary restriction. The instructive power of such tests is somewhat limited, however, for the following reasons. If metabolic rate were the sole determinant of longevity in *C. elegans*, then one should see a reduction in metabolic rate in long-lived nematodes, although such an observation would give no indication of causality. If long-lived nematodes show no change in metabolic rate, or even a small increase, this does not demonstrate that metabolic rate is not a determinant of aging, because several mechanisms may contribute to mutant longevity.

In general, reductions in metabolic rate in long-lived *C. elegans* have not been detected in insulin/IGF-1 signaling mutants or animals subjected to dietary restriction, but they have been in some strains with mitochondrial defects. In *age-1* and *daf-2* mutants, oxygen consumption rate shows no reduction and even slight increases [32, 100, 131, 140, 141]. Lower levels of heat production and higher levels of ATP also were seen in *daf-2* mutants, which might imply a higher level of mitochondrial coupling in this mutant. Consistent with this, levels of $O_2^{\cdot-}$ production in isolated mitochondria are higher in *daf-2(e1370)* mutants than in wild type [37]. One study reported a decline in metabolic rate in *age-1* and *daf-2* mutants, and concluded that the rate of living theory is supported [139]. This study measured CO_2 production, which may explain the discrepancy with other studies (for further discussion of methodological issues in metabolic studies of *C. elegans*, see [140-142]).

The effects of dietary restriction (DR) on metabolic rate in *C. elegans* depends on how DR is exerted. DR by bacterial dilution has no effect on oxygen consumption rate, whereas DR by means of axenic culture or an *eat-2* mutation increases oxygen consumption rate and heat production [143, 144]. Strains with alterations in mitochondrial function vary in terms of metabolic rate. As mentioned, *clk-1* has little effect on metabolic rate [31, 114, 121]. However, metabolic rate was reduced in *isp-1* mutants [109] and animals subjected to RNAi knockdown of several ETC genes [103]. Thus, it is possible that reduced metabolic rate somehow contributes to the increased longevity of long-lived forms of *C. elegans* in some cases. However, the effect of metabolic rate on aging in *C. elegans* really remains unknown.

6.4   Differences in Energy Metabolism between C. elegans and Vertebrates

One worry when using *C. elegans* as a model organism to investigate links between metabolism and aging is that its metabolism is different in several respects from that of higher animals. For example, *C. elegans* possess the glyoxylate pathway, absent in higher animals, that allows the conversion of acetyl CoA to glucose. Expression of the main glyoxylate enzyme, which has both malate synthase and isocitrate lyase activity, is up-regulated in *daf-2* and *age-1* mutant adults, and in dauer larvae [15, 145]. *Caenorhabditis elegans* also synthesize the disaccharide trehalose, also lacking in higher animals.

Nematodes are also capable of anaerobic respiration using an alternative electron acceptor, rhodoquinone [146], and the malate dismutation pathway [147]. When cultured under anoxic conditions, *C. elegans* excrete lactate, acetate, succinate, and propionate [148]. It has been suggested that such anaerobic respiration might reduce $O_2^{\cdot-}$ production levels, thereby increasing life span [149]. Transcript profile studies suggest that this pathway might be up-regulated in dauer larvae and *daf-2* mutants [145, 150]. However, increased anaerobic respiration in *daf-2* mutants would be expected to generate heat, which would increase their calorimetric/respirometric (C/R) ratio. In fact, the C/R ratio is reduced in mutants with reduced insulin/IGF-1 signaling [131].

6.5   Conclusions

Rate-of-living effects can occur in *C. elegans*, but the biochemical processes whose rates are so strongly determinative of aging remain unclear. The evidence for the importance of O_2 consumption is weak, to say the least. One alternative aging rate-determining process that is affected by temperature is protein synthesis. Several studies have recently shown that reduction of function of a number of genes linked to protein biosynthesis increases life span in *C. elegans* [68, 133–135].

7   Overall Conclusions

Tests of the various elements of the metabolic and oxidative damage theories of aging have in many cases failed to support these theories. In particular, there is little clear evidence that metabolic rate is a determinant of aging, although the possibility of effects of metabolic rate on aging have not been excluded. Molecular damage clearly accumulates with age, but it remains uncertain whether this is a primary cause of aging, and the mechanisms that determine the rate of damage accumulation remain unclear.

Trying to draw clear conclusions from work on these topics can sometimes be an exasperating occupation. Clearly, some aspects of the rate-of-living and oxidative damage theories of aging are wrong; yet, others seem to be supported—but usually weakly. It is as if these theories are somewhere near the truth, but not actually there: We are clearly missing something. For example, molecular damage seems to be important in aging; yet, the importance of atmospheric oxygen and $O_2^{\cdot-}$ is highly uncertain. Likewise, mitochondria do seem to be important in aging, yet it is not clear that this has anything to do with $O_2^{\cdot-}$. Mitochondria play many roles in the cell beyond oxidative phosphorylation and ATP production, including calcium homeostasis, steroid biogenesis, pyrimidine biosynthesis, and fatty acid metabolism. The effects of CoQ on aging are particularly hard to interpret, because it also affects many processes both in mitochondria and elsewhere, and is Janus faced in its effect on $O_2^{\cdot-}$, acting either as a prooxidant or antioxidant, depending on the cellular context [124].

A potential problem with the use of model organisms to study aging is the possibility that mechanisms of aging may differ across taxa, i.e., involve private rather than public mechanisms [151]. In the context of metabolic theories, there are a number of reasons for being suspicious that private mechanisms may be at play in *C. elegans*. For example, it is odd that disruption of the electron transport chain usually extends life span instead of causing death, raising a worry that this is a nematode peculiarity; in *C. elegans* O_2 consumption and $O_2^{\cdot-}$ production declines with age; age accumulation of protein oxidation is largely restricted to the mitochondria; and the pattern of age increase of blue fluorescence does seem simply to reflect general age increases in molecular damage.

This survey clearly supports one particular conclusion: That more research in this field is needed. In particular, we need more direct testing of mechanisms by means of reverse genetic approaches, exemplified by the studies by Zimniak and co-workers on the glutathione *S*-transferase GST-10.

Acknowledgements We are very grateful to J. R. Vanfleteren and F. Matthijssens for communication of unpublished information, and for their careful reading of this review in draft form. Any errors that might remain are the fault of the authors. This work was supported by the European Union and the Wellcome Trust.

References

1. Balaban RS, Nemoto S, Finkel T. Mitochondria, oxidants, and aging. Cell 2005;120:483–95.
2. Raha S, Robinson BH. Mitochondria, oxygen free radicals, disease and ageing. Trends Biochem Sci 2000;25:502–8.
3. Wood WB. The nematode *Caenorhabditis elegans*. Plainview, New York: Cold Spring Harbor Press; 1988.
4. Riddle DL, Blumenthal T, Meyer BJ, Priess JR. *C. elegans* II. Plainview, New York: Cold Spring Harbor Laboratory Press; 1997.
5. Fraser A, Kamath R, Zipperlen P, Martinez-Campos M, Sohrmann M, Ahringer J. Functional genomic analysis of *C. elegans* chromosome I by systematic RNA interference. Nature 2000;408:325–30.
6. Garigan D, Hsu A, Fraser A, Kamath R, Ahringer J, Kenyon C. Genetic analysis of tissue aging in *Caenorhabditis elegans*: a role for heat-shock factor and bacterial proliferation. Genetics 2002;161:1101–12.
7. Gerstbrein B, Stamatas G, Kollias N, Driscoll M. In vivo spectrofluorimetry reveals endogenous biomarkers that report healthspan and dietary restriction in *Caenorhabditis elegans*. Aging Cell 2005;4:127–37.
8. Ishii N, Fujii M, Hartman PS et al. A mutation in succinate dehydrogenase cytochrome b causes oxidative stress and ageing in nematodes. Nature 1998;394:694–7.
9. Ishii N, Takahashi K, Tomita S et al. A methyl viologen-sensitive mutant of the nematode *Caenorhabditis elegans*. Mutat Res 1990;237:165–71.
10. Senoo-Matsuda N, Yasuda K, Tsuda M et al. A defect in the cytochrome b large subunit in complex II causes both superoxide anion overproduction and abnormal energy metabolism in *Caenorhabditis elegans*. J Biol Chem 2001;276:41553–8.

11. Ishii N, Ishii T, Hartman PS. The role of the electron transport gene SDHC on lifespan and cancer. Exp Gerontol 2006;41:952–6.

12. Kenyon C. The plasticity of aging: insights from long-lived mutants. Cell 2005;120:449–60.

13. Larsen PL. Aging and resistance to oxidative stress in *Caenorhabditis elegans*. Proc Natl Acad Sci U S A 1993;90:8905–9.

14. Vanfleteren JR. Oxidative stress and ageing in *Caenorhabditis elegans*. Biochem J 1993;292:605–8.

15. Vanfleteren JR, De Vreese A. The gerontogenes age-1 and daf-2 determine metabolic rate potential in aging *Caenorhabditis elegans*. FASEB J 1995;9:1355–61.

16. Sun J, Tower J. FLP recombinase-mediated induction of Cu/Zn-superoxide dismutase transgene expression can extend the life span of adult *Drosophila melanogaster* flies. Mol Cell Biol 1999;19:216–28.

17. Parkes TL, Elia AJ, Dickinson D, Hilliker AJ, Phillips JP, Boulianne GL. Extension of Drosophila lifespan by overexpression of human SOD1 in motorneurons. Nat Genet 1998;19:171–4.

18. Orr W, Sohal R. Does overexpression of Cu,Zn-SOD extend life span in *Drosophila melanogaster*? Exp Gerontol 2003;38:227–30.

19. Adachi H, Fujiwara Y, Ishii N. Effects of oxygen on protein carbonyl and aging in *Caenorhabditis elegans* mutants with long (age-1) and short (mev-1) life spans. J Gerontol 1998;53A:B240–B4.

20. Yasuda K, Ishii T, Suda H et al. Age-related changes of mitochondrial structure and function in *Caenorhabditis elegans*. Mech Ageing Dev 2006;127:763–70.

21. Nakamura A, Yasuda K, Adachi H, Sakurai Y, Ishii N, Goto S. Vitellogenin-6 is a major carbonylated protein in aged nematode, *Caenorhabditis elegans*. Biochem Biophys Res Commun 1999;264:580–3.

22. Herndon L, Schmeissner P, Dudaronek J et al. Stochastic and genetic factors influence tissue-specific decline in ageing *C. elegans*. Nature 2002;419:808–14.

23. Shibata Y, Branicky R, Landaverde IO, Hekimi S. Redox regulation of germline and vulval development in *Caenorhabditis elegans*. Science 2003;302:1779–82.

24. Klass M, Nguyen PN, Dechavigny A. Age correlated changes in the DNA template in the nematode *Caenorhabditis elegans*. Mech Ageing Dev 1983;22:253–63.

25. Melov S, Lithgow GJ, Fischer DR, Tedesco PM, Johnson TE. Increased frequency of deletions in the mitochondrial genome with age of *Caenorhabditis elegans*. Nucleic Acids Res 1995;23:1419–25.

26. Yin D. Biochemical basis of lipofuscin, ceroid, and age pigment-like fluorophores. Free Radic Biol Med 1996;21:871–88.

27. Terman A, Brunk UT. Oxidative stress, accumulation of biological 'garbage', and aging. Antioxid Redox Signal 2006;8:197–204.

28. Sitte N, Huber M, Grune T et al. Proteasome inhibition by lipofuscin/ceroid during postmitotic aging of fibroblasts. FASEB J 2000;14:1490–8.

29. Davis BO, Anderson GL, Dusenbery DB. Total luminescence spectroscopy of fluorescence changes during aging in *Caenorhabditis elegans*. Biochemistry 1982;21:4089–95.

30. Klass MR. Aging in the nematode *Caenorhabditis elegans*: major biological and environmental factors influencing life span. Mech Ageing Dev 1977;6:413–29.

31. Braeckman BP, Houthoofd K, Brys K et al. No reduction of energy metabolism in Clk mutants. Mech Ageing Dev 2002;123:1447–56.

32. Houthoofd K, Braeckman BP, Lenaerts I et al. DAF-2 pathway mutations and food restriction in aging *Caenorhabditis elegans* differentially affect metabolism. Neurobiol Aging 2005;26:689–96.

33. Hosokawa H, Ishii N, Ishida H, Ichimori K, Nakazawa H, Suzuki K. Rapid accumulation of fluorescent material with ageing in an oxygen-sensitive mutant mev-1 of *Caenorhabditis elegans*. Mech Ageing Dev 1994;74:161–70.

34. Clokey GV, Jacobsen LA. The autofluorescent 'lipofuscin' granules in the intestinal cells of *Caenorhabditis elegans* are secondary lysosomes. Mech Ageing Dev 1986;35:79–94.

35. Ishii N, Goto S, Hartman PS. Protein oxidation during aging of the nematode *Caenorhabditis elegans*. Free Radic Biol Med 2002;33:1021–5.
36. Yasuda K, Adachi H, Fujiwara Y, Ishii N. Protein carbonyl accumulation in aging dauer formation-defective (daf) mutants of *Caenorhabditis elegans*. J Gerontol 1999;54A: B47–51; discussion B2–3.
37. Brys K, Vanfleteren JR, Braeckman BP. Testing the rate-of-living / oxidative damage theory of aging in the nematode model *Caenorhabditis elegans*. Exp Gerontol 2007;42:845–51.
38. Hartman P, Ponder R, Lo HH, Ishii N. Mitochondrial oxidative stress can lead to nuclear hypermutability. Mech Ageing Dev 2004;125:417–20.
39. Honda S, Ishii N, Suzuki K, Matsuo M. Oxygen-dependent perturbation of life span and aging rate in the nematode. J Gerontol 1993;48A:B57–B61.
40. Senoo-Matsuda N, Hartman PS, Akatsuka A, Yoshimura S, Ishii N. A complex II defect affects mitochondrial structure, leading to ced-3- and ced-4-dependent apoptosis and aging. J Biol Chem 2003;278:22031–6.
41. Henderson ST, Johnson TE. daf-16 integrates developmental and environmental inputs to mediate aging in the nematode *Caenorhabditis elegans*. Curr Biol 2001;11:1975–80.
42. Keaney M, Matthijssens F, Sharpe M, Vanfleteren JR, Gems D. Superoxide dismutase mimetics elevate superoxide dismutase activity in vivo but do not retard aging in the nematode *Caenorhabditis elegans*. Free Radic Biol Med 2004;37:239–50.
43. Ayyadevara S, Engle MR, Singh SP et al. Lifespan and stress resistance of *Caenorhabditis elegans* are increased by expression of glutathione transferases capable of metabolizing the lipid peroxidation product 4-hydroxynonenal. Aging Cell 2005;4:257–71.
44. Arkblad EL, Tuck S, Pestov NB et al. A *Caenorhabditis elegans* mutant lacking functional nicotinamide nucleotide transhydrogenase displays increased sensitivity to oxidative stress. Free Radic Biol Med 2005;38:1518–25.
45. Blum J, Fridovich I. Superoxide, hydrogen peroxide, and oxygen toxicity in two free-living nematode species. Arch Biochem Biophys 1983;222:35–43.
46. Halliwell B, Gutteridge JMC. Free radicals in biology and medicine. Oxford, UK: Oxford Science Publications; 1999.
47. Mathers J, Fraser JA, McMahon M, Saunders RD, Hayes JD, McLellan LI. Antioxidant and cytoprotective responses to redox stress. Biochem Soc Symp 2004:157–76.
48. Fridovich I. Superoxide radical and superoxide dismutases. Annu Rev Biochem 1995;64:97–112.
49. Giglio AM, Hunter T, Bannister JV, Bannister WH, Hunter GJ. The copper/zinc superoxide dismutase gene of *Caenorhabditis elegans*. Biochem Mol Biol Int 1994;33:41–4.
50. Jensen LT, Culotta VC. Activation of CuZn superoxide dismutases from *Caenorhabditis elegans* does not require the copper chaperone CCS. J Biol Chem 2005;280:41373–9.
51. Giglio M-P, Hunter T, Bannister JV, Bannister WH, Hunter GJ. The manganese superoxide dismutase gene of *Caenorhabditis elegans*. Biochem Mol Biol Int 1994;33:37–40.
52. Hunter T, Bannister WH, Hunter GJ. Cloning, expression, and characterization of two manganese superoxide dismutases from *Caenorhabditis elegans*. J Biol Chem 1997;272:28652–9.
53. Suzuki N, Inokuma K, Yasuda K, Ishii N. Cloning, sequencing and mapping of a manganese superoxide dismutase gene of the nematode *Caenorhabditis elegans*. DNA Res 1996;3:171–4.
54. Honda Y, Honda S. The daf-2 gene network for longevity regulates oxidative stress resistance and Mn-superoxide dismutase gene expression in *Caenorhabditis elegans*. FASEB J 1999;13:1385–93.
55. Wang J, Kim S. Global analysis of dauer gene expression in *Caenorhabditis elegans*. Development 2003;130:1621–34.
56. Fujii M, Ishii N, Joguchi A, Yasuda K, Ayusawa D. Novel superoxide dismutase gene encoding membrane-bound and extracellular isoforms by alternative splicing in *Caenorhabditis elegans*. DNA Res 1998;5:25–30.
57. Petriv OI, Rachubinski RA. Lack of peroxisomal catalase causes a progeric phenotype in *Caenorhabditis elegans*. J Biol Chem 2004;279:19996–20001.
58. Taub J, Lau JF, Ma C et al. A cytosolic catalase is needed to extend lifespan in C. elegans daf-C and clk-1 mutants. Nature 1999;399:162–6.

59. Togo SH, Maebuchi M, Yokota S, Bun-Ya M, Kawahara A, Kamiryo T. Immunological detection of alkaline-diaminobenzidine-negative peroxisomes of the nematode *Caenorhabditis elegans* purification and unique pH optima of peroxisomal catalase. Eur J Biochem 2000;267:1307–12.

60. Yanase S, Yasuda K, Ishii N. Adaptive responses to oxidative damage in three mutants of *Caenorhabditis elegans* (age-1, mev-1 and daf-16) that affect life span. Mech Ageing Dev 2002;123:1579–87.

61. McElwee J, Bubb K, Thomas J. Transcriptional outputs of the *Caenorhabditis elegans* forkhead protein DAF-16. Aging Cell 2003;2:111–21.

62. McElwee JJ, Schuster E, Blanc E, Thomas JH, Gems D. Shared transcriptional signature in *C. elegans* dauer larvae and long-lived daf-2 mutants implicates detoxification system in longevity assurance. J Biol Chem 2004;279:44533–43.

63. Murphy CT, McCarroll SA, Bargmann CI et al. Genes that act downstream of DAF-16 to influence the lifespan of *C. elegans*. Nature 2003;424:277–84.

64. Houthoofd K, Braeckman B, Johnson T, Vanfleteren J. Life extension via dietary restriction is independent of the Ins/IGF-1 signalling pathway in *Caenorhabditis elegans*. Exp Gerontol 2003;38:947–54.

65. Anderson GL. Superoxide dismutase activity in dauer larvae of *Caenorhabditis elegans* (Nematoda: Rhabditidae). Can J Zool 1982;60:288–91.

66. Houthoofd K, Braeckman B, Lenaerts I et al. Ageing is reversed, and metabolism is reset to young levels in recovering dauer larvae of *C. elegans*. Exp Gerontol 2002;37:1015–21.

67. Taub J, Lau J, Ma C et al. A cytosolic catalase is needed to extend adult lifespan in *C. elegans* daf-C and clk-1 mutants. Nature 2003;421:764.

68. Henderson ST, Bonafe M, Johnson TE. daf-16 protects the nematode *Caenorhabditis elegans* during food deprivation. J Gerontol 2006;61A:444-60.

69. Hsu A, Murphy C, Kenyon C. Regulation of aging and age-related disease by DAF-16 and heat shock factor. Science 2003;300:1142–5.

70. Sampayo JN, Olsen A, Lithgow GJ. Oxidative stress in *Caenorhabditis elegans*: protective effects of superoxide dismutase/catalase mimetics. Aging Cell 2003;2:319–26.

71. Melov S, Ravenscroft J, Malik S et al. Extension of life span with superoxide dismutase/catalase mimetics. Science 2000;289:1567–9.

72. Keaney M, Gems D. No increase in lifespan in *Caenorhabditis elegans* upon treatment with the superoxide dismutase mimetic EUK-8. Free Radic Biol Med 2003;34:277–82.

73. Magwere T, West M, Riyahi K, Murphy MP, Smith RA, Partridge L. The effects of exogenous antioxidants on lifespan and oxidative stress resistance in *Drosophila melanogaster*. Mech Ageing Dev 2006;127:356–70.

74. Bayne AC, Sohal RS. Effects of superoxide dismutase/catalase mimetics on life span and oxidative stress resistance in the housefly, *Musca domestica*. Free Radic Biol Med 2002;32:1229–34.

75. Gourley BL, Parker SB, Jones BJ, Zumbrennen KB, Leibold EA. Cytosolic aconitase and ferritin are regulated by iron in *Caenorhabditis elegans*. J Biol Chem 2003;278:3227–34.

76. Barsyte D, Lovejoy D, Lithgow G. Longevity and heavy metal resistance in daf-2 and age-1 long-lived mutants of *Caenorhabditis elegans*. FASEB J 2001;15:627–34.

77. Golden TR, Melov S. Microarray analysis of gene expression with age in individual nematodes. Aging Cell 2004;3:111–24.

78. Gems D, McElwee JJ. Broad spectrum detoxification: the major longevity assurance process regulated by insulin/IGF-1 signaling? Mech Ageing Dev 2005;126:381–7.

79. Gibson GG, Skett P. Introduction to drug metabolism. 3rd edn. Bath, UK: Nelson Thornes; 2001.

80. McElwee JJ, Schuster E, Blanc E et al. Evolutionary conservation of regulated longevity assurance mechanisms. Genome Biol 2007;8:R132.

81. Hayes JD, McLellan LI. Glutathione and glutathione-dependent enzymes represent a co-ordinately regulated defence against oxidative stress. Free Radic Res 1999;31:273–300.

82. Tawe W, Eschbach M, Walter R, Henkle-Duhrsen K. Identification of stress-responsive genes in *Caenorhabditis elegans* using RT-PCR differential display. Nucleic Acids Res 1998;26:1621–7.

83. Leiers B, Kampkotter A, Grevelding C, Link C, Johnson T, Henkle-Duhrsen K. A stress-response glutathione S-transferase confers resistance to oxidative stress in *Caenorhabditis elegans*. Free Radic Biol Med 2003;34:1405–15.
84. Ayyadevara S, Dandapat A, Singh SP et al. Lifespan extension in hypomorphic daf-2 mutants of *Caenorhabditis elegans* is partially mediated by glutathione transferase CeGSTP2-2. Aging Cell 2005;4:299–307.
85. Engle MR, Singh SP, Nanduri B, Ji X, Zimniak P. Intertebrate glutathione transferases conjugating 4-hydroxynonenal: CeGST 5.4 from *Caenorhabditis elegans*. Chemicobiol Int 2001;133:244–8.
86. Ayyadevara S, Dandapat A, Singh SP et al. Life span and stress resistance of *Caenorhabditis elegans* are differentially affected by glutathione transferases metabolizing 4-hydroxynon-2-enal. Mech Ageing Dev 2007;128:196–205.
87. Epstein J, Gershon D. Studies on ageing in nematodes IV. The effect of anti-oxidants on cellular damage and life span. Mech Ageing Dev 1972;1:257–64.
88. Zuckerman BM, Geist MA. Effects of vitamin E on the nematode *Caenorhabditis elegans*. Age (Omaha, Nebr) 1983;6:1–4.
89. Harrington LA, Harley CB. Effect of vitamin E on lifespan and reproduction in *Caenorhabditis elegans*. Mech Ageing Dev 1988;43:71–8.
90. Adachi H, Ishii N. Effects of tocotrienols on life span and protein carbonylation in *Caenorhabditis elegans*. J Gerontol 2000;55A:B280–5.
91. Ishii N, Senoo-Matsuda N, Miyake K et al. Coenzyme Q10 can prolong C. elegans lifespan by lowering oxidative stress. Mech Ageing Dev 2004;125:41–6.
92. Boveris A. Mitochondrial production of superoxide radical and hydrogen peroxide. In: Reivich M, Coburn R, Lahiri S, Chance B, eds. Tissue hypoxia and ischemia. New York: Plenum Press; 1977.67–82.
93. St-Pierre J, Buckingham JA, Roebuck SJ, Brand MD. Topology of superoxide production from different sites in the mitochondrial electron transport chain. J Biol Chem 2002;277:44784–90.
94. Staniek K, Nohl H. Are mitochondria a permanent source of reactive oxygen species? Biochim Biophys Acta 2000;1460:268–75.
95. Imlay JA, Fridovich I. Assay of metabolic superoxide production in *Escherichia coli*. J Biol Chem 1991;266:6957–65.
96. Nohl H, Hegner D. Do mitochondria produce oxygen radicals in vivo? Eur J Biochem 1978;82:563–7.
97. Murfitt R, Vogel K, Sanadi D. Characterization of the mitochondria of the free-living nematode *Caenorhabditis elegans*. Comp Biochem Physiol 1976;53B:423–30.
98. Okimoto R, Macfarlane JL, Clary DO, Wolstenholme DR. The mitochondrial genomes of two nematodes, *Caenorhabditis elegans* and *Ascaris suum*. Genetics 1992;130:471–98.
99. De Cuyper C, Vanfleteren JR. Oxygen consumption during development and aging of the nematode *Caenorhabditis elegans*. Comp Biochem Physiol 1982;73A(2):283–9.
100. Vanfleteren JR, De Vreese A. Rate of aerobic metabolism and superoxide production rate potential in the nematode *Caenorhabditis elegans*. J Exp Zool 1996;274:93–100.
101. Suda H, Shouyama T, Yasuda K, Ishii N. Direct measurement of oxygen consumption rate on the nematode *Caenorhabditis elegans* by using an optical technique. Biochem Biophys Res Commun 2005;330:839–43.
102. Anson RM, Hansford RG. Mitochondrial influence on aging rate in *Caenorhabditis elegans*. Aging Cell 2004;3:29–34.
103. Dillin A, Hsu A, Arantes-Oliveira N et al. Rates of behavior and aging specified by mitochondrial function during development. Science 2002;298:2398–401.
104. Lee S, Lee R, Fraser A, Kamath R, Ahringer J, Ruvkun G. A systematic RNAi screen identifies a critical role for mitochondria in C. elegans longevity. Nat Genet 2003;33:40–8.
105. Hansen M, Hsu AL, Dillin A, Kenyon C. New genes tied to endocrine, metabolic, and dietary regulation of lifespan from a *Caenorhabditis elegans* genomic RNAi screen. PLoS Genet 2005;1:119–28.

106. Hamilton B, Dong Y, Shindo M et al. A systematic RNAi screen for longevity genes in *C. elegans*. Genes Dev 2005;19:1544–55.
107. Rea SL. Metabolism in the *Caenorhabditis elegans* Mit mutants. Exp Gerontol 2005;40:841–9.
108. Yoneda T, Benedetti C, Urano F, Clark SG, Harding HP, Ron D. Compartment-specific perturbation of protein handling activates genes encoding mitochondrial chaperones. J Cell Sci 2004;117:4055–66.
109. Feng J, Bussiere F, Hekimi S. Mitochondrial electron transport is a key determinant of life span in *Caenorhabditis elegans*. Dev Cell 2001;1:633–44.
110. Tsang WY, Sayles LC, Grad LI, Pilgrim DB, Lemire BD. Mitochondrial respiratory chain deficiency in *Caenorhabditis elegans* results in developmental arrest and increased life span. J Biol Chem 2001;276:32240–6.
111. Tsang WY, Lemire BD. Mitochondrial genome content is regulated during nematode development. Biochem Biophys Res Commun 2002;291:8–16.
112. Ewbank JJ, Barnes TM, Lakowski B, Lussier M, Bussey H, Hekimi S. Structural and functional conservation of the *Caenorhabditis elegans* timing gene clk-1. Science 1997;275:980–3.
113. Stenmark P, Grunler J, Mattsson J, Sindelar PJ, Nordlund P, Berthold DA. A new member of the family of di-iron carboxylate proteins. Coq7 (clk-1), a membrane-bound hydroxylase involved in ubiquinone biosynthesis. J Biol Chem 2001;276:33297–300.
114. Felkai S, Ewbank JJ, Lemieux J, Labbe J-C, Brown GG, Hekimi S. CLK-1 controls respiration, behavior and aging in the nematode *Caenorhabditis elegans*. EMBO J 1999;18:1783–92.
115. Stepanyan Z, Hughes B, Cliche DO, Camp D, Hekimi S. Genetic and molecular characterization of CLK-1/mCLK1, a conserved determinant of the rate of aging. Exp Gerontol 2006;41:940–51.
116. Miyadera H, Amino H, Hiraishi A et al. Altered quinone biosynthesis in the long-lived clk-1 mutants of *Caenorhabditis elegans*. J Biol Chem 2001;276:7713–6.
117. Wong AE, Boutis P, Hekimi S. Mutations in the clk-1 gene of *Caenorhabditis elegans* affect developmental and behavioral timing. Genetics 1995;139:1247–59.
118. Larsen P, Clarke, CF. Extension of life-span in *Caenorhabditis elegans* by a diet lacking coenzyme Q. Science 2002;295:120–3.
119. Hartman PS, Ishii N, Kayser EB, Morgan PG, Sedensky MM. Mitochondrial mutations differentially affect aging, mutability and anesthetic sensitivity in *Caenorhabditis elegans*. Mech Ageing Dev 2001;122:1187–201.
120. Ichimiya H, Huet RG, Hartman P, Amino H, Kita K, Ishii N. Complex II inactivation is lethal in the nematode *Caenorhabditis elegans*. Mitochondrion 2002;2:191–8.
121. Braeckman BP, Houthoofd K, De Vreese A, Vanfleteren JR. Apparent uncoupling of energy production and consumption in long-lived Clk mutants of *Caenorhabditis elegans*. Curr Biol 1999;9:493–6.
122. Lass A, Sohal RS. Effect of coenzyme Q(10) and alpha-tocopherol content of mitochondria on the production of superoxide anion radicals. FASEB J 2000;14:87–94.
123. Kwong LK, Kamzalov S, Rebrin I et al. Effects of coenzyme Q(10) administration on its tissue concentrations, mitochondrial oxidant generation, and oxidative stress in the rat. Free Radic Biol Med 2002;33:627–38.
124. Miyadera H, Kano K, Miyoshi H, Ishii N, Hekimi S, Kita K. Quinones in long-lived clk-1 mutants of *Caenorhabditis elegans*. FEBS Lett 2002;512:33–7.
125. Jonassen T, Davis DE, Larsen PL, Clarke CF. Reproductive fitness and quinone content of *Caenorhabditis elegans* clk-1 mutants fed coenzyme Q isoforms of varying length. J Biol Chem 2003;278:51735–42.
126. Branicky R, Nguyen PA, Hekimi S. Uncoupling the pleiotropic phenotypes of clk-1 with tRNA missense suppressors in *Caenorhabditis elegans*. Mol Cell Biol 2006;26:3976–85.
127. Lee SS, Kennedy S, Tolonen AC, Ruvkun G. DAF-16 target genes that control *C. elegans* life-span and metabolism. Science 2003;300:644–7.
128. Murakami S, Johnson TE. A genetic pathway conferring life extension and resistance to UV stress in *Caenorhabditis elegans*. Genetics 1996;143:1207–18.

129. Brand MD. Uncoupling to survive? The role of mitochondrial inefficiency in ageing. Exp Geront 2000;35:811–20.

130. Iser WB, Kim D, Bachman E, Wolkow C. Examination of the requirement for ucp-4, a putative homolog of the mammalian uncoupling proteins, for stress tolerance and longevity in *C. elegans*. Mech Ageing Dev 2005;126;1090–1096.

131. Houthoofd K, Fidalgo MA, Hoogewijs D et al. Metabolism, physiology and stress defense in three aging Ins/IGF-1 mutants of the nematode *Caenorhabditis elegans*. Aging Cell 2005;4:87–95.

132. Kenyon C, Chang J, Gensch E, Rudener A, Tabtiang R. A C. elegans mutant that lives twice as long as wild type. Nature 1993;366:461–4.

133. Pan KZ, Palter JE, Rogers AN et al. Inhibition of mRNA translation extends lifespan in *Caenorhabditis elegans*. Aging Cell 2007;6:111–9.

134. Hansen M, Taubert S, Crawford D, Libina N, Lee SJ, Kenyon C. Lifespan extension by conditions that inhibit translation in *Caenorhabditis elegans*. Aging Cell 2007;6:95–110.

135. Syntichaki P, Troulinaki K, Tavernarakis N. eIF4E function in somatic cells modulates ageing in *Caenorhabditis elegans*. Nature 2007;445:922–6.

136. Beckman KB, Ames BN. The free radical theory of aging matures. Physiol Rev 1998;78:547–81.

137. Finkel T, Holbrook NJ. Oxidants, oxidative stress and the biology of ageing. Nature 2000;408:239–47.

138. Gems D, Sutton AJ, Sundermeyer ML et al. Two pleiotropic classes of daf-2 mutation affect larval arrest, adult behavior, reproduction and longevity in *Caenorhabditis elegans*. Genetics 1998;150:129–55.

139. Van Voorhies W, Ward S. Genetic and environmental conditions that increase longevity in *Caenorhabditis elegans* decrease metabolic rate. Proc Natl Acad Sci USA 1999;96:11399–403.

140. Braeckman B, Houthoofd K, De Vreese A, Vanfleteren J. Assaying metabolic activity in ageing *Caenorhabditis elegans*. Mech Ageing Dev 2002;123:105–19.

141. Braeckman B, Houthoofd K, Vanfleteren J. Assessing metabolic activity in aging *Caenorhabditis elegans*: concepts and controversies. Aging Cell 2002;1:82–8.

142. Van Voorhies W. The influence of metabolic rate on longevity in the nematode *Caenorhabditis elegans*. Aging Cell 2002;1:91–101.

143. Houthoofd K, Braeckman B, Lenaerts I et al. Axenic growth up-regulates mass-specific metabolic rate, stress resistance, and extends life span in Caenorhabditis elegans. Exp Gerontol 2002;37:1371–8.

144. Houthoofd K, Braeckman B, Lenaerts I et al. No reduction of metabolic rate in food restricted *Caenorhabditis elegans*. Exp Gerontol 2002;37:1359–69.

145. McElwee JJ, Schuster E, Blanc E, Gems D. Partial reiteration of dauer larva metabolism in long lived daf-2 mutant adults in *Caenorhabditis elegans*. Mech Ageing Dev 2006;127:458–72.

146. Takamiya S, Matsui T, Taka H, Murayama K, Matsuda M, Aoki T. Free-living nematodes *Caenorhabditis elegans* possess in their mitochondria an additional rhodoquinone, an essential component of the eukaryotic fumarate reductase system. Arch Biochem Biophys 1999;371:284–9.

147. Tielens A, Rotte C, van Hellemond J, Martin W. Mitochondria as we don't know them. Trends Biochem Sci 2002; 27:564–72.

148. Foll R, Pleyers A, Lewandovski G, Wermter C, Hegemann V, Paul R. Anaerobiosis in the nematode *Caenorhabditis elegans*. Comp Biochem Physiol B Biochem Mol Biol 1999;124:269–80.

149. Rea S, Johnson TE. A metabolic model for lifespan determination in *Caenorhabditis elegans*. Dev Cell 2003; 5:197–203.

150. Holt S, Riddle D. SAGE surveys *C. elegans* carbohydrate metabolism: evidence for an anaerobic shift in the long-lived dauer larva. Mech Ageing Dev 2003;124:779–800.

151. Martin GM, Austad SN, Johnson TE. Genetic analysis of ageing: role of oxidative damage and environmental stresses. Nat Genet 1996;13:25–34.

7
Roles of Oxidative Stress in the Aging Process of *Drosophila melanogaster*

Robin J. Mockett, Rajindar S. Sohal, and William C. Orr

Summary The oxidative stress hypothesis of aging predicts that progression of the aging process could be retarded and the life spans of animals could be extended by decreases in oxidant production, enhancement of antioxidant defenses, or augmentation of repair capabilities. Some of the results from studies of *Drosophila* support this idea, but much of the existing evidence seems to be at odds with the most straightforward predictions of the hypothesis. In fact, the most conservative interpretation of the existing studies is that the predictions of the hypothesis need to be revised, in recognition of the physiological roles of oxidant production, the limited access of antioxidants to some sites of oxidant production, and the influence of confounding factors, such as altered rates of metabolism, in studies of life spans in poikilotherms, including *Drosophila*. Indeed, effects on life span alone do not provide a sufficient basis to infer the efficacy of any experimental treatment on mechanisms of aging in the poikilotherm. Consideration of such issues suggests that despite its wide appeal, further lines of investigation are necessary to verify or falsify the oxidative stress hypothesis.

Keywords Aging, oxidants, free radicals, antioxidants, repair, rate of living theory, *Drosophila*.

1 Introduction

1.1 The Free Radical/Oxidative Stress Hypothesis of Aging

The key feature of the free radical hypothesis of aging, first proposed by Harman [1], is that oxidative molecular damage arising from the production of free radicals in biological systems is the primary cause of their physiological deterioration, which results in increasing incidence of mortality as a function of time. The oxidative stress hypothesis represents an extension of this theory to include oxidants that do not necessarily contain unpaired electrons, and to emphasize that there is an inherent gap between oxidants and antioxidants [2]. Rather than structural damage alone,

From: *Aging Medicine: Oxidative Stress in Aging: From Model Systems to Human Diseases*   111
Edited by: S. Miwa, K.B. Beckman, and F.L. Muller © Humana Press, Totowa, NJ

this hypothesis also emphasizes the role of the redox state, with its attendant effects on a broad range of redox-sensitive functions, including cell signaling and gene regulatory mechanisms. The term "nitrosative stress" also has been used to distinguish the effects of reactive nitrogen species from those of reactive oxygen species (ROS). In this chapter, the terms "oxidative stress" and "oxidant" are used to encompass all of the physiologically relevant agents, without restriction based on their electron pairing status or chemical elements responsible for the oxidation.

As stated above, the phrase "oxidative stress" implies that there is an imbalance between the production of oxidants, which cause molecular damage, and antioxidant defenses and repair processes, which either prevent or reverse damage. The existence of an imbalance is inferred from the presence of steady-state amounts of products of reactions between oxidants and various cellular macromolecules, even in young, healthy animals. The hypothesis that oxidative stress increases in aging *Drosophila melanogaster* is supported by observations that the rate of mitochondrial hydrogen peroxide (H_2O_2) production increases as a function of age [3], and that the redox state becomes more prooxidizing, as indicated by the reduced glutathione (GSH): oxidized glutathione (GSSG) ratio and amount of protein mixed disulfides [4].

The oxidative stress hypothesis of aging gives rise to at least three straightforward predictions, which are amenable to experimental verification: (1) prevention of oxidant production would diminish the accumulation of damage, retard the aging process, and extend the life span of the organism; 2) enhancement of antioxidant levels would result in neutralization of an increased fraction of oxidants, and likewise extend the life span; and 3) enhancement of repair capability should also slow the accumulation of damage and increase longevity. An attractive feature of such experimental approaches, in contrast to studies demonstrating correlations between oxidative stress and longevity, is that, in addition to testing the hypothesis, they also may give rise to interventions to diminish the effects of aging.

The majority of this chapter (see Subsections 3–5) is concerned with the results and interpretation of experiments intended to test the aforementioned three predictions of the hypothesis, with primary emphasis on approaches involving transgenic modifications in the *Drosophila* model system. Subsection 2 identifies several caveats germane to the interpretation of longevity studies in all poikilothermal organisms, and Subsection 1.2., below, raises issues that must be considered regardless of the model system.

1.2   Interpretation of Experimental Tests of the Oxidative Stress Hypothesis

A number of factors need to be considered in the interpretation of experimental tests of the oxidative stress hypothesis. For example, a decrease in oxidant production or enhancement of antioxidant levels could be predicted to extend the life spans of animals, but contrary results do not automatically falsify the hypothesis, for the following reasons:

1. Augmentation of the level of an individual antioxidant may not necessarily affect the overall balance between oxidants and antioxidants. Cellular antioxidant defenses are provided by numerous enzymes and nonenzymatic mechanisms, which are often functionally overlapping and interdependent. Consequently, alterations in one or two components of the defense network may not have a significant, physiological impact. Another possibility is that there might be a compensatory down-regulation of expression of some antioxidants in response to the enhancement of others, or of endogenous antioxidants in response to exogenous antioxidant supplementation, as was observed for superoxide dismutase and glutathione in houseflies supplemented with ascorbate or β-carotene [5].

2. Elevation of antioxidant levels may not prevent oxidative damage, particularly when oxidants are produced at inaccessible sites. Such a situation can arise because individual antioxidants are generally restricted either to the aqueous or membrane compartments of the cell, and because enzymatic antioxidants in particular are much bulkier than the oxidants they detoxify. In situations where a potent oxidant acts at its site of generation, as in the metal-catalyzed oxidation of the metal-nucleotide binding pocket of glutamine synthetase [6], an enhancement of bulk antioxidant levels in the cell may not have any meaningful, preventive effect.

A related issue is that oxidants cause a multiplicity of structural modifications in proteins, lipids and nucleic acids, not all of which have equal functional ramifications. For example, the oxidation of several methionine residues in some proteins has no serious effect, whereas additional oxidations lead to enzyme inactivation and degradation [7]. Carbonylation of amino acid residues at sites other than the active site may have no effect on catalytic activity [8]. Additionally, the measurement of various kinds of adducts is complicated by issues such as in vitro oxidation during sample preparation, particularly for DNA [9]. Thus, predicting and determining the specific effects of antioxidant overexpression on levels of oxidative damage can be complex, even if oxidative damage is indeed a causal factor in aging.

3. As is the case in humans, it must be recognized that aging may not be the only cause of death in flies. Aside from accidents, death also may be caused by adverse conditions (e.g., infection, malnourishment, environmental toxins, desiccation, sudden changes in temperature, and somatic mutations in essential genes), which are mechanistically distinct from normal aging. A treatment that simultaneously attenuated the aging process and caused death by an age-independent mechanism would not necessarily increase longevity. Thus, if a purported antiaging treatment does not extend the life span, or even decreases survivorship, then the relevant pathway is excluded as the sole determinant of life span, but not as a contributor or even a primary causal factor in the aging process. For example, if oxidative stress were a causal factor in aging, then a treatment diminishing oxidant production could attenuate oxidative damage that normally leads to death as a result of aging. Simultaneously, the same treatment could cause pathology resulting in death unrelated to the aging process, e.g., it could interfere with cellular signaling pathways mediated by oxidants, or with the immune response to microbial infection. The pathological effects would

mask the anticipated extension of life span resulting from the decreased rate of aging, which could lead to a mistaken conclusion that oxidation was not a causal factor in the aging process. Such possibilities illustrate that identification of the proximate causes of death in insect populations, particularly in the laboratory, is of immediate importance to remove some of the uncertainties surrounding the interpretation of mortality data.

2 Physiological Adaptation

One particular issue, which has a critical bearing on the interpretation of studies involving experimental manipulations of life span, is that various types of organisms differ fundamentally in their capacities to respond to suboptimal conditions, either by becoming relatively inactive (e.g., dormant, torpid, or hibernatory) or by reducing their fecundity, thereby increasing their likelihood of survival until the restoration of conditions conducive to successful reproduction [10]. A counterintuitive effect of these adaptive responses is that adverse conditions can actually increase the length of life. Such conditions could consequently be regarded as beneficial, if longevity were the sole criterion of benefit, whereas an opposite conclusion would be suggested by the fact that they diminish levels of physical, metabolic, and/or reproductive activity.

A related issue is that experimental manipulations of the environment or genotype, which are intended to test various theories of aging by decreasing its rate, may in fact increase the life span by triggering a common adaptive mechanism of resource conservation. Thus, the extension of life span may be accompanied by trade-off effects, which decrease the quality of life. For example, exposure of *Drosophila* to either X-rays or ethidium bromide can extend the life span, but the longer-lived flies have decreased rates of egg production or oxygen consumption, respectively [11, 12]. Consequently, life span prolongation by a particular dietary or genetic modification is not sufficient to establish either the desirability or likely effectiveness of the corresponding treatment as a human antiaging therapy, or to conclude that truly novel (previously unknown) mechanisms of aging have been identified.

The varying capacities of poikilotherms and homeotherms to exhibit adaptive effects, involving decreases in metabolic rate, fertility, or both, cannot be overlooked in any meaningful analysis of mortality data from flies or other poikilothermal organisms used as models of human aging. Failure to do so has the potential to implicate various unrelated biochemical processes in the causation of aging, when in fact such processes are involved only tangentially, as a consequence of their influence on the adaptive response to stress. These conclusions do not invalidate or undercut poikilothermal model systems, which are of great value because of their short life spans, ease of mass culture, and genetic tractability (e.g., accessibility for mutagenesis and transgenesis), and because it is plausible that they do share mechanisms of aging in common with homeotherms. Instead, the presence or absence of trade-offs can be regarded as a criterion to distinguish which among the wide range of life-extending treatments in poikilotherms are most likely to be conserved in homeotherms, without adverse effects, and which are most likely to reveal fundamental, causal mechanisms of aging.

2.1   The Rate-of-Living Hypothesis

A decrease in the rate of metabolism is among the various manifestations of entry into a quiescent state, which results secondarily in an increased length of life. Although there is no necessary requirement that the lifetime energy expenditure, also referred to as the metabolic potential, should be constant, independent of the genotype or environmental conditions encountered by an organism, such a postulate was integral to early hypotheses about the relationship between rates of aging and metabolism [13]. An inverse relationship between longevity and metabolic rate was initially suggested by Rubner [13], who demonstrated that the lifetime energy expenditure per gram of body weight was roughly the same in several mammalian species. These species differed considerably in their basal metabolic rates and maximum life spans, such that those with faster rates of energy expenditure tended to have shorter life spans. Rubner's postulate was modified in Pearl's rate-of-living hypothesis [14], which also proposed that the metabolic potential was comparatively fixed and that the duration of life and rate of metabolism were inversely related. However, Pearl additionally stated that the genetically determined capacity for metabolic activity would not be realized in all environments, and in a notable departure from Rubner's proposal, the rate-of-living hypothesis referred explicitly and repeatedly to the individual, rather than different genotypes within or between species, suggesting "that duration of life is a function of two variables.

a. The constitution (organization or pattern) of the individuals, genetically determined.
b. The average rate of metabolism or rate of energy expenditure during life." (see pp. 139–140 in [14]).

Pearl also stated "that the relation between rate of growth and duration of life in the same individual is an inverse one … in general the duration of life varies inversely as the rate of energy expenditure during life." (see p. 145 in [14]).

2.2   Departures from the Rate-of-Living Model

It is necessary periodically to emphasize [2, 10, 15] that the ideas of Pearl and Rubner are routinely and mistakenly conflated and that the rate-of-living hypothesis concerning the *individual organism* cannot be disproven by comparisons of metabolic rates amongst widely divergent genera, or even different strains within a single species. Criticisms of the rate-of-living hypothesis are often made on the basis of predictions that are not implicit in the hypothesis, as originally postulated by Pearl.

For example, a recent comparison of basal metabolic rate and maximum life span data across a wide range of taxa convincingly refutes Rubner's idea, except insofar as it might be applied within some small groups of closely related species [16], but it is a misrepresentation of the hypothesis as stated by Pearl.

Comparisons of metabolic rates among genotypes within a single species or strain are more directly relevant to the predictions of the rate-of-living hypothesis, but not ideal. Strictly speaking, the hypothesis can be tested only in genetically identical organisms exposed to different environmental conditions, within a range normally associated with differences in metabolic rates, but no other physiological effects. The primary experimental problem is that any variation in conditions affecting the rate of metabolism and longevity would be expected to have additional, direct effects on fitness. Thus, the relationship between the metabolic rate and length of life should *tend* to be inverse, as stated by Pearl, but it should not be expected to fit a hyperbolic equation.

Notwithstanding such caveats, recent comparisons among outbred MF1 mice [17] and recombinant inbred (RI) *Drosophila* lines under normal conditions of maintenance [18] have shown that rates of oxygen consumption and carbon dioxide production are not always negatively correlated with longevity, and that they may even be positively correlated in mice. Similarly, there was no significant correlation between either oxygen consumption or heat production and survival times of individual wild-type flies from the Dahomey stock, although the life span of this stock is very short [19]. Despite the existence of some genetic heterogeneity among the individual mice and RI lines, and even individual wild-type flies, it is reasonable to conclude from these studies that an inverse relationship with the rate of metabolism is not the sole or primary determinant of longevity in all cases. This conclusion gives rise to three questions: (1) why should metabolic rates and other putative indices of adaptation continue to be measured in poikilotherms with extended life spans; (2) what common mechanisms, other than decreased metabolism, can explain the fact that poikilothermal life spans are extended by a wide range of proximate causes; and (3) what are the ramifications for the oxidative stress hypothesis?

2.3   Ramifications for Longevity Studies in Poikilotherms

The falsification of the postulate of a fixed, realized metabolic potential in widely divergent genotypes has prompted a suggestion that the attention given to metabolic rates in comparative studies of aging seems to be unjustified [16]. Although a decrease in attention to metabolic rates might be tenable for comparative studies among eutherians, a similar approach where experimental manipulations lead to extensions of life span within an otherwise constant genetic background, particularly in poikilothermal model systems, would be a disastrous error. The fact that metabolic rate is not the only variable affecting longevity under all conditions, including physiologically relevant conditions, does not indicate that it is never a factor under any conditions. Indeed, it remains altogether plausible that large differences in rates of metabolism are not merely a correlate but a causal factor underlying differences in longevity of flies at different temperatures [20], or in cages that permit varying levels of physical activity or flight [15]. The fact that exposure to the potent mutagen, ethidium bromide, extends the life span of *Drosophila*, and could logically be proposed to slow the rate of aging if its inhibitory effect on oxygen consumption were unknown [12], should be sufficient to demonstrate that poikilothermal life extension must be interpreted in the context of the presence or absence of adaptive responses. The primary value of continued measurements of potential trade-off variables, including fertility, locomotor activity, and flight activity, in addition to metabolic rates, is to distinguish between life-extending manipulations where the pursuit of some other mechanistic basis is more or less likely to be fruitful. As described in Subsection 2.4, the existence of life span variations that could not be explained by offsetting changes in metabolic activity has already led to the identification of putative underlying mechanisms, which might explain some apparent discrepancies between experimental results and the predictions of the oxidative stress hypothesis.

2.4 Ramifications for the Oxidative Stress Hypothesis

The rate-of-living and oxidative stress hypotheses have been linked, because the rate of oxygen consumption was used as an index of the rate of living [20], and because ROS production was assumed to constitute a constant fraction of oxygen consumption, regardless of the rate at which oxygen was consumed [15]. If the latter postulate were true, then evidence supporting or opposing the rate-of-living hypothesis would indirectly support or oppose the oxidative stress hypothesis, in addition to its implications for the interpretation of life span data in general. In fact, the implications of data that conflict with the rate-of-living model depend on the underlying mechanisms. Two plausible mechanisms have been suggested, each of which is compatible with oxidative stress as a causal factor in aging. Both mechanisms are based on the fact that ROS production is not always closely correlated with the rate of oxygen consumption in different genotypes and environments.

1. The "uncoupling to survive" hypothesis recognizes that the fraction of electrons diverted to ROS production is significantly impacted by membrane potential and the rate of electron flow through the electron transport chain [17, 21, 22]. When proton pumping associated with electron transport exceeds the reverse flow of protons across the inner mitochondrial membrane via F_0F_1-ATPase and adenine nucleotide translocase, the presence of uncoupling proteins permits some dissipation of the membrane potential. Consequently, the flow of electrons (and consumption of oxygen) continues, thereby decreasing the fraction of upstream electron carriers which are in a reduced state capable of univalent reduction of oxygen. As discussed in Subsection 3.2., this theory provides a potential mechanism to attenuate ROS production, thereby permitting a direct test of the oxidative stress hypothesis of aging.

2. The "membrane pacemaker" hypothesis proposes that the extent of membrane lipid oxidation resulting from mitochondrial respiration is dependent upon the proportions of various fatty acyl chains that are present, because the fatty acids differ in their susceptibility to peroxidation [23]. Differences in membrane lipid composition in long- versus short-lived species, and in response to life-extending treatments such as caloric restriction, and possibly decreased insulin signaling, are consistent with the membrane pacemaker hypothesis [23, 24]. Thus, this hypothesis serves not only to uncouple the rate-of-living and oxidative stress hypotheses of aging, but also suggests a common mechanism whereby various, seemingly disparate, treatments can increase longevity. Nevertheless, the extension of this idea to *Drosophila* is problematic for several reasons. First, it is opposed by the absence (at least in the head) of fatty acids with the highest relative peroxidizability in this short-lived species [25]. Conceivably, the absence of peroxidizable lipids could represent an evolutionary adaptation compensating for high ambient oxygen concentrations in tissues served by the flies' tracheolar network. Second, houseflies do contain low levels of highly peroxidizable fatty acids, but the total polyunsaturated fatty acid content is higher and life span is longer at 18 °C than at 25 °C [26]. Third, an additional variable affecting the membrane lipid composition is homeoviscous adaptation, whereby membrane viscosity is maintained by changes in the proportions of saturated and unsaturated fatty acids within a given species acclimated to different temperatures (see pp. 422–436 in [27]). Such variations within species can complicate efforts to correlate differences between species with differences in life span. Finally, the fatty acid composition of membrane lipids in *Drosophila* heads is strongly influenced by diet [25]. The latter observation could, however, suggest a mechanism whereby diets rich in nutrients other than simple carbohydrates decrease the life span [28].

3 Experimental Manipulations of Oxidant Production

Treatments that increase oxidative stress and diminish life span show the correlation predicted by the oxidative stress hypothesis [29, 30]; however, support of this kind is indirect, because decreased life span is not proof of acceleration of normal aging. In contrast, the limits imposed by normal aging cannot be exceeded without affecting the underlying causes of this process. Thus, the questions whether diminution of oxidant production could be accomplished experimentally, and whether the result would be an extension of life span, were considered for many years to be of crucial importance to the validation of the oxidative stress hypothesis [31].

3.1 Mitochondrial Catalase

An initial strategy to diminish oxidant generation was the ectopic expression of catalase in the mitochondrial matrix, which is a major site of H_2O_2 production [32, 33].

Although catalase is an antioxidant, which eliminates H_2O_2 rather than preventing H_2O_2 production, the net effect of targeted catalase activity would be to diminish the exposure of extramitochondrial cellular constituents to mitochondrially generated H_2O_2, and not merely to increase the level of defense. Indeed, mitochondrial expression of catalase essentially abolished the release of H_2O_2 under in vitro conditions. Nevertheless, in *Drosophila*, there was no positive effect on longevity [32], and further augmentation of mitochondrial catalase levels, particularly in the presence of excess Mn superoxide dismutase (SOD), resulted in a substantial (up to 43%) shortening of life span [33]. It is plausible, but unproven, that the effect on life span emerged only beyond levels of catalase sufficient to eliminate H_2O_2 production in vitro because of differences between rates of oxidant release by mitochondria in vivo versus in vitro. In either case, the results suggest that a severe disruption of mitochondrial oxidant release has adverse effects in *Drosophila*, which preclude any extension of life span. These results do not rule out the possibility that diminishing oxidant production has a beneficial effect on the aging process, which is masked by pathological effects on other physiological processes, but they demonstrate that alternative experimental strategies would be needed to reveal the hypothetical relationship between oxidant production and aging.

Although the focus of this chapter is on results obtained in *Drosophila*, a similar experiment in transgenic mice should be mentioned, because it seemed initially to demonstrate a difference between the species. Ectopic expression of catalase in the mitochondria of two founder lines was reported to extend the median life span by 17–21%, although the effect was diminished in a >99% B6 background [34]. A difference between flies and mice could arise from tissue-specific enhancement of catalase activity, but the same group indicated recently that the original results may not be reproducible [35]. Until this issue is resolved, any attempt to explain the apparent differences between species will be premature.

3.2 Uncoupling Proteins (UCPs)

The negative effect of mitochondrial catalase expression on the life span of *Drosophila*, in conjunction with the multiplicity of effects of oxidants on gene expression [36], suggests that a moderate decrease in oxidant production may be more likely to have a net positive effect than a complete abolition of H_2O_2 release. This idea, along with the "uncoupling to survive" hypothesis, was supported by the effects of targeted expression of human UCP 2 (hUCP2) in the neurons of adult *Drosophila*. hUCP2 expression increased mitochondrial state 4 respiration in fly heads, but decreased H_2O_2 production by approximately 40% and extended the median life span of female flies by 28% [37]. The effect on male life span was less consistent, and there was no effect on maximum life span; however, the effect on median life span was not offset by any decrease in fecundity or locomotor activity. Additionally, the accumulation of 4-hydroxy-2-nonenal, a marker of peroxidative damage, was decreased by 32% in middle-aged flies.

These results seem to support the hypothesis that oxidative stress is a causal factor in the aging process of flies, and they suggest that this process can be attenuated by decreasing the rate of oxidant production. The same study also indicated limits to the scope of life span extension by this method: ubiquitous expression of hUCP2 was lethal during development, and hUCP2 expression in muscle had no positive effect on life span, suggesting (1) that a decrease but not complete abolition of oxidant production is beneficial, and (2) that the rate of aging is most strongly influenced by oxidant production in the nervous system. However, it should be noted that the life spans of mice overexpressing UCP2 exclusively in hypocretin neurons also were extended, and this effect was accompanied by an elevation of local temperature and decrease in core body temperature [38]. The tissue-specific effect of UCP2 overexpression on temperature, and the inverse relationship between temperature and metabolic rate in *Drosophila*, suggest that measurements of whole-fly respiration and mitochondrial respiration outside the head will be necessary to establish the mechanism by which hUCP2 expression extends the life span of flies.

3.3   *Iron Metabolism*

Loosely-bound transition metals, particularly iron, have long been recognized as a potent source of ROS as a result of their catalysis of hydroxyl free radical production via the Fenton reaction [39]. The total iron content of flies is known to increase as a function of age, and supplementation with dietary tea extracts both inhibited the accumulation of iron and prolonged the life span of flies [40]. A plausible hypothesis for future investigation is that maintenance of iron stores at young adult levels would not interfere with oxidant production required for various physiological processes, but it would forestall surplus oxidant production and slow the accumulation of damage. Ideally, the hypothesis would be tested by supplementation with an iron chelator, or overexpression of an iron-binding protein, such as ferritin, with more precisely defined properties than tea extracts. An investigation of the normal tissue distribution and sequestration status of the accumulated iron in aging flies also would be necessary, to target a chelator or overexpressed protein with the appropriate spatial and temporal pattern to avoid iron deficiency.

3.4   *Cytochrome c Oxidase (COX)*

An additional mechanism for the attenuation of ROS generation, without disrupting basal levels required for signaling, is suggested by the finding that COX, the terminal oxidoreductase of the mitochondrial electron transport chain, exhibits an age-associated decline in activity in *Drosophila* [41]. There was no similar decline in the activities of other mitochondrial oxidoreductases, but there was a decline in state 3 and uncoupled respiration rates. Prevention or reversal of the decline in COX activity would therefore be predicted to relieve a bottleneck impeding the flow of electrons, and consequently to ameliorate the decline in energy production. By decreasing the fraction of upstream electron carriers in a reduced state, an enhancement of COX activity in aging flies would be expected additionally to attenuate the age-associated increase in univalent oxygen reduction leading to oxidative stress. Studies in HeLa cells have shown that a specific increase in the abundance of one regulatory subunit, COX Vb, is associated with an increase in COX enzymatic activity, decrease in ROS production, and enhanced resistance to hyperoxia [42]. These findings raise the possibility that the age-related decline in COX activity and increase in mitochondrial oxidant production in *Drosophila* could be attenuated by the over-expression of one or a few COX subunits, thus providing a direct test of the oxidative stress hypothesis of aging.

4 Overexpression of Antioxidants

The relationship between aging and perturbations of antioxidant activities has been examined more extensively than the corresponding relationships involving either oxidant production or repair processes. This situation reflects the relative ease of antioxidant supplementation, because correlations between antioxidant levels, aging and the maximum life spans of different genotypes (within or among species) are weaker than the correlations involving oxidant production, oxidative damage, and repair [31].

4.1 Nonenzymatic Antioxidants

Dietary supplementation with nonenzymatic antioxidants has sometimes been reported to increase survivorship in *Drosophila*, but the effects were small in magnitude [43] and/or accompanied by a decreased metabolic rate [44], or they were irreproducible ([45], R. S. Sohal, unpublished results), or they were observed only in relatively short-lived backgrounds [46, 47]. In some cases, the life span was not extended [48]. In other cases, metabolic rates and fertility were not measured [49, 50]. There are no clear examples in *Drosophila* of an antioxidant, or indeed of any other compound administered in the diet, which substantially extended mean/median life span beyond levels normally observed in robust control populations (typically 50–75 days at 25 °C [14, 48, 51–53]), and for which both of the most obvious trade-off mechanisms, involving metabolism and reproduction, were ruled out.

4.2 Enzymatic Antioxidants

The majority of the work on antioxidative enzymes has been reviewed elsewhere, from several different perspectives [10, 54–56], and it is only summarized here. In general, the complete absence of SODs or catalase has severe, adverse effects [57,

58], but dramatic decreases in activity short of a bona fide null condition are well tolerated [59, 60]. Conversely, overexpressions of Cu-Zn and Mn SODs, catalase, and thioredoxin reductase in the native patterns had no positive effect on life span in long-lived genetic backgrounds [61–64]. Likewise, overexpressing various combinations of these enzymes had no beneficial effect [52].

Overexpression of Cu-Zn SOD has been reported to extend the life span of *Drosophila*, either (1) in combination with catalase in the native pattern [65], (2) in a spatially restricted pattern in motor neurons [66], (3) in a temporally restricted, inducible manner in adult flies [51], or (4) in adult flies in combination with Mn SOD [67], which had been reported to extend life span in its own right [68]. The conclusions of these studies were, however, undercut by, respectively: (1) insufficient control sample sizes and irreproducibility [52, 55]; (2) limited reproducibility in long-lived, wild-caught genetic backgrounds [53]; (3) lack of reproducibility in long-lived backgrounds within the same study [51]; and (4) generally short life spans and/or evidence of diminished oxygen consumption in several (but not all) of the genotypes examined. Collectively, the evidence suggests that SOD activity may have a small effect on longevity, mainly in short-lived backgrounds, but it has not as yet been demonstrated that an effect can be observed in a long-lived background, independent of a trade-off mechanism involving respiration, locomotion, or fertility. Additionally, the fact that enhancement of antioxidant activities sometimes increases survival times under severe, experimental oxidative stress, without exhibiting any beneficial effect in the same genotypes under unstressed conditions [32, 33, 63], suggests that the extension of life span in short-lived populations may reflect increased resistance to an unidentified source of stress, rather than an effect on the aging process.

The hypothesis that enhancement of antioxidant levels in the nervous system can retard the aging process was supported by a recent study of glutamate-cysteine ligase (GCL) overexpression [69]. Neuronal overexpression of the catalytic subunit, GCLc, increased mean and maximum life spans up to 50% in a long-lived background (84 *vs* 56-61 days for driver and responder controls in a GAL4-UAS binary transgenic system). Global overexpression of the modulatory subunit also extended the life span by up to 24%. There was no decrease in oxygen consumption, at least among young adults overexpressing GCLc. Thus, enhancement of the glutathione biosynthetic capacity and possibly superoxide dismutase activity in the nervous system can extend the life span of flies. Furthermore, the superoxide-glutathione sink hypothesis for free radical detoxification [70] leads to a prediction that simultaneous overexpression of both Cu-Zn SOD and GCL in the nervous system could have synergistic effects on life span. The simultaneous overexpression of both subunits of GCL, of additional enzymes affecting the redox state, and of antioxidants restricted to tissues outside the nervous system, are other topics worthy of further investigation.

5   Repair

In general, tests of the effects of reversal of oxidative damage have been less comprehensive, involving fewer repair proteins and pathways, than the corresponding investigations of damage prevention. Given that oxidant production is now recognized to be physiologically necessary, and that at least some damage is unavoidable under such conditions, the repair or removal of damage by-products should be regarded as an area of high priority for future testing of the oxidative stress hypothesis of aging.

5.1 Methionine Sulfoxide Reductase

Overexpression of a single repair protein, the peptide methionine sulfoxide reductase A (MSRA), had the most pronounced positive effect on life span so far reported for a single enzyme associated with oxidative stress [71]. MSRA overexpression in the nervous system increased the median life span by 56–73% in female *Drosophila* and by 60–83% in male flies. A similar result was obtained when MSRA was overexpressed globally, and a smaller extension of life span was observed in flies over-expressing MSRA primarily in the eye. Although the control life spans in the majority of these experiments were not particularly long (typically 35–50 days), the life span was extended in some cases beyond levels normally observed in *Drosophila* at 25 °C (95 days in females overexpressing MSRA in the nervous system *vs* 55–61 days for the control flies). Metabolic rates were not measured in these flies, but physical activity and reproductive output were consistently greater in the MSRA overexpressors throughout most of the adult life span. Testing the effects of MSRA overexpression in longer-lived, isogenic backgrounds would be necessary to establish reproducibility, but the existing results do support the hypothesis that oxidative stress is a causal factor in the aging process of *Drosophila*.

5.2 Protein Carboxyl Methyltransferase (PCMT)

In addition to MSRA, overexpression of PCMT, the first enzyme in another protein repair pathway, has been reported to extend the life span of flies [72]. Ubiquitous overexpression of PCMT increased survivorship at 29 °C, indicating that modified protein isoaspartyl residues, the substrate of PCMT, may limit survival under conditions of thermal stress. Overexpression of PCMT did not extend the life span at 25 °C, suggesting that an increased capacity to repair this substrate does not postpone the aging process under unstressed conditions.

5.3 Small Heat Shock Proteins (Hsp22)

An alternative strategy for the prevention or repair of damage is the overexpression of Hsp, which are normally induced by conditions associated with protein damage, such as thermal or oxidative stress and aging [73]. Hsp22, a small mitochondrial heat shock protein, was overexpressed in long-lived *Drosophila* backgrounds by two independent groups, with strikingly divergent results. Morrow et al. [74] detected a 32% increase in mean life span, resulting from either ubiquitous or motor neuron specific overexpression of Hsp22 with the GAL4-UAS expression system. The control life spans were at least 60 days in each case, and flies overexpressing Hsp22 in motor neurons had a mean life span of 90 days [74]. Locomotor activity and resistance to experimental oxidative stress were increased, but there were no measurements of metabolic rates or fertility. In contrast, Bhole et al. [73] reported that ubiquitous Hsp22 overexpression using the "tet-on" system either decreased or had no significant effect on survivorship at 25 or 29 °C, and resistance to experimental oxidative stress was also diminished. The control life spans at 25 °C were ~60–80 days, but there were no measurements of metabolic rates or fertility. The reasons for these divergent results are unknown, but they could involve differences in the magnitude, timing or tissue specificity of Hsp22 overexpression with the GAL4-UAS *vs* tet-on system [73].

6   Conclusions (Perspective)

Several main points have been emphasized in this chapter. First, poikilothermal animals constitute valuable model systems in which to test various hypotheses of aging, but careful consideration should be given to the substantial physiological differences between poikilotherms and homeotherms. To avoid potentially erroneous interpretations, extrapolations, and predictions of effects in higher animals, it is essential to test for decreases in metabolic, reproductive, or locomotor activity, and other measures of fitness, which might underlie extensions of life span in simple model organisms. Second, although the oxidative stress hypothesis of aging gives rise to straightforward predictions, organismal physiology is sufficiently complex that the actual effects of a single experimental manipulation may be difficult to predict. Consequently, the hypothesis cannot be verified or falsified conclusively based on any single line of investigation. Third, experimental decreases in rates of in vivo oxidant production, such as those induced by mitochondrial catalase, which had been predicted to provide the most convincing evidence for or against this hypothesis, have in fact met with rather limited success, for reasons which are independent of the relationship between oxidative stress and aging. Specifically, the physiological roles of oxidants may be disrupted by preventing oxidant production, so that any beneficial effect on the aging process is limited or completely masked by unrelated, adverse effects. Fourth, enhancement of antioxidant defenses, which has been the most widely used approach to test the oxidative stress hypothesis and prevent age-related oxidative damage, has had little or no positive effect on the life span of *Drosophila* in most studies, especially when long-lived strains were examined. To date, the most notable beneficial effects in a long-lived background have been elicited by neuronal overexpression of GCLc, underscoring the importance of maintaining glutathione levels. Fifth, where beneficial effects on life span have been observed, they have tended to implicate the nervous system as a site particularly vulnerable to oxidative stress. Finally, overexpression studies involving repair enzymes have yielded some evidence in support of the oxidative stress hypothesis, but more extensive work in this area is necessary. An overarching conclusion is that the currently available evidence is suggestive of a causal role for oxidative stress in the aging process of *Drosophila*, but this evidence is not yet solid or definitive.

Acknowledgments This work was supported by National Institutes of Health-National Institute on Aging grants R01 AG7657 (to R.S.S.) and R01 AG15122 (to W.C.O.).

References

1. Harman D. Aging: a theory based on free radical and radiation chemistry. J Gerontol 1956;11:298–300.
2. Sohal RS. The free radical hypothesis of aging: an appraisal of the current status. Aging Clin Exp Res 1993;5:3–17.
3. Sohal RS, Agarwal A, Agarwal S, Orr WC. Simultaneous overexpression of copper- and zinc-containing superoxide dismutase and catalase retards age-related oxidative damage and increases metabolic potential in *Drosophila melanogaster*. J Biol Chem 1995;270:15671–15674.
4. Rebrin I, Bayne AC, Mockett RJ, Orr WC, Sohal RS. Free aminothiols, glutathione redox state and protein mixed disulphides in aging *Drosophila melanogaster*. Biochem J 2004;382: 131–136.
5. Sohal RS, Allen RG, Farmer KJ, Newton RK, Toy PL. Effects of exogenous antioxidants on the levels of endogenous antioxidants, lipid-soluble fluorescent material and life span in the housefly, *Musca domestica*. Mech Ageing Dev 1985;31:329–336.
6. Climent I, Levine RL. Oxidation of the active site of glutamine synthetase: conversion of arginine-344 to γ-glutamyl semialdehyde. Arch Biochem Biophys 1991;289:371 375.
7. Levine RL, Berlett BS, Moskovitz J, Mosoni L, Stadtman ER. Methionine residues may protect proteins from critical oxidative damage. Mech Ageing Dev 1999;107:323–332.
8. Yarian CS, Rebrin I, Sohal RS. Aconitase and ATP synthase are targets of malondialdehyde modification and undergo an age-related decrease in activity in mouse heart mitochondria. Biochem Biophys Res Commun 2005;330:151–156.
9. Halliwell B. Can oxidative DNA damage be used as a biomarker of cancer risk in humans? Problems, resolutions and preliminary results from nutritional supplementation studies. Free Radic Res 1998;29:469–486.
10. Sohal RS, Mockett RJ, Orr WC. Mechanisms of aging: an appraisal of the oxidative stress hypothesis. Free Radic Biol Med 2002;33:575–586.
11. Lamb MJ. The effects of radiation on the longevity of female *Drosophila subobscura*. J Insect Physiol 1964;10:487–497.
12. Fleming JE, Leon HA, Miquel J. Effects of ethidium bromide on development and aging of *Drosophila*: implications for the free radical theory of aging. Exp Gerontol 1981;16:287–293.
13. Rubner M. Das problem der lebensdauer und seine beziehungen zu wachstum und ernahrung. Munich, Germany: Oldenburg, 1908.
14. Pearl R. The rate of living. New York: Alfred A. Knopf, Inc., 1928.
15. Sohal RS. The rate of living theory: a contemporary interpretation. In: Collatz K-G, Sohal RS, eds. Insect aging. Berlin, Germany: Springer-Verlag, 1986:23–44.
16. de Magalhães JP, Costa J, Church GM. An analysis of the relationship between metabolism, developmental schedules, and longevity using phylogenetic independent contrasts. J Gerontol Biol Sci 2007;62A:149–160.

17. Speakman JR, Talbot DA, Selman C et al. Uncoupled and surviving: individual mice with high metabolism have greater mitochondrial uncoupling and live longer. Aging Cell 2004;3:87–95.

18. Van Voorhies WA, Khazaeli AA, Curtsinger JW. Testing the "rate of living" model: further evidence that longevity and metabolic rate are not inversely correlated in *Drosophila melanogaster*. J Appl Physiol 2004;97:1915–1922.

19. Hulbert AJ, Clancy DJ, Mair W, Braeckman BP, Gems D, Partridge L. Metabolic rate is not reduced by dietary-restriction or by lowered insulin/IGF-1 signalling and is not correlated with individual lifespan in *Drosophila melanogaster*. Exp Gerontol 2004;39:1137–1143.

20. Miquel J, Lundgren PR, Bensch KG, Atlan H. Effects of temperature on the life span, vitality and fine structure of *Drosophila melanogaster*. Mech Ageing Dev 1976;5:347–370.

21. Nicholls DG. Mitochondrial function and dysfunction in the cell: its relevance to aging and aging-related disease. Int J Biochem Cell Biol 2002;34:1372–1381.

22. Miwa S, St-Pierre J, Partridge L, Brand MD. Superoxide and hydrogen peroxide production by *Drosophila* mitochondria. Free Radic Biol Med 2003;35:938–948.

23. Hulbert AJ. On the importance of fatty acid composition of membranes for aging. J Theor Biol 2005;234:277–288.

24. Pamplona R, Barja G, Portero-Otín M. Membrane fatty acid unsaturation, protection against oxidative stress, and maximum life span. Ann NY Acad Sci 2002;959:475–490.

25. Stark WS, Lin T-N, Brackhahn D, Christianson JS, Sun GY. Fatty acids in the lipids of *Drosophila* heads: effects of visual mutants, carotenoid deprivation and dietary fatty acids. Lipids 1993;28:345–350.

26. Sohal RS, Müller A, Koletzko B, Sies H. Effect of age and ambient temperature on *n*-pentane production in adult housefly, *Musca domestica*. Mech Ageing Dev 1985;29:317–326.

27. Hochachka PW, Somero GN. Biochemical adaptation. Princeton, New Jersey: Princeton University Press, 1984.

28. Mair W, Piper MDW, Partridge L. Calories do not explain extension of life span by dietary restriction in *Drosophila*. PLoS Biol 2005;3:e223.

29. Arking R, Buck S, Berrios A, Dwyer S, Baker III GT. Elevated paraquat resistance can be used as a bioassay for longevity in a genetically based long-lived strain of *Drosophila*. Dev Genet 1991;12:362–370.

30. Baret P, Fouarge A, Bullens P, Lints FA. Life-span of *Drosophila melanogaster* in highly oxygenated atmospheres. Mech Ageing Dev 1994;76:25–31.

31. Mockett RJ, Sohal RS. Oxidative stress may be a causal factor in senescence of animals. In: Robine J-M, Vaupel JW, Jeune B, Allard M, eds. Longevity: to the limits and beyond. Berlin, Germany: Springer-Verlag, 1997:139–154.

32. Mockett RJ, Bayne A-CV, Kwong LK, Orr WC, Sohal RS. Ectopic expression of catalase in *Drosophila* mitochondria increases stress resistance but not longevity. Free Radic Biol Med 2003;34:207–217.

33. Bayne A-CV, Mockett RJ, Orr WC, Sohal RS. Enhanced catabolism of mitochondrial superoxide/hydrogen peroxide and aging in transgenic *Drosophila*. Biochem J 2005;391:277–284.

34. Schriner SE, Linford NJ, Martin GM et al. Extension of murine life span by overexpression of catalase targeted to mitochondria. Science 2005;308:1909–1911.

35. Choi CQ. Old mice hard to replicate. Scientist 2007;21:64.

36. Allen RG, Tresini M. Oxidative stress and gene regulation. Free Radic Biol Med 2000;28:463–499.

37. Fridell Y-WC, Sánchez-Blanco A, Silvia BA, Helfand SL. Targeted expression of the human uncoupling protein 2 (hUCP2) to adult neurons extends life span in the fly. Cell Metab 2005;1:145–152.

38. Conti B, Sanchez-Alavez M, Winsky-Sommerer R et al. Transgenic mice with a reduced core body temperature have an increased life span. Science 2006;314:825–828.

39. Herbert V, Shaw S, Jayatilleke E, Stopler-Kasdan T. Most free-radical injury is iron-related: it is promoted by iron, hemin, holoferritin and vitamin C, and inhibited by desferoxamine and apoferritin. Stem Cells 1994;12:289–303.

40. Massie HR, Aiello VR, Williams TR. Inhibition of iron absorption prolongs the life span of *Drosophila*. Mech Ageing Dev 1993;67:227–237.
41. Ferguson M, Mockett RJ, Shen Y, Orr WC, Sohal RS. Age-associated decline in mitochondrial respiration and electron transport in *Drosophila melanogaster*. Biochem J 2005;390:501–511.
42. Campian JL, Gao X, Qian M, Eaton JW. Cytochrome *c* oxidase activity and oxygen tolerance. J Biol Chem 2007;282:12430–12438.
43. Sagi O, Wolfson M, Utko N, Muradian K, Fraifeld V. p66[ShcA] and ageing: modulation by longevity-promoting agent aurintricarboxylic acid. Mech Ageing Dev 2005;126:249–254.
44. Miquel J, Fleming J, Economos AC. Antioxidants, metabolic rate and aging in *Drosophila*. Arch Gerontol Geriatr 1982;1:159–165.
45. Brack C, Bechter-Thüring E, Labuhn M. *N*-Acetylcysteine slows down ageing and increases the life span of *Drosophila melanogaster*. Cell Mol Life Sci 1997;53:960–966.
46. Bonilla E, Medina-Leendertz S, Díaz S. Extension of life span and stress resistance of *Drosophila melanogaster* by long-term supplementation with melatonin. Exp Gerontol 2002;37:629–638.
47. Wood JG, Rogina B, Lavu S, Howitz K, Helfand SL, Tatar M, Sinclair D. Sirtuin activators mimic caloric restriction and delay ageing in metazoans. Nature 2004;430:686–689.
48. Massie HR, Shumway ME, Whitney SJP, Sternick SM, Aiello VR. Ascorbic acid in *Drosophila* and changes during aging. Exp Gerontol 1991;26:487–494.
49. Cui X, Dai X-G, Li W-B, Zhang B-L, Fang Y-Z. Effects of lu-duo-wei capsule on prolonging life span of housefly and *Drosophila melanogaster*. Am J Chin Med. 1999;27:407–413.
50. Driver C, Georgiou A. How to re-energise old mitochondria without shooting yourself in the foot. Biogerontology 2002;3:103–106.
51. Sun J, Tower J. FLP recombinase-mediated induction of Cu/Zn-superoxide dismutase transgene expression can extend the life span of adult *Drosophila melanogaster* flies. Mol Cell Biol 1999;19:216–228.
52. Orr WC, Mockett RJ, Benes JJ, Sohal RS. Effects of overexpression of copper-zinc and manganese superoxide dismutases, catalase, and thioredoxin reductase genes on longevity in *Drosophila melanogaster*. J Biol Chem 2003;278:26418 26422.
53. Spencer CC, Howell CE, Wright AR, Promislow DEL. Testing an 'aging gene' in long-lived *Drosophila* strains: increased longevity depends on sex and genetic background. Aging Cell 2003;2:123–130.
54. Tower J. Aging mechanisms in fruit flies. Bioessays 1996;18:799–807.
55. Tatar M. Transgenes in the analysis of life span and fitness. Am Nat 1999;154:S67–S81.
56. Haenold R, Wassef DM, Heinemann SH, Hoshi T. Oxidative damage, aging and anti-aging strategies. Age 2005;27:183–199.
57. Phillips JP, Campbell SD, Michaud D, Charbonneau M, Hilliker AJ. Null mutation of copper/zinc superoxide dismutase in *Drosophila* confers hypersensitivity to paraquat and reduced longevity. Proc Natl Acad Sci U S A 1989;86:2761–2765.
58. Duttaroy A, Paul A, Kundu M, Belton A. A *Sod2* null mutation confers severely reduced adult life span in *Drosophila*. Genetics 2003;165:2295–2299.
59. Mackay WJ, Bewley GC. The genetics of catalase in *Drosophila melanogaster*: isolation and characterization of acatalasemic mutants. Genetics 1989;122:643–652.
60. Mockett RJ, Radyuk SN, Benes JJ, Orr WC, Sohal RS. Phenotypic effects of familial amyotrophic lateral sclerosis mutant *Sod* alleles in transgenic *Drosophila*. Proc Natl Acad Sci USA 2003;100:301–306.
61. Orr WC, Sohal RS. The effects of catalase gene overexpression on life span and resistance to oxidative stress in transgenic *Drosophila melanogaster*. Arch Biochem Biophys 1992;297:35–41.
62. Orr WC, Sohal RS. Effects of Cu-Zn superoxide dismutase overexpression on life span and resistance to oxidative stress in transgenic *Drosophila melanogaster*. Arch Biochem Biophys 1993;301:34–40.
63. Mockett RJ, Sohal RS, Orr WC. Overexpression of glutathione reductase extends survival in transgenic *Drosophila melanogaster* under hyperoxia but not normoxia. FASEB J 1999;13:1733–1742.

64. Mockett RJ, Orr WC, Rahmandar JJ, Benes JJ, Radyuk SN, Klichko VI, Sohal RS. Overexpression of Mn-containing superoxide dismutase in transgenic *Drosophila melanogaster*. Arch Biochem Biophys 1999;371:260–269.
65. Orr WC, Sohal RS. Extension of life-span by overexpression of superoxide dismutase and catalase in *Drosophila melanogaster*. Science 1994;263:1128–1130.
66. Parkes TL, Elia AJ, Dickinson D, Hilliker AJ, Phillips JP, Boulianne GL. Extension of *Drosophila* lifespan by overexpression of human *SOD1* in motorneurons. Nat Genet 1998;19:171–174.
67. Sun J, Molitor J, Tower J. Effects of simultaneous over-expression of Cu/ZnSOD and MnSOD on *Drosophila melanogaster* life span. Mech Ageing Dev 2004;125:341-349.
68. Sun J, Folk D, Bradley TJ, Tower J. Induced overexpression of mitochondrial Mn-superoxide dismutase extends the life span of adult *Drosophila melanogaster*. Genetics 2002;161:661–672.
69. Orr WC, Radyuk SN, Prabhudesai L et al. Overexpression of glutamate-cysteine ligase extends life span in *Drosophila melanogaster*. J Biol Chem 2005;280:37331-37338.
70. Winterbourn CC. Superoxide as an intracellular radical sink. Free Radic Biol Med 1993;14:85–90.
71. Ruan H, Tang XD, Chen M-L et al. High-quality life extension by the enzyme peptide methionine sulfoxide reductase. Proc Natl Acad Sci U S A 2002;99:2748–2753.
72. Chavous DA, Jackson FR, O'Connor CM. Extension of the *Drosophila* lifespan by overexpression of a protein repair methyltransferase. Proc Natl Acad Sci U S A 2001;98:14814–14818.
73. Bhole D, Allikian MJ, Tower J. Doxycycline-regulated over-expression of hsp22 has negative effects on stress resistance and life span in adult *Drosophila melanogaster*. Mech Ageing Dev 2004;125:651–663.
74. Morrow G, Samson M, Michaud S, Tanguay RM. Overexpression of the small mitochondrial Hsp22 extends *Drosophila* life span and increases resistance to oxidative stress. FASEB J 2004;18:598–599.

8
Does Oxidative Stress Limit Mouse Life Span?

Florian L. Muller

Summary The free radical theory of aging, in its strongest form, holds that oxidative stress is *the* determinant of animal life span. However, evolutionary studies suggest that it is highly unlikely that life span is determined by a single biochemical process. Rather, one can restate the free radical radical theory in a more modest form, i.e., that oxidative stress is one of several biochemical processes that limit life span. An obvious prediction of this hypothesis is that knocking out antioxidant genes should result in shortened life span (assuming that knocking-out these genes in fact result in increased oxidative stress). In the best-case scenario, one would expect a segmental progeroid syndrome. In this chapter, I summarize recent studies examining the life span, pathology, and oxidative stress status of antioxidant and oxidative damage repair knockout mice (*Sod1*, *Sod2*, *Sod3*, *Gpx1*, *Prdx1*, *Prdx2*, and *MsrA*). Data from these studies indicate that whether or not oxidative stress is limiting to life span depends on what markers of oxidative stress are assayed. Specifically, the DNA damage product, 8-oxodeoxyguanosine (8-oxo-dG), is elevated in several antioxidant knockout mice that show no life span reduction. Whether elevations in 8-oxo-dG genuinely represent increased in situ oxidative DNA damage or arise during sample preparation is debatable. Other markers that suffer from minimal isolation artifacts, such as F_2-isoprostanes, are only elevated in antioxidant knockout mice that actually have a shortened life span (the $Sod1^{-/-}$ to be precise). I suggest that measuring the expression of genes responsive to oxidative damage circumvents the issue of isolation artifacts that one way or another affects all biochemical assays of oxidative damage. By these criteria, $Sod1^{-/-}$ but not $Gpx1^{-/-}$, $Sod2^{+/-}$, or $Gpx4^{+/-}$ mice show an elevation in oxidative stress. If one takes as reference the standard set by DNA repair deficient mice, a strong case can be made that $Sod1^{-/-}$ exhibit "accelerated aging." Thus, I conclude that the present data do not disprove the hypothesis that oxidative stress is limiting to life span in mice.

Keywords Superoxide, life span, oxidative stress, F_2-isoprostanes, antioxidant enzyme, aging.

From: *Aging Medicine: Oxidative Stress in Aging: From Model Systems to Human Diseases* 129
Edited by: S. Miwa, K.B. Beckman, and F.L. Muller © Humana Press, Totowa, NJ

1 Introduction

The origins of the free radical theory of aging and the biochemistry of reactive oxygen species (ROS) have been reviewed in Chapter 1. In a landmark review, Beckman and Ames made an important addition to the field by dividing the theory into "strong" and "weak" versions [1]. The strong version of the free radical theory states that oxidative stress determines life span, whereas the weaker version postulates that oxidative stress is "associated" with age-related disease.

I want to suggest a further distinction. Evolutionary biologists have by enlarge concluded that aging is not adaptive and that it is the result of lack of selective pressure at advanced age, because of high extrinsic mortality reduces cohort seize over time independently of aging [2, 3]. In practice, this means that it is unlikely that one single biochemical process has been evolutionarily selected to "kill-off" the organism, but rather, multiple biochemical processes (although finite in number) are likely to cause deterioration more or less simultaneously (a similar idea has previously been put forward by Shmookler Reis [4]). I suggest that one can further divide the free radical theory of aging into an "ambitious" version, i.e., that oxidative stress is *the* life span determinant (reducing oxidative stress is necessary and sufficient for life span extension) and a "modest" version, i.e., oxidative stress is life span-limiting, even though it is not life span-determining (reducing oxidative stress is necessary but not sufficient for life span extension). This distinction may sound trivial, but in practice these two versions differ in one critical point: only the "ambitious" version predicts that reducing oxidative stress by overexpression of antioxidant enzymes or supplementation with antioxidants will result in increased life span (something that by enlarge has not been observed). The modest version would argue that reducing oxidative stress is necessary for extending life span, even though, in and of itself, it is not sufficient. But both the ambitious and the modest versions predict that increasing oxidative stress, by, for example, ablating antioxidant enzymes, will result in shortened life span.

In this chapter, I review the evidence concerning the question of whether oxidative damage can be life span-limiting. This may sound like a "duh," but in fact, it is not. Other biochemical mechanisms at one time thought to be limiting to life span, such as telomere shortening or even mitochondrial DNA (mtDNA) point mutations, have not passed this test (first, second, and third generation telomerase knockout mice do not show a decreased life span [5–7]), and PolG mutant mice, with a 500-fold increase in mtDNA point mutation load, show a minimally altered life span [8]). I review the literature on oxidative stress status and life span of antioxidant knockout mice. In several instances, the knockout or deficiency of an antioxidant enzyme does not result in shortening of life span. This could be taken as evidence against the idea that oxidative stress is life span-limiting; however, the critical question is whether in these instances, oxidative stress is truly elevated. As I show, the answer to this question is not trivial, and it depends on the maker of oxidative damage under consideration.

2 The Model: *Mus musculus "laboratorienscis"*

Lab mice originate from mouse fanciers of the late nineteenth century, and they have undergone considerable transformations compared with their wild cousins [9]. For one, they are tame enough to be handled without restrainers. Lab mice weight between 20 and 50 g, and they can be housed in relatively small containers. As such, housing the animals is relatively cheap, in most institutions, costing around $1 per cage (one to six animals) per day. Typically, precautions are taken to avoid infectious diseases (so-called barrier conditions), and the animals are tested for the presence of typical mouse pathogens (specific pathogen free conditions). Sentinel animals are sacrificed and analyzed for the presence of such pathogens at regular intervals. Although this means that never during their lifetime will these mice experience, say, viral influenza, they are not germ-free and bacterial infections readily occur. It could be argued that this favorable environment likely minimizes the deleteriousness of the phenotype of antioxidant knockout mice in particular and even knockout mice in general. For example, it has been shown that $Gpx1^{-/-}$ (which as I describe below, are essentially normal) develop myocarditis after coxsackie virus infection, which is harmless in control mice [10]. Regardless of this consideration, not using barrier conditions for aging experiments means adding a further variable (unidentified epidemic diseases) to an already complicated experiment. Although simply "aging" a colony of mice is trivial, doing a properly controlled survival experiment is not.

Under optimal conditions, the life span of the commonly used C57B6/J mouse is ~30 months average and 40 months maximum [11, 12]. C57B/6J is considered a long-lived inbred strain. Dr. David Harrison maintains an excellent website at The Jackson Laboratory (Bar Harbor, ME), which contains the most up to date information on the life span of specific inbred, F2, and outbred mouse strains (http://www.jax. org/staff/harrison/labsite/tablea.html). Outbred strains such as CD-1 or 4-way crosses are not as long-lived, although F1 crosses of C57B6/J with DBA/J or CBA/J or C3He/J are noticeably longer lived than pure C57B6/J. Although it is scientifically acceptable to perform life span experiments on mixed backgrounds, this will add considerable variability, requiring more animals. Power calculations from the Richardson lab indicate that to reliably observe a 10% alteration in life span, ~80 animals per group are necessary.

Mice differ substantially from humans with regard to age-related morbidity. Although the major cause of age-related deaths in humans is cerebrovascular or cardiovascular incidents, in mice, this is not the case. Rather, the major cause of death is cancer [13, 14], the predominant type, varies depending on the inbred strain: for example in C57B6/J mice, it is lymphoma, whereas in C3He it is hepatocellular carcinoma [11, 13]. Another key difference is that mice do not spontaneously develop diabetes the way humans do. Despite these and other imperfections, one of the great advantages of mice as model systems is the ease by which recombination-deletions, gene targeting, is achieved (this is much more difficult even in rats). As such, mice have become the most popular mammalian model for aging and other physiological studies; given this and simple cost consideration, it is likely to stay that way for a long time to come.

3   Life Span of Antioxidant and Oxidative Damage Repair Knockout Mice

As in most areas of biology, gene targeting [15] has revolutionized the study of the free radical theory of aging. Several antioxidant knockout mice have been generated, although the actual life span has only been determined in a handful. I only discuss in depth those antioxidant knockouts for which an actual life span is available, but a few key points are worth summarizing first. Completely knocking out any component of the thiol redox system, either glutathione biosynthesis, thioredoxins, or thioredoxin reductases, results in early embryonic lethality [16–18]. Regrettably, the effect on life span of heterozygotes of any of these is not known. The oxidative stress status of even fewer antioxidant knockout mice is known for sure. Discussed below are those antioxidant knockout mice for which a life span curve is available and those that have been investigated with respect to oxidative stress status.

3.1 $Sod1^{-/-}$ or CuZnSOD Knockout

CuZnSOD is the major superoxide scavenger in the cytoplasm [19] and in the mitochondrial intermembrane space [20, 21]. This latter localizaton is likely of great importance because the mitochondrial electron transport chain can release superoxide into that compartment [22–24]. It was a surprise to the field, given the putative importance of CuZnSOD, that $Sod1^{-/-}$ mice seemed healthy at birth and that they were initially reported to be phenotypically normal [25]. This result would have proven highly inconsistent with the idea that excess superoxide and oxidative stress is somehow involved in aging. However, although $Sod1^{-/-}$ mice are essentially normal at birth, it is now known that they exhibit high levels of oxidative stress, multiple age-dependent pathologies, and have a 30% reduction in life span [26–36]. One could perhaps even say that $Sod1^{-/-}$ mice exhibit "acceleration" of selected age-related pathologies, akin to a segemental progeroid syndrome [37–40]. $Sod1^{-/-}$ females have very low fertility, or more accurately, $Sod1^{-/-}$ females exhibit a rapid, age-dependent decline in fertility ([35], see table in [36]), with a loss in female fertility being characteristic of aging in both mice and humans. $Sod1^{-/-}$ mice exhibit acceleration of age-related hearing loss [41–43] macular degeneration [44], and early appearance of cataracts [30], all of which also occur, albeit at a later age, in normal popluations. In a recent study from our laboratory, we reported that $Sod1^{-/-}$ mice exhibit an acceleration of age related muscle atrophy [27], most likely the consequence of the subtle neuromuscular alterations described previously [25, 45, 46].

Indeed, age-related muscle mass loss "sarcopenia" is a hallmark of aging in vertebrate species (discussed extensively in [27, 47]). In that same study, we also observed a decrease in bone mineral density in $Sod1^{-/-}$ animals, a characteristic of osteoporosis. $Sod1^{-/-}$ mice also exhibit thymic involution (a critical piece of data buried in [48]), which is a hallmark age-related pathology in diverse mammals, including humans and mice [49].

Thus, several typical age-related pathologies are accelerated in $Sod1^{-/-}$ mice, and I argue that, if the precedent set by the DNA damage repair knockout mice is used, an even better case could be made that $Sod1^{-/-}$ mice indeed exhibit accelerated aging (assuming one accepts such a thing exists; [39, 40, 50]). However, although several age-related pathologies are clearly accelerated in these mice, the apparent cause of death in 50% of females and 80% of males is hepatocellular carcinoma (which is a classic age-related pathology in the CD-1 but highly atypical in the C57BL6 genetic background). Nevertheless, the remaining mice that show no evidence of hepatocellular carcinoma still have a reduced life span, and it is striking to note that carbonic anhydrase III, an age-related marker that decreases with age in both mice and rats [51], was found to be decreased earlier in the life span of $Sod1^{-/-}$ mice compared with wild-type (WT) mice [28]. So, is accelerated aging or the hepatocellular carcinoma responsible for shorter life span? The Richardson lab is currently tackling this question by rescuing the liver cancer in $Sod1^{-/-}$ mice by ectopic expression of SOD1 under a hepatocyte specific promoter.

The other question that arises is whether the $Sod1^{+/-}$ also exhibit a shortened life span, i.e., whether there is a dose response between CuZnSOD levels and life span. Based on a small N size study by Elchuri et al., the answer would seem to be "no." However, the survival curve of the $Sod1^{+/-}$ seems shorter, and anything but large changes (bigger than 15%) would have been detected by such small numbers (intriguingly, a small increase in hepatocellular carcinoma was also observed in $Sod1^{+/-}$ mice compared with +/+ controls). The Richardson lab has done a preliminary survival curve of $Sod1^{+/-}$ in a mixed genetic background, which did in fact show a shorter life span than our typical C57B6 controls. The Richardson lab is currently exploring this question in more depth by performing a larger scale cohort survival and end of life pathology on the $Sod1^{+/-}$ mice.

3.2 MnSOD Knockout

MnSOD ($Sod2$) is the only scavenger of superoxide in the mitochondrial matrix [19]. Genetic ablation of this enzyme in mice ($Sod2^{-/-}$ mice) results in neonatal lethality between 1 to 24 days after birth, depending on the genetic background [52–54]. The cause of death in the longest-lived genetic background (at ~24 days after birth), is neurodegeneration [52]. These mice have ~4-fold increase in DNA oxidative damage as measured by 8-OH-guanine, 8-OH-adenine, and 5-OH-cytosine [55], and they also have severe enzymatic deficits consistent with massive oxidative stress [55]. The neonatal lethality of $Sod2^{-/-}$ is consistent with the notion that mitochondrial superoxide is limiting to life span. But again, the more interesting question is whether there is a dose–response relationship between MnSOD levels and life span, i.e., whether the life span of the $Sod2^{+/-}$ mice is also decreased. To address this, the Richardson lab measured oxidative damage and life span in $Sod2^{+/-}$ mice, mice with one WT and one deleted allele of $Sod2$ [56–58]. In all tissues assayed, the activity of MnSOD was decreased by around 50%, and no compensatory increase in the activity of Gpx1 or CuZnSOD was observed [57–59]. Respiratory function and aconitase activity were compromised in mitochondria isolated from $Sod2^{+/-}$ mice [56], and both nuclear and mitochondrial DNA oxidative damage (8-oxodeoxyguanosine [8-oxo-dG]) was increased by approximately 30 to 80%) [57–59]. Thus, in $Sod2^{+/-}$ mice, the increase in oxidative damage would predict a shortened life span if oxidative stress is indeed limiting to life span. Against expectations, the natural life span of $Sod2^{+/-}$ mice was statistically indistinguishable from that of WT mice [59]. If anything, the life span of $Sod2^{+/-}$ mice tended to be longer (median 894 vs 918, maximum 1189 vs 1239 days) in $Sod2^{+/-}$ compared with $Sod2^{+/+}$ littermates. In addition, various aging biomarkers (e.g., cataracts, immune response, advanced glyco-oxidation end products) also failed to show any differences between $Sod2^{+/-}$ and $Sod2^{+/+}$ mice. Thus, these results would seem incompatible with the idea that oxidative stress is limiting to lifespan. But is oxidative stress really increased? Different markers of oxidative damage yield different answers. For example, mtDNA deletions [60], F_2-isoprostanes, protein carbonyls (H. Van Remmen and A. Chaudhuri, unpublished data) and methionine sulfoxide levels [61] were not significantly elevated in $Sod2^{+/-}$ mice compared with WT mice (this is particularly critical regarding F_2-isoprostanes, because these are typically the most responsive parameter to experimental oxidative stress [62, 63]). Regarding the in vitro mitochondrial defects (e.g., decreased aconitase activity, respiratory function [56]): it is well known that mitochondria deteriorate (e.g., respiratory control ratio drops, aconitase function decreases) rapidly once the tissue is broken, and this decline continues as mitochondria are stored on ice [64], and it is likely that autooxidation (continuous exposure to 21% vs in vivo 3% O_2 tension)-driven oxidative stress is critical in the functional decline of isolated mitochondria [64, 65]. For this reason, it is conceivable that the differences in mitochondrial function observed between the $Sod2^{+/-}$ and WT mitochondria may in fact arise as a consequence of oxidative stress during isolation, rather than being endogenously present.

To some degree, all biochemical markers of oxidative damage are subject to artefactual oxidation during isolation, and it is therefore difficult to rule out that any differences observed between an antioxidant knockout mouse and its control were in fact endogenously present as opposed to arising to differences in protection from oxidative stress during isolation. What is needed is a marker of oxidative stress that is not influenced by artefactual oxidation during isolation. The answer may lay in the use of oxidative stress-dependent gene expression changes. Prolla's lab performed gene expression studies on $Sod2^{+/-}$ mice; if microarrays of mice treated with paraquat (an artificial superoxide generator) are taken as reference [66], the gene expression pattern of $Sod2^{+/-}$ mice does not suggest elevated oxidative damage [67]. These data would indicate that oxidative stress in vivo in $Sod2^{+/-}$ mice is not elevated.

If the deficiency of MnSOD in *Sod2*[+/−] mice does not result in elevated oxidative stress in vivo, the normal lifespan of the *Sod2*[+/−] does not directly challenge the notion that oxidative stress is limiting to lifespan.

What is clear though, is that when *Sod2*[+/−] mice are placed in pathophysiological situations thought to cause increased mitochondrial superoxide production, pathology is worsened [68–70]. For example, *Sod2*[+/−] mice are more susceptible to ischemia-reperfusion injury [68, 70–72], and when they are crossed to the transgenic SOD1 G93A ALS mouse model, result in accelerated disease course and shortened lifespan [69]. Conversely, when crossed to the SOD1 H46RH48Q ALS mouse model, which does not exhibit obvious mitochondrial dysfunction, *Sod2*[+/−] does not worsen pathology or shorten lifespan (F. L. Muller, Ph.D. thesis). My integrated interpretation of the above-mentioned data is that a 50% reduction MnSOD is sufficient to control the levels of superoxide produced during "normal" metabolism (there is no elevation in oxidative stress under basal conditions, and no shortening of lifespan), but not during pathophysiological situations that result in elevated mitochondrial superoxide production.

3.3   *Sod3*[−/−]*/EC-SOD Knockout*

The third and least abundant SOD enzyme is EC-SOD, encoded by the *Sod3* gene. EC-SOD is located in the extracellular space [73]. *Sod3*[−/−] mice lacking EC-SOD are viable and fertile, and they do not show a decreased lifespan [74, 75]. Although the lack of lifespan shortening would seem to be evidence against the notion and oxidative stress is lifespan-limiting, it is unclear whether these mice actually have increased levels of oxidative stress: young *Sod3*[−/−] mice (3 months old) do not show an increase in urinary F_2-isoprostanes; in contrast, at 3 months of age mice lacking CuZnSOD (*Sod1*[−/−] mice) show increased levels of F_2-isoprostanes [75]. It is important to realize that mice differ in certain critical aspects in terms of aging from humans, i.e., the cardiovascular disease and hypertension, so prominent in human aging, are not age-related causes of death in mice. Hence, *Sod3* might be much more important for human lifespan and age-related pathology.

3.4   *Glutathione Peroxidase-1 Knockout*

Glutathione peroxidase 1 is abundant in almost all mouse tissues, and it is traditionally thought to be the main cellular scavenger of H_2O_2 [76]. Against expectations, a complete knockout of this enzyme fails to generate a dramatic phenotype [77]. The only major pathological finding to date, is that *Gpx1*[−/−] mice develop cataracts at a considerably younger age than WT mice [78, 79]. There was an initial report that *Gpx1*[−/−] mice had reduced body mass [80], but we have been unable to confirm this in our colony, even in *Gpx1*[−/−]*Sod2*[+/−] double knockout mice [81]. Preliminary data from our laboratory indicate that there is no difference in the mortality curves between $Gpx1^{-/-}$, and WT mice to at least 36 months (Van Remmen et al., unpublished). This would seem to conflict with the notion that oxidative stress is limiting to lifespan, but again the question remains whether oxidative damage is actually increased by the absence of Gpx1. Although it is clear that $Gpx1^{-/-}$ mice are sensitive to a host of (exogenous) oxidative stressors (e.g., paraquat, diquat, infection), whether baseline levels of oxidative damage are elevated remains unclear (the same arguments as for the $Sod2^{+/-}$ apply here). Measures of DNA oxidative damage (8-oxo-dG) exhibit a modest increase (between 10 and 35%). F_2-isoprostanes, perhaps currently the most robust marker of in situ oxidative damage [62, 63], do not show an elevation in $Gpx1^{-/-}$ mice compared with WT (H. Van Remmen and A. Richardson, unpublished). Mitochondrial H_2O_2 release measured using Amplex Red is not increased in skeletal muscle mitochondria from $Gpx1^{-/-}$ mice (Muller et al., unpublished). Microarray data indicate that $Gpx1^{-/-}$ mice do not exhibit the signature of oxidative stress (taken as WT mice treated with diquat) and that they fail to show a significant compensatory antioxidant up-regulation (Han et al., in preparation). Thus, the unaltered lifespan in the $Gpx1^{-/-}$ mice does not necessarily conflict with the notion that oxidative stress is limiting to lifespan, because it is not clear whether oxidative damage is actually elevated in the $Gpx1^{-/-}$ mice. The lack of elevated oxidative stress despite the lack of compensatory antioxidant up-regulation does suggest that Gpx1 is not the major scavenger of endogenously produced H_2O_2, even though it is necessary to protect the animal from high levels of H_2O_2 produced during stress events. My interpretation of the available data is the low levels of endogenously produced H_2O_2 are scavenged by peroxiredoxins [82].

3.5 Peroxiredoxin Knockout

The peroxiredoxins, initially known as thiol-specific antioxidants [83], are a family of cysteine-dependent thioredoxin peroxidases [84]. The substrates for these enzymes include (but are not limited to) hydrogen peroxide, organic peroxides, alkly peroxides, and peroxynitrite [84, 85]. Six peroxiredoxins have been identified in the mammalian genome [84]. Based on genetic studies in yeast and mice and recent biochemical reevaluation, it is evident to me that peroxiredoxins rather than Gpx1 are the main scavengers of endogenously produced H_2O_2 [82, 86–89]. Knockout mouse models lacking peroxiredoxin 1, 2, and 6 have been generated, although the lifespan of only $Prdx1$ mice has been reported [90–92]. Peroxiredoxin 1 is a thioredoxin peroxidase found in most tissues, and it is especially abundant in erythrocytes. Mice null for peroxiredoxin 1 ($Prdx1^{-/-}$) have hemolytic anemia, an increase in cancer incidence, and a reduced lifespan [92]. An increase in oxidative damage to both DNA and protein has been demonstrated in $Prdx1^{-/-}$ mice [92, 93]. Interestingly, the $Prdx1^{+/-}$ heterozygous knockout mice show an increased tumor burden but no decrease in lifespan compared with WT mice [92], a result similar to the lifespan and pathology study in $Sod2^{+/-}$ mice [59]. Peroxiredoxin 2 is also found in most tissues, and it is especially abundant in erythrocytes [91]. Mice null for *Prxd2* also suffer from hemolytic anemia [91], but apparently in a milder form than observed in the *Prxd1*−/− mice. *Prxd2*−/− mice also show decreased circulating erythrocyte half-life. In addition, senescence in mouse embryonic fibroblasts has been shown to be accelerated in *Prx2*−/− mice [94]. No data on lifespan are yet available for *Prxd2*−/− mice [91]. The strong phenotype of peroxiredoxin knockout mice stands in contrast to the lack of phenotype of *Gpx1* knockout mice, underscoring the greater importance of peroxiredoxins in scavenging endogenously produced hydrogen peroxide.

3.6 MsrA Knockout

MsrA selectively reduces (using thioredoxin as an electron source) the sulfoxide of methionine back to its thioether. It is found in the cytoplasm and in mitochondria [95]. Overexpression of one isoform of this enzyme (methionine-*S*-sulfoxide reductase, *MsrA*) [96] leads to an increase in lifespan in flies [97] and yeast [98]. In mice, deletion of *MsrA* shortens lifespan [99], although the survival data are based on a small number of animals. The cause of death was not analyzed in this study, nor was general pathology reported. The Richardson lab is currently replicating the lifespan studies (preliminary survival curves, indicating no difference in lifespan between WT and *MsrA*−/− mice were presented at the 2007 meeting of the American Aging Association).

3.7 Knockouts of the Ogg1/Myh 8-oxo-dG Control System

Ogg1 and *Myh* act in concert to prevent the mutagenic effects of the oxidatively damaged nucleotide 8-oxo-dG (for a more detailed description, see Chapter 12). Knocking out *Ogg1* in mice does not result in any gross phenotypical abnormalities [100], despite the fact that 8-oxo-dG levels are elevated severalfold (but see below regarding methodology) in both nuclear [101, 102] and in mitochondrial DNA (~20-fold higher in mtDNA [103]; but see also [104]). However, it has been noted that a residual 8-oxo-dG repair capacity exists in *Ogg1*−/− mice, both in nuclear and mitochondrial DNA [105, 106]. This residual repair activity is due to the cockayne syndrome B protein (Csb) [106], but is not dependent on transcription coupled repair [105]. A double knockout of both *Ogg1*−/− and *Csb*−/− has been bred: the increase in 8-oxo-dG was even more dramatic than *Ogg1*−/−, and Fapy-glycosylase–sensitive sites were increased in all cell types examined, not just in the liver. As with *Ogg1*−/−, the relative difference between *Csb*−/− and *Ogg1*−/− *Csb*−/− became more pronounced with advancing age [105]. Thus, at 9 to 14 months, there was a fivefold increase in Fapy-glycosylase–sensitive sites in hepatocytes compared with *Csb*−/− and *Csb*−/− *Ogg1*−/−. Despite this, *Ogg1*−/−*Csb1*−/− mice are viable to at least 22 months [105]. No formal survival studies have been reported with either *Ogg1*−/− or *Ogg1*−/−*Csb*−/− mice. However, that a 20-fold increase in 8-oxo-dG in mtDNA allows survival to at least 27 months does not bode well for any version of the free radical theory, which argues that the DNA oxidative damage product 8-oxo-dG is central to aging. Although it is yet unclear whether mtDNA (or even nuclear DNA) mutagenesis is actually elevated in $Ogg1^{-/-}$ mice (one would think it would). Nor should it be taken as evidence that oxidative damage is not involved in cancer—the hepatocarcinogenesis present in $Sod1^{-/-}$ would dispute this claim. However, how much 8-oxo-dG is increased really depends on how it is measured [104]. There are profound disagreements across different labs and methodologies even when chemical standards are used [107, 108]. Thus, although some investigators report dramatic (>20-fold) elevations in 8-oxo-dG in mtDNA of $Ogg1^{-/-}$ mice, others find essentially no change [104].

The most likely explanation for the lack of cancer or lifespan reduction is that even if 8-oxo-dG is elevated, its mutagenic effects are minimized by the action of Myh, which prevents 8-oxo-dG dependent mutagenesis by removing mispaired dA opposite of 8-oxo-dG (see Chapter 12 for a more detailed mechanistic description). Consistent with this idea, it was recently reported that $Ogg1/Myh$ double knockout mice do indeed have a shortened lifespan due to a massive increase in carcinogenesis [109, 110].

3.8   Combinations of Antioxidant Knockouts

The question of redundancy and complementation in the antioxidant system is often raised. For example, crossing $Gpx1^{-/-}$ and $Gpx2^{-/-}$ (neither of which have a particularly strong deleterious phenotype on their own) results in a lethal intestinal inflammation ([111], although this does not happen in a C57B6 background). This result seems to be the exception rather than the rule. For example, crossing $Sod3^{-/-}$ and $Sod1^{-/-}$ mice does not yield a shorter lifespan than is observed in $Sod1^{-/-}$ mice alone [75]. Similarly, we find that $Sod1^{-/-} \times Sod2^{+/-}$ mice also do not have a shorter lifespan than $Sod1^{-/-}$ mice alone [26]. The lack of "interaction" of various SOD knockouts is somewhat expected considering that anionic superoxide is highly membrane-impermeable (reviewed in [22, 24]) and reaffirms that the pools of superoxide are highly segregated. In fact, we have bred multiple antioxidant knockout mice involving $Gpx1$, $Gpx4$, $Sod1$, and $Sod2$, and we have found no synthetic aggravation of phenotype. Indeed, we have generated $Sod1^{-/-} Gpx1^{-/-} Sod2^{+/-}$ mice, which in terms of lifespan and general phenotype are essentially identical to $Sod1^{-/-}$ mice alone [26]. However, one interesting combination that has yet to be described is a double $Prdx1 Prdx2$ knockout mouse. Based on the similar distribution, close homology and similar function one would predict that a double knockout of these two enzymes be extra-deleterious, perhaps even lethal.

3.9   Gene Expression Changes in Antioxidant Knockout Mice: An Independent Measure of In Vivo Oxidative Stress and Compensatory Antioxidant Up-Regulation

Oxidative damage dependent gene expression changes provide an alternative way of assessing oxidative stress in antioxidant knockout mice. Foremost, they bypass the problems of artefactual oxidation during isolation that affect to varying extents all biochemical measures of oxidative damage. But first, what genes are altered by oxidative damage? To answer this question, the Richardson lab (Han, Muller, Perez et al., in preparation) performed a large-scale unbiased expression analysis in mice injected with diquat (a superoxide producer). This resulted in thousands of gene expression changes: we compared the pattern of gene expression with previous array studies in cell culture (human cells treated with H_2O_2) and in mice hearts injected with paraquat [66, 112]. Although many genes are altered by oxidative stress in a tissue-specific manner, only a handful of these alterations are conserved across different tissues and different species. To put it succinctly, the conserved gene "expression signature" of oxidative stress is an up-regulation of p53-target genes (p21, *Atf3, Gadd45α, Ddit4, Gdf15,* and *Plk3*) and metallothioneins. Interestingly, this is also one of the expression signatures of aging [67].

Using this information, is the expression signature of oxidative stress status present in antioxidant knockout mice? In the $Gpx4^{+/-}$, $Sod2^{+/-}$ (muscle) and $Gpx1^{-/-}$ (liver) the answer is an unambiguous "no." Alternatively, in the $Sod1^{-/-}$ liver, metallothioneins are indeed induced as are a number of p53 target genes. Thus, by this measure, only $Sod1^{-/-}$ mice (which do show a reduced lifespan) actually are under oxidative stress.

One of the caveats of interpreting phenotypes in antioxidant knockout mice is the possibility that other antioxidant enzymes are up-regulated in compensation. As such, compensatory up-regulations may mask the true extent of the phenotypic consequences, and hence the physiological importance, of the knocked out enzyme. Gene expression arrays provide the technology to address this issue. From our investigations of the gene expression changes in $Sod1^{-/-}$ and $Gpx1^{-/-}$ mice (liver), the simple answer is that there is not one antioxidant gene upregulation in the $Gpx1^{-/-}$ mice. In the $Sod1^{-/-}$ mice (liver), there is indeed an up-regulation of antioxidant genes, most notably a sixfold increase in sulfiredoxin and metallothioneins (Han, Muller, Perez, in preparation); stated more broadly, there is a general up-regulation of thiol antioxidants. Work in the yeast model indicates that absence of $Sod1^{-/-}$ could be partially rescued by giving thiol antioxidants in the media [113, 114]. Thus, compensatory antioxidant up-regulations may minimize the deleteriousness of the lack of CuZnSOD, but such an explanation cannot be invoked for Gpx1. Huang's lab did not observe a robust antioxidant upregulation in $Sod2^{-/-}$ brains [115].

4 Conclusions

The data accumulated so far allow a rough answer to the question of whether oxidative stress is limiting to lifespan in mice. The tremendous deleterious phenotypes of $Sod2^{-/-}$ and $Gpx4^{-/-}$ mice make it clear that controlling the levels of oxygen free radicals is indispensable for life and that very high levels of oxidative damage are indeed lifespan-limiting. The shortened lifespan and acceleration of diverse age-related pathologies in the $Sod1^{-/-}$ mice indicate that even a less extreme elevation in oxidative damage is lifespan-limiting. The question now remains, How high does oxidative stress need to be to be lifespan-limiting? That is, what is the shape of the

Table 8.1 Summary of lifespan and oxidative damage parameters in antioxidant knockout mice

Gene symbol	Gene name	Life span	F$_2$-isoprostanes	8-oxo-dG	p53 targets
Sod1$^{-/-}$	CuZn superoxide dismutase	Reduced by 30%	↑↑	↑↑	↑↑
Sod2$^{-/-}$	Mn superoxide dismutase	Neonatal lethal	N/D	N/D	↑
Sod2$^{+/-}$	Mn superoxide dismutase	Normal	–	↑	–
Sod3$^{-/-}$	Extracellular superoxide dismutase	Normal	–	N/D	N/D
Gpx1$^{-/-}$	Glutathione peroxidase 1	"Normal," not shortened	–	↑	–
Gpx4$^{-/-}$	Phospholipid glutathione peroxidase	Embryonic lethal	N/D	N/D	N/D
MsrA$^{-/-}$	Methionine *R*-sulfoxide reductase	Shortened?	N/D	N/D	N/D
Ogg1$^{-/-}$	8-oxo-dG glycosylase	Probably unchanged	–	↑↑↑	N/D

The table shows lifespan of antioxidant knockout mice, limited to those in which it has been accurately determined, a question mark is placed after MsrA$^{-/-}$ mice, because of the Richardson's labs initial inability to replicate the shortened lifespan (see text for details). Changes in F$_2$-isoprostanes and 8-oxo-dG are taken from the literature and some unpublished data from the Van Remmen and Richardson labs (see text for details). Changes in p53 target gene expression come from the literature as well as unpublished data from the Van Remmen/Richardson labs (see text for details). N/D not determined.

dose–response curve between lifespan and elevated oxidative damage? Data from the *Sod2*$^{+/-}$ mice indicate certain markers of oxidative stress (8-oxo-dG) are elevated even though these animals show no decrease in lifespan. In fact, 8-oxo-dG levels are even more dramatically elevated in the *Ogg1*$^{-/-}$ mice, despite the fact that these mice do not show a reduced lifespan. However, there are some major issues surrounding the measurement of 8-oxo-dG. Other measures of oxidative damage, such as the F$_2$-isoprostanes, are not increased in *Sod2*$^{+/-}$, *Gpx1*$^{-/-}$, or *Ogg1*$^{-/-}$ mice (see summary in Table 8.1) and gene expression changes indicative of elevated oxidative stress are only observed in the *Sod1*$^{-/-}$ not *Gpx1*$^{-/-}$, *Sod2*$^{+/-}$, or *Gpx4*$^{+/-}$ mice.

My own integrated assessment of the evidence is that no instance of unambiguous elevation of oxidative stress, without also observing a decrease in lifespan, has yet been reported in mice. Thus, I conclude that the hypothesis that oxidative stress is limiting to lifespan has (so far; with the caveat that the lifespan and oxidative stress status of few antioxidant knockouts has been explored in detail) not been disproved and that the modest version of the free radical theory may well prove to hold true.

References

1. Beckman KB, Ames BN. The free radical theory of aging matures. Physiol Rev 1998;78(2):547–81.
2. Austad SN. Is aging programed? Aging Cell 2004;3(5):249–51.

3. Martin GM, Austad SN, Johnson TE. Genetic analysis of ageing: role of oxidative damage and environmental stresses. Nat Genet 1996;13(1):25–34.

4. Shmookler Reis RJ. Model systems for aging research: syncretic concepts and diversity of mechanisms. Genome 1989;31(1):406–12.

5. Rudolph KL, Chang S, Lee HW et al. Longevity, stress response, and cancer in aging telomerase-deficient mice. Cell 1999;96(5):701–12.

6. Lee HW, Blasco MA, Gottlieb GJ, Horner JW 2nd, Greider CW, DePinho RA. Essential role of mouse telomerase in highly proliferative organs. Nature 1998;392(6676):569–74.

7. Blasco MA, Lee HW, Hande MP et al. Telomere shortening and tumor formation by mouse cells lacking telomerase RNA. Cell 1997;91(1):25–34.

8. Vermulst M, Bielas JH, Kujoth GC et al. Mitochondrial point mutations do not limit the natural lifespan of mice. Nat Genet 2007;39(4):540–3.

9. Peters LL, Robledo RF, Bult CJ, Churchill GA, Paigen BJ, Svenson KL. The mouse as a model for human biology: a resource guide for complex trait analysis. Nat Rev Genet 2007;8(1):58–69.

10. Beck MA, Esworthy RS, Ho YS, Chu FF. Glutathione peroxidase protects mice from viral-induced myocarditis. FASEB J 1998;12(12):1143–9.

11. Mewissen DJ, Rust JH, Haren J, Cluten MJ. [Epidemiology of C57 black/6M mouse strain: statistical findings in a control population]. C R Seances Soc Biol Fil 1977;171(5):1140–4.

12. Kunstyr I, Leuenberger HG. Gerontological data of C57BL/6J mice. I. Sex differences in survival curves. J Gerontol 1975;30(2):157–62.

13. Storer JB. Longevity and gross pathology at death in 22 inbred mouse strains. J Gerontol 1966;21(3):404–9.

14. Festing MF, Blackmore DK. Lifespan of specified-pathogen-free (MRC category 4) mice and rats. Lab Anim 1971;5(2):179–92.

15. Capecchi MR. Gene targeting in mice: functional analysis of the mammalian genome for the twenty-first century. Nat Rev Genet 2005;6(6):507–12.

16. Jakupoglu C, Przemeck GK, Schneider M et al. Cytoplasmic thioredoxin reductase is essential for embryogenesis but dispensable for cardiac development. Mol Cell Biol 2005;25(5):1980–8.

17. Conrad M, Jakupoglu C, Moreno SG et al. Essential role for mitochondrial thioredoxin reductase in hematopoiesis, heart development, and heart function. Mol Cell Biol 2004;24(21):9414–23.

18. Shi ZZ, Osei-Frimpong J, Kala G et al. Glutathione synthesis is essential for mouse development but not for cell growth in culture. Proc Natl Acad Sci U S A 2000;97(10):5101–6.

19. Fridovich I. Fundamental aspects of reactive oxygen species, or what's the matter with oxygen? Ann N Y Acad Sci 1999;893:13–8.

20. Okado-Matsumoto A, Fridovich I. Subcellular distribution of superoxide dismutases (sod) in rat liver. Cu,Zn-sod in mitochondria. J Biol Chem 2001;276(42):38388–93.

21. Sturtz LA, Diekert K, Jensen LT, Lill R, Culotta VC. A fraction of yeast Cu,Zn-superoxide dismutase and its metallochaperone, ccs, localize to the intermembrane space of mitochondria. A physiological role for sod1 in guarding against mitochondrial oxidative damage. J Biol Chem 2001;276(41):38084–9.

22. Muller F. The nature and mechanism of superoxide production by the electron transport chain: its relevance to aging. J Am Aging Assoc 2000;23:227–53.

23. Han D, Williams E, Cadenas E. Mitochondrial respiratory chain-dependent generation of superoxide anion and its release into the intermembrane space. Biochem J 2001;353(2):411–6.

24. Muller FL, Liu Y, Van Remmen H. Complex III Releases superoxide to both sides of the inner mitochondrial membrane. J Biol Chem 2004;279(47):49064–73.

25. Reaume AG, Elliott JL, Hoffman EK et al. Motor neurons in Cu/Zn superoxide dismutase-deficient mice develop normally but exhibit enhanced cell death after axonal injury. Nat Genet 1996;13(1):43–7.

26. Muller FL, Lustgarten MS, Jang Y, Richardson A, Van Remmen H. Trends in oxidative aging theories. Free Radic Biol Med 2007;43(4):477–503.

27. Muller FL, Song W, Liu Y et al. Absence of CuZn superoxide dismutase leads to elevated oxidative stress and acceleration of age-dependent skeletal muscle atrophy. Free Radic Biol Med 2006;40(11):1993–2004.

28. Elchuri S, Oberley TD, Qi W et al. CuZnSOD deficiency leads to persistent and widespread oxidative damage and hepatocarcinogenesis later in life. Oncogene 2005;24(3):367–80.
29. Kessova IG, Ho YS, Thung S, Cederbaum AI. Alcohol-induced liver injury in mice lacking Cu, Zn-superoxide dismutase. Hepatology 2003;38(5):1136–45.
30. Reddy VN, Kasahara E, Hiraoka M, Lin LR, Ho YS. Effects of variation in superoxide dismutases (SOD) on oxidative stress and apoptosis in lens epithelium. Exp Eye Res 2004;79(6):859–68.
31. Busuttil RA, Garcia AM, Cabrera C et al. Organ-specific increase in mutation accumulation and apoptosis rate in CuZn-superoxide dismutase-deficient mice. Cancer Res 2005;65(24):11271–5.
32. Keithley EM, Canto C, Zheng QY, Wang X, Fischel-Ghodsian N, Johnson KR. Cu/Zn superoxide dismutase and age-related hearing loss. Hear Res 2005;209(1–2):76–85.
33. Kostrominova TY, Pasyk KA, Van Remmen H, Richardson AG, Faulkner JA. Adaptive changes in structure of skeletal muscles from adult Sod1 homozygous knockout mice. Cell Tissue Res 2006.
34. Hadjur S, Ung K, Wadsworth L et al. Defective hematopoiesis and hepatic steatosis in mice with combined deficiencies of the genes encoding Fancc and Cu/Zn superoxide dismutase. Blood 2001;98(4):1003–11.
35. Matzuk MM, Dionne L, Guo Q, Kumar TR, Lebovitz RM. Ovarian function in superoxide dismutase 1 and 2 knockout mice. Endocrinology 1998;139(9):4008–11.
36. Ho YS, Gargano M, Cao J, Bronson RT, Heimler I, Hutz RJ. Reduced fertility in female mice lacking copper-zinc superoxide dismutase. J Biol Chem 1998;273(13):7765–9.
37. Martin GM. Genetics and aging; the Werner syndrome as a segmental progeroid syndrome. Adv Exp Med Biol 1985;190:161–70.
38. Martin GM. Syndromes of accelerated aging. Natl Cancer Inst Monogr 1982;60:241–7.
39. Hasty P, Vijg J. Accelerating aging by mouse reverse genetics: a rational approach to understanding longevity. Aging Cell 2004;3(2):55–65.
40. Hasty P, Campisi J, Hoeijmakers J, van Steeg H, Vijg J. Aging and genome maintenance: lessons from the mouse? Science 2003;299(5611):1355–9.
41. McFadden SL, Ding D, Reaume AG, Flood DG, Salvi RJ. Age-related cochlear hair cell loss is enhanced in mice lacking copper/zinc superoxide dismutase. Neurobiol Aging 1999;20(1):1–8.
42. McFadden SL, Ding D, Burkard RF et al. Cu/Zn SOD deficiency potentiates hearing loss and cochlear pathology in aged 129,CD-1 mice. J Comp Neurol 1999;413(1):101–12.
43. Ohlemiller KK, McFadden SL, Ding DL et al. Targeted deletion of the cytosolic Cu/Zn-superoxide dismutase gene (Sod1) increases susceptibility to noise-induced hearing loss. Audiol Neurootol 1999;4(5):237–46.
44. Imamura Y, Noda S, Hashizume K et al. Drusen, choroidal neovascularization, and retinal pigment epithelium dysfunction in SOD1-deficient mice: a model of age-related macular degeneration. Proc Natl Acad Sci U S A 2006;103(30):11282–7.
45. Flood DG, Reaume AG, Gruner JA et al. Hindlimb motor neurons require Cu/Zn superoxide dismutase for maintenance of neuromuscular junctions. Am J Pathol 1999;155(2):663–72.
46. Shefner JM, Reaume AG, Flood DG et al. Mice lacking cytosolic copper/zinc superoxide dismutase display a distinctive motor axonopathy. Neurology 1999;53(6):1239–46.
47. Muller FL, Song W, Jang Y et al. Denervation-Induced skeletal muscle atrophy is associated with increased mitochondrial ROS production. Am J Physiol 2007; 293: R1159–1168.
48. Iuchi Y, Okada F, Onuma K et al. Elevated oxidative stress in erythrocytes due to a SOD1 deficiency causes anaemia and triggers autoantibody production. Biochem J 2007;402(2):219–27.
49. Mitchell WA, Meng I, Nicholson SA, Aspinall R. Thymic output, ageing and zinc. Biogerontology 2006;7(5–6):461–70.
50. Miller RA. 'Accelerated aging': a primrose path to insight? Aging Cell 2004;3(2):47–51.
51. Cabiscol E, Levine RL. Carbonic anhydrase III. Oxidative modification in vivo and loss of phosphatase activity during aging. J Biol Chem 1995;270(24):14742–7.
52. Huang TT, Carlson EJ, Kozy HM et al. Genetic modification of prenatal lethality and dilated cardiomyopathy in Mn superoxide dismutase mutant mice. Free Radic Biol Med 2001;31(9):1101–10.

53. Li Y, Huang TT, Carlson EJ et al. Dilated cardiomyopathy and neonatal lethality in mutant mice lacking manganese superoxide dismutase. Nat Genet 1995;11(4):376–81.
54. Lebovitz RM, Zhang H, Vogel H et al. Neurodegeneration, myocardial injury, and perinatal death in mitochondrial superoxide dismutase-deficient mice. Proc Natl Acad Sci USA 1996;93(18):9782–7.
55. Melov S, Coskun P, Patel M et al. Mitochondrial disease in superoxide dismutase 2 mutant mice. Proc Natl Acad Sci U S A 1999;96(3):846–51.
56. Williams MD, Van Remmen H, Conrad CC, Huang TT, Epstein CJ, Richardson A. Increased oxidative damage is correlated to altered mitochondrial function in heterozygous manganese superoxide dismutase knockout mice. J Biol Chem 1998;273(43):28510–5.
57. Van Remmen H, Salvador C, Yang H, Huang TT, Epstein CJ, Richardson A. Characterization of the antioxidant status of the heterozygous manganese superoxide dismutase knockout mouse. Arch Biochem Biophys 1999;363(1):91–7.
58. Van Remmen H, Williams MD, Guo Z et al. Knockout mice heterozygous for Sod2 show alterations in cardiac mitochondrial function and apoptosis. Am J Physiol 2001;281(3): H1422–32.
59. Van Remmen H, Ikeno Y, Hamilton M et al. Life-long reduction in MnSOD activity results in increased DNA damage and higher incidence of cancer but does not accelerate aging. Physiol Genomics 2003;16(1):29–37.
60. Lynn S, Van Remmen H, Epstein CJ, Huang TT. Investigation of mitochondrial DNA deletions in post-mitotic tissues of the heterozygous superoxide dismutase 2 knockout mouse: effect of ageing and genotype on the tissue-specific accumulation. Free Radic Biol Med 2001;31:S58 S.
61. Stadtman ER, Van Remmen H, Richardson A, Wehr NB, Levine RL. Methionine oxidation and aging. Biochim Biophys Acta 2005;1703(2):135–40.
62. Kadiiska MB, Gladen BC, Baird DD et al. Biomarkers of oxidative stress study III. Effects of the nonsteroidal anti-inflammatory agents indomethacin and meclofenamic acid on measurements of oxidative products of lipids in CCl4 poisoning. Free Radic Biol Med 2005;38(6):711–8.
63. Kadiiska MB, Gladen BC, Baird DD et al. Biomarkers of oxidative stress study II: are oxidation products of lipids, proteins, and DNA markers of CCl4 poisoning? Free Radic Biol Med 2005;38(6):698–710.
64. Chan SH, Higgins E Jr. Uncoupling activity of endogenous free fatty acids in rat liver mitochondria. Can J Biochem 1978;56(2):111–6.
65. Brewer GJ, Jones TT, Wallimann T, Schlattner U. Higher respiratory rates and improved creatine stimulation in brain mitochondria isolated with anti-oxidants. Mitochondrion 2004;4(1):49–57.
66. Edwards MG, Sarkar D, Klopp R, Morrow JD, Weindruch R, Prolla TA. Age-related impairment of the transcriptional responses to oxidative stress in the mouse heart. Physiol Genomics 2003;13(2):119–27.
67. Edwards MG, Anderson RM, Yuan M, Kendziorski CM, Weindruch R, Prolla TA. Gene expression profiling of aging reveals activation of a p53-mediated transcriptional program. BMC Genomics 2007;8:80.
68. Asimakis GK, Lick S, Patterson C. Postischemic recovery of contractile function is impaired in SOD2 but not SOD1 mouse hearts. Circulation 2002;105(8):981–6.
69. Andreassen OA, Ferrante RJ, Klivenyi P et al. Partial deficiency of manganese superoxide dismutase exacerbates a transgenic mouse model of amyotrophic lateral sclerosis. Ann Neurol 2000;47(4):447–55.
70. Murakami K, Kondo T, Kawase M et al. Mitochondrial susceptibility to oxidative stress exacerbates cerebral infarction that follows permanent focal cerebral ischemia in mutant mice with manganese superoxide dismutase deficiency. J Neurosci 1998;18(1):205–13.
71. Lewen A, Matz P, Chan PH. Free radical pathways in CNS injury. J Neurotrauma 2000;17(10):871–90.

72. Kim GW, Kondo T, Noshita N, Chan PH. Manganese superoxide dismutase deficiency exacerbates cerebral infarction after focal cerebral ischemia/reperfusion in mice: implications for the production and role of superoxide radicals. Stroke 2002;33(3):809–15.
73. Marklund SL. Human copper-containing superoxide dismutase of high molecular weight. Proc Natl Acad Sci U S A 1982;79(24):7634–8.
74. Carlsson LM, Jonsson J, Edlund T, Marklund SL. Mice lacking extracellular superoxide dismutase are more sensitive to hyperoxia. Proc Natl Acad Sci U S A 1995;92(14):6264–8.
75. Sentman ML, Granstrom M, Jakobson H, Reaume A, Basu S, Marklund SL. Phenotypes of mice lacking extracellular superoxide dismutase and copper- and zinc-containing superoxide dismutase. J Biol Chem 2006;281(11):6904–9.
76. Halliwell B, Gutteridge J. Free radicals in biology and medicine. Third ed. New York: Oxford University Press; 1999.
77. Ho YS, Magnenat JL, Bronson RT et al. Mice deficient in cellular glutathione peroxidase develop normally and show no increased sensitivity to hyperoxia. J Biol Chem 1997;272(26):16644–51.
78. Wolf N, Penn P, Pendergrass W et al. Age-related cataract progression in five mouse models for anti-oxidant protection or hormonal influence. Exp Eye Res 2005;81(3):276–85.
79. Reddy VN, Giblin FJ, Lin LR et al. Glutathione peroxidase-1 deficiency leads to increased nuclear light scattering, membrane damage, and cataract formation in gene-knockout mice. Invest Ophthalmol Vis Sci 2001;42(13):3247–55.
80. Esposito LA, Kokoszka JE, Waymire KG, Cottrell B, MacGregor GR, Wallace DC. Mitochondrial oxidative stress in mice lacking the glutathione peroxidase-1 gene. Free Radic Biol Med 2000;28(5):754–66.
81. Van Remmen H, Qi W, Sabia M et al. Multiple deficiencies in antioxidant enzymes in mice result in a compound increase in sensitivity to oxidative stress. Free Radic Biol Med 2004;36(12):1625–34.
82. Low FM, Hampton MB, Peskin AV, Winterbourn CC. Peroxiredoxin 2 functions as a non-catalytic scavenger of low level hydrogen peroxide in the erythrocyte. Blood 2006.
83. Chae HZ, Kim IH, Kim K, Rhee SG. Cloning, sequencing, and mutation of thiol-specific antioxidant gene of *Saccharomyces cerevisiae*. J Biol Chem 1993;268(22):16815–21.
84. Wood ZA, Schroder E, Robin Harris J, Poole LB. Structure, mechanism and regulation of peroxiredoxins. Trends Biochem Sci 2003;28(1):32–40.
85. Dubuisson M, Vander Stricht D, Clippe A et al. Human peroxiredoxin 5 is a peroxynitrite reductase. FEBS Lett 2004;571(1–3):161–5.
86. Smith S, Hwang JY, Banerjee S, Majeed A, Gupta A, Myung K. Mutator genes for suppression of gross chromosomal rearrangements identified by a genome-wide screening in *Saccharomyces cerevisiae*. Proc Natl Acad Sci U S A 2004;101(24):9039–44.
87. Huang ME, Kolodner RD. A biological network in *Saccharomyces cerevisiae* prevents the deleterious effects of endogenous oxidative DNA damage. Mol Cell 2005;17(5):709–20.
88. Ragu S, Faye G, Iraqui I, Masurel-Heneman A, Kolodner RD, Huang ME. Oxygen metabolism and reactive oxygen species cause chromosomal rearrangements and cell death. Proc Natl Acad Sci U S A 2007;104(23):9747–52.
89. Inoue Y, Matsuda T, Sugiyama K, Izawa S, Kimura A. Genetic analysis of glutathione peroxidase in oxidative stress response of *Saccharomyces cerevisiae*. J Biol Chem 1999;274(38):27002–9.
90. Wang X, Phelan SA, Forsman-Semb K et al. Mice with targeted mutation of peroxiredoxin 6 develop normally but are susceptible to oxidative stress. J Biol Chem 2003;278(27):25179–90.
91. Lee TH, Kim SU, Yu SL et al. Peroxiredoxin II is essential for sustaining lifespan of erythrocytes in mice. Blood 2003;101(12):5033–8.
92. Neumann CA, Krause DS, Carman CV et al. Essential role for the peroxiredoxin Prdx1 in erythrocyte antioxidant defence and tumour suppression. Nature 2003;424(6948):561–5.
93. Egler RA, Fernandes E, Rothermund K et al. Regulation of reactive oxygen species, DNA damage, and c-Myc function by peroxiredoxin 1. Oncogene 2005;24(54):8038–50.

94. Han YH, Kim HS, Kim JM, Kim SK, Yu DY, Moon EY. Inhibitory role of peroxiredoxin II (Prx II) on cellular senescence. FEBS Lett 2005;579(21):4897–902.

95. Hansel A, Kuschel L, Hehl S et al. Mitochondrial targeting of the human peptide methionine sulfoxide reductase (MSRA), an enzyme involved in the repair of oxidized proteins. FASEB J 2002;16(8):911–3.

96. Kim HY, Gladyshev VN. Methionine sulfoxide reduction in mammals: characterization of methionine-R-sulfoxide reductases. Mol Biol Cell 2004;15(3):1055–64.

97. Ruan H, Tang XD, Chen ML et al. High-quality life extension by the enzyme peptide methionine sulfoxide reductase. Proc Natl Acad Sci U S A 2002;99(5):2748–53.

98. Koc A, Gasch AP, Rutherford JC, Kim HY, Gladyshev VN. Methionine sulfoxide reductase regulation of yeast lifespan reveals reactive oxygen species-dependent and -independent components of aging. Proc Natl Acad Sci U S A 2004;101(21):7999–8004.

99. Moskovitz J, Bar-Noy S, Williams WM, Requena J, Berlett BS, Stadtman ER. Methionine sulfoxide reductase (MsrA) is a regulator of antioxidant defense and lifespan in mammals. Proc Natl Acad Sci U S A 2001;98(23):12920–5.

100. Klungland A, Rosewell I, Hollenbach S et al. Accumulation of premutagenic DNA lesions in mice defective in removal of oxidative base damage. Proc Natl Acad Sci USA 1999;96(23):13300–5.

101. Osterod M, Hollenbach S, Hengstler JG, Barnes DE, Lindahl T, Epe B. Age-related and tissue-specific accumulation of oxidative DNA base damage in 7,8-dihydro-8-oxoguanine-DNA glycosylase (Ogg1) deficient mice. Carcinogenesis 2001;22(9):1459–63.

102. Minowa O, Arai T, Hirano M et al. Mmh/Ogg1 gene inactivation results in accumulation of 8-hydroxyguanine in mice. Proc Natl Acad Sci U S A 2000;97(8):4156–61.

103. de Souza-Pinto NC, Eide L, Hogue BA et al. Repair of 8-oxodeoxyguanosine lesions in mitochondrial DNA depends on the oxoguanine DNA glycosylase (OGG1) gene and 8-oxo-guanine accumulates in the mitochondrial DNA of OGG1-defective mice. Cancer Res 2001;61(14):5378–81.

104. Trapp C, McCullough AK, Epe B. The basal levels of 8-oxoG and other oxidative modifications in intact mitochondrial DNA are low even in repair-deficient (Ogg1$^{-/-}$/Csb$^{-/-}$) mice. Mutat Res 2007.

105. Osterod M, Larsen E, Le Page F et al. A global DNA repair mechanism involving the Cockayne syndrome B (CSB) gene product can prevent the in vivo accumulation of endogenous oxidative DNA base damage. Oncogene 2002;21(54):8232–9.

106. Stevnsner T, Nyaga S, de Souza-Pinto NC et al. Mitochondrial repair of 8-oxoguanine is deficient in Cockayne syndrome group B. Oncogene 2002;21(57):8675–82.

107. Riis B. Comparison of results from different laboratories in measuring 8-oxo-2~-deoxyguanosine in synthetic oligonucleotides. Free Radic Res 2002;36(6):649–59.

108. Collins AR, Cadet J, Moller L, Poulsen HE, Vina J. Are we sure we know how to measure 8-oxo-7,8-dihydroguanine in DNA from human cells? Arch Biochem Biophys 2004;423(1):57–65.

109. Russo MT, De Luca G, Degan P et al. Accumulation of the oxidative base lesion 8-hydroxy-guanine in DNA of tumor-prone mice defective in both the Myh and Ogg1 DNA glycosylases. Cancer Res 2004;64(13):4411–4.

110. Xie Y, Yang H, Cunanan C et al. Deficiencies in mouse Myh and Ogg1 result in tumor predisposition and G to T mutations in codon 12 of the K-ras oncogene in lung tumors. Cancer Res 2004;64(9):3096–102.

111. Esworthy RS, Aranda R, Martin MG, Doroshow JH, Binder SW, Chu FF. Mice with combined disruption of Gpx1 and Gpx2 genes have colitis. Am J Physiol Gastrointest Liver Physiol 2001;281(3):G848–55.

112. Desaint S, Luriau S, Aude JC, Rousselet G, Toledano MB. Mammalian antioxidant defenses are not inducible by H_2O_2. J Biol Chem 2004;279(30):31157–63.

113. Zyracka E, Zadrag R, Koziol S, Krzepilko A, Bartosz G, Bilinski T. Yeast as a biosensor for antioxidants: simple growth tests employing a *Saccharomyces cerevisiae* mutant defective in superoxide dismutase. Acta Biochim Pol 2005;52(3):679–84.

114. Tamai KT, Gralla EB, Ellerby LM, Valentine JS, Thiele DJ. Yeast and mammalian metal-lothioneins functionally substitute for yeast copper-zinc superoxide dismutase. Proc Natl Acad Sci U S A 1993;90(17):8013–7.
115. Lynn S, Huang EJ, Elchuri S et al. Selective neuronal vulnerability and inadequate stress response in superoxide dismutase mutant mice. Free Radic Biol Med 2005;38(6):817–28.

Part II-B
The Comparative Approach

9
Mitochondrial Free Radical Production and Caloric Restriction: Implications in Vertebrate Longevity and Aging

Mónica López-Torres and Gustavo Barja

Summary In this chapter, studies focusing on the relationship between oxidative stress and aging in different vertebrate species and in calorie-restricted animals are reviewed. Endogenous antioxidants inversely correlate with species maximum longevity, and experiments modifying their levels can increase survival and mean life span but not maximum life span. Evidence shows that long-lived vertebrates consistently have low mitochondrial free radical generation rates and also a low fatty acid unsaturation of cellular membranes, two crucial factors determining their aging rate. Oxidative damage to mitochondrial DNA is also lower in long-lived vertebrates than in short-lived vertebrates. Conversely, caloric restriction, the best described experimental manipulation that consistently increases mean and maximum life span, also decreases mitochondrial reactive oxygen species (ROS) generation and oxidative damage to mitochondrial DNA. Recent data suggest that the decrease in mitochondrial ROS generation would be due to protein restriction rather than calories, pointing out a key role for dietary methionine. Longevity would be achieved in part due to a low endogenous oxidative damage generation rate, but also due to a macromolecular composition highly resistant to oxidative modification, as it is the case for lipids and proteins.

Keywords Longevity, aging, maximum life span potential (MLSP), free radicals, ROS production, DNA damage, mitochondria, caloric restriction, protein restriction, methionine restriction, fatty acid unsaturation.

1 Introduction

Aging is an endogenous and progressive process, probably due to more than one factor. So, even if animals are exposed to optimum conditions throughout their lives, they will still age at a rate characteristic for their species. Thus, their maximum life span potentials (MLSPs) are mainly determined by their genes, not by the environment. Conversely, mean life span is mainly determined by the environment and to a lesser extent by the genotype. Nowadays, oxygen-free radical

From: *Aging Medicine: Oxidative Stress in Aging: From Model Systems to Human Diseases*   149
Edited by: S. Miwa, K.B. Beckman, and F.L. Muller © Humana Press, Totowa, NJ

generation is widely accepted among the main causes of aging [1–4]. According to the theory, reactive oxygen species (ROS), endogenously produced during tissue respiration, continuously and progressively lead to tissue damage and finally to aging. In healthy tissues, mitochondria are the main source of ROS, and the only cellular organelles containing DNA. ROS can damage lipids, proteins, and most importantly, DNA. Evidence supports that the mitochondrial ROS generation rate can be one of the main factors determining the aging rate and thus the MLSP of each species. Many reports have shown that mitochondrial ROS (mtROS) generation is lower in long-lived animal species than in short-lived species (3–6). Interestingly, caloric restriction, the best known experimental manipulation that decreases the aging rate and increases MLSP in many species, consistently decreases mtROS generation rate [7, 8]. In both types of approaches, comparative models and dietary restriction, lower rates of mtROS generation relate to lower levels of oxidative damage to mitochondrial DNA (mtDNA) and proteins [9, 10] and to slow aging rate. Although the effects of caloric restriction are well established, the mechanisms underlying them are not fully understood. Recent studies on the role of specific dietary components are beginning to clarify this point [11]. Protein restriction, and more precisely methionine restriction, seems to be the factor determining the effect of dietary restriction, independently of calories, on mtROS generation and oxidative stress.

Alternatively, comparative studies also indicate that the aging rate and thus the MLSP are not only related to mtROS generation rate but also to the macromolecular composition. Long-lived species show low endogenous damage generation rates and at the same time a macromolecular composition highly resistant to oxidative modification. Long-lived mammals and birds have cellular membranes with low degrees of unsaturation, mainly due to a minimal presence of highly unsaturated fatty acids. These membranes are less sensitive to oxidative stress and show lower levels of lipid peroxidation and lipoxidation-derived protein modification [12–14]. In the same line of evidence, it has been recently reported that protein methionine content is negatively correlated with MLSP in mammals [15]; so, long-lived species would have proteins less sensitive to oxidative modification due to their comparative lower levels of methionine residues.

2 Mitochondrial ROS Generation Rate: Comparative Studies

As mentioned, mitochondria are considered the major source of ROS in healthy tissues because the main generator, the electron transport chain, is located at the inner mitochondrial membrane. Because mitochondria are also the only cellular organelles with their own DNA, they would be an important target for oxidative damage. Tissue levels of oxidative stress can be controlled either by the rate of ROS generation or scavenging. After intense analyses years ago, it was clarified that antioxidants, although probably important for protection against different age-related diseases, do not control the rate of aging. Different types of studies support this idea. The initial hypotheses that aging could be the result of a decrease in antioxidants at old age was soon discarded, because it was observed that changes in tissue endogenous antioxidant levels as a function of age did not follow a consistent pattern and that they did not necessarily decrease in old animals [3, 16, 17]. The slow rate of aging of long-lived animals could theoretically be due to a constitutively higher antioxidant defense system. But when comparative studies were performed, the opposite situation was found. Long-lived vertebrates did not show higher but instead lower tissue levels of endogenous antioxidant enzymes and low-molecular-weight antioxidants [6, 18, 19] indicating, that their in vivo ROS generation rate must also be low. Experimentally increasing tissue antioxidant levels through either dietary supplementation, pharmacological induction, or transgenic techniques moderately increases mean life span sometimes, but it does not change MLSP [3, 20–24]. Only marginal or very small increases in MLSP have been rarely reported [25]. And finally, studies in which genes encoding for certain antioxidants are knocked out show that animals can develop different pathologies but that their aging rates do not seem to be affected [26, 27]. In summary, the data consistently indicate that antioxidants do not control aging and maximum longevity, although they can protect against many pathologies and different causes of early death due to their capacity to decrease oxidative stress in different situations leading to an increase in survival (mean life span) but not in MLSP.

Although endogenous tissue antioxidants do not determine the aging rate, their negative correlation with maximum longevity strongly indicates that the endogenous rate of free radical generation in tissues in vivo under normal conditions must be lower in long-lived than in short-lived species [6, 18]. If having low endogenous antioxidant levels, long-lived animals had high ROS production rates, they would not be able to cope with oxidative damage, and oxidative stress balance and homeostasis would be lost. The strategy of decreasing mtROS generation instead of increasing either antioxidant capacity or repair systems seems to be a better mechanism to increase longevity during evolution. To generate large amounts of ROS and then try to neutralize them before they reach important targets like DNA, or even worse, try to repair DNA after heavily damaging it, would be very inefficient. This is even clearer taking into account (1) the high energetic cost of continuously maintaining high antioxidant and repair molecules levels in tissues of long-lived animals, (2) the capacity of short- and long-lived animals to temporarily induce these protective molecules when needed, and (3) the proximity or even contact between main ROS sources and macromolecular targets of paramount importance for aging, such as mtDNA. Instead, lowering ROS generation rate near its main aging-related target decreases its damage much more efficiently and at a much lower cost. Indeed, almost every investigation performed to date on the subject has shown that mtROS generation rate is lower in tissues of long-lived than short-lived animals [3, 5, 6, 18, 28–32]. This is the case for all kinds of long-lived homeothermic vertebrates independently of their mass-adjusted rates of O_2 consumption (low in animals of large body size and high in small animals) and explains why endogenous tissue antioxidants negatively correlates with MLSP across species (long-lived animals have constitutively low levels of endogenous antioxidants because they generate ROS at a low rate).

The studies performed in birds (exceptionally long-lived homeotherms) are particularly illustrative, because these animals, like the other two groups, bats and primates, live much longer than mammals of similar body size or metabolic rate, in contradiction with the rate of living theory of aging. In spite of their high rates of O_2 consumption, the three bird species studied to date, from different families (parakeets, canaries, and pigeons, with MLSPs of 21, 24, and 35 years, respectively) have lower rates of mtROS generation than mice and rats (MLSPs of 3.5–4 years) [5, 29–33]. In most cases, this is possible because the percentage of total electron flow in the respiratory chain directed to ROS production (percentage of free radical leak [FRL]) is lower in birds [3, 5, 30, 31]. Their respiratory chain transports electrons more efficiently, avoiding univalent electron leaks to oxygen upstream of cytochrome oxidase. Lower mtROS generation rates in human than in rat brain also have been reported recently [34]. Primates (especially humans) and bats also live longer than expected for their body size and metabolic rate. So, both small birds and large mammals have low ROS production rates and slow aging rates, whereas metabolic rate is low in large mammals but high in small birds. Thus, mtROS generation rate correlates better with MLSP than metabolic rate.

The presence of different FRL values in the mitochondrial respiratory chain from different species shows that the amount of electrons directed to ROS generation is not fixed and that it can be regulated in each species as it would be expected for a parameter controlling the endogenous aging rate. A role for mtROS generation in the control of aging also is supported by current studies in invertebrates [35], fungi [36], and cultivated cells [37]. Conversely, in most of the performed comparative studies, the difference in mtROS production between species was usually smaller than the difference in MLSP [5, 29–31]. This would be consistent with the idea that aging is caused by more than one single major factor.

The site in the respiratory chain where ROS generation is lower in long-lived species has also been studied. ROS production at the respiratory chain was classically attributed to complex III ubisemiquinone [38]. However, as it was initially described in submitochondrial particles [39, 40], complex I also contains an important ROS generator [29, 30, 34, 41–45] today accepted as part of mainstream basic knowledge [46]. Moreover, available studies indicate that the respiratory complex responsible for the lower mtROS generation in long-lived species is complex I [29–31, 33]. Concerning the identity of the ROS generator inside complex I, flavin mononucleotide [34, 47], ubisemiquinones [43, 44], or iron-sulfur clusters [41, 42, 48] have been proposed. Both flavin and FeS clusters are situated in the hydrophilic complex I domain facing the mitochondrial matrix compartment where mtDNA is located. On the contrary, ROS generated at complex III seem to be mainly directed to the cytosolic side [33], although recent studies suggest that part of the generation also could take place toward the matrix side [49].

3 Mitochondrial DNA Oxidative Damage: Comparative Studies

Tissue antioxidants help to control oxidative stress in vivo. They are a first line of defense contributing to the maintenance of cellular oxidative stress homeostasis. But ROS neutralization and scavenging by antioxidants is not 100% efficient. There is always a certain level of oxidative damage to macromolecules even in healthy animals. Such oxidative damage affects cellular lipids, proteins, and DNA. Although ROS can damage different macromolecules, damage to DNA must be of crucial importance for aging because it can result in permanently altered or lost genetic information. ROS can attack DNA directly at the sugar-phosphate backbone or at the bases, producing many different oxidatively modified purines and pyrimidines, like the most commonly used marker of DNA oxidative damage, 8-oxo-7,8-dihydro-2′-deoxyguanosine (8-oxodG), and single- and double-strand breaks and DNA mutations. Although increases in oxidative damage to nuclear and mitochondrial DNA have sometimes [50–53] but not always [53, 54] been described during aging, continuous oxidative damage to DNA throughout life could have a more important long-term consequence due to its capacity to generate DNA mutations.

A general compensation between ROS production and scavenging must take place in both short- and long-lived animals, allowing tissue homeostasis and thus survival in both cases. The cellular ROS turnover must be high in the first and low in the second group of animals. But if short-lived species have high ROS production rates and high antioxidant levels, whereas long-lived species have low ROS production rates and low antioxidant levels, the final balance could be similar in both situations. This is not the case because in short-lived animals, the high rates of ROS production would lead to a higher local concentration of ROS near the sites of ROS generation. At those sites, the source of ROS and their targets are in such proximity that antioxidants cannot effectively intercept ROS before they damage the targets. The local concentration of ROS near targets relevant to aging (such as mtDNA) situated close to the places of ROS generation (the inner mitochondrial membrane) or even in contact with them is expected to be lower in long-lived species due to their lower rates of ROS production. This would result in a lower steady-state level of oxidative damage and accumulation of somatic mutations in their mtDNA and a slow rate of aging [6]. Supporting this idea, heart and brain 8-oxodG levels in mtDNA negatively correlate with MLSP in mammals [9], and this is in agreement with the lower urinary excretion of 8-oxo-7,8-dihydroguanine in long-lived animals than in short-lived animals [55]. Long-lived birds also generally show lower oxidative damage to mtDNA than short-lived rodents of similar body size [56]. The accumulation rate of mtDNA mutations with age is slower in humans than in mice, too [57]. Furthermore, the 8-oxo-dG concentration is around 10-fold higher in mtDNA than in nuclear DNA (nDNA) in the heart and brain of all the 11 species of mammals and birds studied [9, 56], a similar difference to that observed for spontaneous mutations when comparing both DNAs [58]. On the other hand, 8-oxodG in nDNA does not correlate with MLSP [9]. Although mitochondria seem to lack some forms of DNA repair such as that for pyrimidine dimers, it is known that its capacity to repair 8-oxodG is similar or even greater than that in the nucleus [59]. Data suggest that rates of both ROS attack and DNA repair are higher in mtDNA of short-lived than in long-lived species, and they show that they are higher in mtDNA than in nDNA in all species [60] in agreement with the location of mtDNA very close to the ROS production site at the mitochondria. The higher rate of mtROS production of short-lived animals may be an important cause of their faster accumulation rate of mtDNA mutations during aging [57]. Mutations in mtDNA, both deletions and point mutations, occur with aging in postmitotic tissues and can reach high levels in old individuals [61–63]. The mutated mtDNA clonally expands towards predominance during aging, having detrimental consequences for cellular function.

4 Caloric Restriction, Mitochondrial ROS Production, DNA Oxidative Damage, and Longevity

Comparative studies suggest a causal relationship between mtROS production and aging, but experimental studies are needed to confirm it. Caloric restriction (CR) is the best-known experimental manipulation that decreases the aging rate and increases MLSP in different species, and it has beneficial effects for health in laboratory rodents, other animals and possibly primates including humans [64]. Physiological changes induced by CR have been extensively described in rodents and include, reduced body size, lower body temperature, and decreases in plasma growth hormone and insulin-like growth factor-1 (IGF-1) levels. Restricted animals also show lower glucose and insulin levels in plasma along with an increased sensitivity to insulin compared to ad libitum-fed animals. Delays in sexual maturation and decreases in fertility also have been reported. The modification in the insulin/IGF-1 signalling pathway has particular interest, because it is a highly conserved system that has been proposed to regulate longevity in a variety of animals from nematodes to mammals [65].

A growing body of evidence supports the hypothesis that CR works, at least in part, by decreasing oxidative stress [2, 10]. Because reports had shown that changes in antioxidant levels in caloric-restricted animals were inconsistent and could not explain the increases in MLSP [66], many studies have focused on the effect of CR on mtROS production [7, 8, 10, 66–68]. These studies, usually applying 40% CR (40% of the ad libitum food intake), show that long-term CR significantly decreases the rate of mtROS generation in different rat tissues ([10, 66–68]; Table 9.1), whereas after short-term CR decreases are only sometimes detected depending upon the tissue studied [7, 8, 10]. CR for 6–7 weeks was enough to decrease mtROS production and 8-oxodG in mtDNA and nDNA in rat liver [7]. Furthermore, the decrease in mtROS production in caloric-restricted rats, specifically takes place at complex I in every tissue studied up to date (heart, liver, and brain), together with no changes in mitochondria O_2 consumption and a decrease in %FRL, indicating

Table 9.1 Summary of the effects of CR, PR, and MetR on oxidative stress and MLSP

	CR (40%)	PR (40%)	MetR (40 and 80%)
mtROS	↓(cx I)	↓(cx I)	↓(cx I and III)[a]
mtVO$_2$	=	=	=
%FRL	↓	↓	↓
Ox. mtDNA	↓	↓	↓
Ox. proteins	↓	↓	↓
MLSP	↑↑	↑	↑

mtVO$_2$, mitochondrial oxygen consumption; Ox. mtDNA, mtDNA oxidative damage (8-oxodG); Ox. proteins, specific markers of protein oxidation, lipoxidation, and glycoxidation; cx I, complex I; cx III, complex III.
[a] Effect of MetR was more important on complex I than on complex III based on the magnitude of the decrease in mtROS and on the number of tissues affected.

that mitochondria from caloric-restricted animals are more efficient avoiding ROS production per unit electron flow [7, 10, 67, 68]. These changes also have been found when comparing long-lived and short-lived species. This suggests that these changes could constitute a highly conserved mechanism for MLSP extension both within and among species.

The decrease in mtROS generation in caloric-restricted rats is accompanied by significant decreases in 8-oxodG levels, only in mtDNA, or both in mtDNA and nDNA depending upon the tissue [7, 8, 10, 67, 68, 69], and by decreases in oxidation-, glycoxidation-, and lipoxidation-derived damage to heart mitochondrial proteins ([70]; Table 9.1). Besides, repair of 8-oxodG in mtDNA through the mitochondrial base excision repair pathway does not increase, or even decreases, in CR [71]. In fact, mtROS generation, 8-oxodG steady-state levels in mtDNA, and 8-oxodG repair through mitochondrial base excision repair decrease to a similar extent in 40% CR, around 30–40%. Similarly to what happened with tissue antioxidant levels in long-lived animals, mitochondrial 8-oxodG repair is lower in caloric-restricted rodents, likely because their rate of mtROS production is also lower than in ad libitum-fed animals. Both, endogenous antioxidants and mtDNA repair systems can be temporarily induced when needed, decreasing global energetic costs. Decreasing mtROS generation is less costly and more efficient than continuously maintaining high levels of antioxidants and repair systems. All these data suggest that lowering the mtROS generation rate is a highly conserved mechanism developed during evolution and shared by both long-lived and caloric restricted animals to decrease steady-state oxidative damage to lipids, proteins and especially mtDNA, and thereby mtDNA mutations and the aging rate [3, 4, 8–10]. Recent reports also have found increases in mitochondrial biogenesis and mitochondrial bioenergetic efficiency [72, 73], decreases in cellular steady-state ROS levels [73] and lack of changes or even increases in O_2 consumption [72, 74–76] in CR models, including mammals, nematodes, and yeast.

5 Protein Restriction, Methionine Restriction and Longevity

The antiaging effect of CR has been commonly attributed to the decreased intake of calories themselves rather than to decreases in specific dietary components. Recent reports in *Drosophila melanogaster* challenge this classical consensus [77, 78]. Carbohydrate or lipid dietary restriction does not seem to change longevity in rodents, whereas in studies on protein restriction (PR), increases in MLSP have been found in almost every investigation. Ten of 11 studies of PR in rats or mice (16 of 18 different life-long survival experiments in these studies) reported increases in MLSP (reviewed in [13]), although the magnitude of the increase (19.2% as mean increase in the 16 positive studies) was lower than that usually found in 40% CR studies (about 40% increase in MLSP). Because the mean degree of PR in the 16 positive studies was 66.7%, the MLSP increase expected at 40% PR would be around 11.5% (assuming proportionality between the degree of restriction and the intensity of the life extension effect, as it is the case in CR). Isocaloric methionine restriction (MetR), in contrast, increases MLSP in rats and mice independently of energy restriction [79–81]. Thus, the decreased methionine intake could be responsible for the MLSP increase induced by PR and for part of the life extension effect of CR. A 65% MetR in mice led to a significant MLSP increase of at least 10% [81], in agreement with the mean increase in longevity at 40% PR (11.5% increase). The rat MetR studies reported a 44% [80] or a 11% increase [79] in MLSP at 80% MetR. This would be equivalent to a mean 14% increase when extrapolated to 40% MetR, which is also within the range of the PR effect on MLSP in mice and rats. Therefore, available data suggests that reduced methionine intake can be responsible for all the life extension effect of PR (Table 9.1). The decrease in protein (and methionine) intake can be responsible for around one half of the life extension effect of CR in rodents (about 40% increase in longevity). This lower but significant life extension effect of PR compared with CR would agree with the idea that aging has more than one single cause. Recent studies in *D. melanogaster* support the idea of a CR life extension effect independent from calories [77, 78]. Reducing the amount of dietary casein from 4 to 2% and from 2% to 1 or 0.5% increases *D. melanogaster* longevity [78].

6 Protein Restriction, Methionine Restriction, mtROS Production, and Oxidative Damage

The effect of PR on mtROS generation has been recently analyzed; 40% PR decreases mtROS production specifically at complex I, lowers the %FRL, and decreases 8-oxodG in mtDNA ([11]; Table 9.1), and it also decreases specific markers of protein oxidative modification, fatty acid unsaturation, and complex I content [82] in rat liver mitochondria, very similar results to those found in CR [70]. Concerning the other two main dietary components, neither lipid [83] nor carbohydrate [84] restriction change mtROS production or oxidative damage to mtDNA in rat liver, in agreement with their lack of effect on MLSP (see above). So, proteins are the dietary components responsible for the decreases in mtROS production and oxidative stress and possibly for part of the MLSP increase during CR. Because MetR was responsible for the life extension effect of PR, studies on the MetR effect on mtROS production and oxidative damage were needed, as concerning oxidative stress, only changes in liver and blood glutathione (GSH) had been reported in MetR models [80]. It was found that isocaloric 80% MetR (with dietary glutamate substitution) decreases mtROS production (mainly at complex I), %FRL, 8-oxodG levels in mtDNA, complex I content, protein oxidative modification markers, and fatty acid unsaturation in rat heart and liver mitochondria [85]. These results are very similar to those described in CR and PR, with the only difference that mitochondrial O_2 consumption values where increased in the MetR experiment. Nevertheless, additional experiments of isocaloric 40 and 80% MetR in rats without glutamate substitution (unpublished data) also showed dose-dependent decreases in mtROS production and %FRL without changes in mitochondrial O_2 consumption, strongly suggesting that MetR can be responsible for 100% of the decrease in mtROS generation and oxidative stress that takes place in 40% PR and 40% CR, and possibly for all the life extension effect of PR and part of it in the case of CR.

Other lines of evidence also suggest that methionine may be involved in aging and longevity. It was recently found that the methionine content of tissue proteins strongly and inversely correlates with MLSP in mammals [15]. Methionine content in proteins is also lower in tissues from long-lived birds than from short-lived mammals of similar body size [13]. Therefore, the longer the MLSP of a species, the lower its protein methionine content. Conversely, excessive methionine in the diet damages different vital organs and increases tissue oxidative stress, with similar negative effects to those observed in rats fed high protein diets (reviewed in [13]). Contrarily, recent studies show that MetR not only increases rodent longevity but at the same time slows cataract development, minimizes age-related changes in T cells, and lowers serum glucose, IGF-1 and insulin levels in mice [81]. MetR also decreases visceral fat mass (around 70%), plasma insulin, IGF-1, and insulin response to glucose, and avoids age-related increases in blood cholesterol and triglycerides in rats [86]. This dramatic decrease in visceral fat mass points out that such a change, characteristic of CR animals, could not be necessarily related to the lower intake of calories.

Various possible mechanisms could be responsible for the decrease in mtROS generation in MetR, PR, and CR. The three dietary protocols decrease the amount of complex I and IV [11, 82, 85] in rat mitochondria. The strong decrease in complex I concentration could explain the lower complex I activity detected in CR [87] and could be responsible at least in part for the lower mtROS generation in CR. The decrease in mtROS production during CR, PR, and MetR also could be related to other mechanisms. Dietary methionine could be detrimental due to its conversion to homocysteine in vivo. Homocysteine has a free thiol group that can be oxidized leading to the formation of protein mixed disulfides. Addition of oxidized GSH (GSSG) to mitochondrial complex I increases its superoxide radical generation rate [88]. This effect of GSSG can explain the direct correlation found between GSSG/

GSH ratio and 8-oxodG levels in mtDNA in tissues of aging mice [52]. Thus, mtROS production could be regulated by thiol agents, including homocysteine, offering another plausible molecular mechanism to explain MetR and supplementation effects on mitochondrial oxidative stress, tissue oxidative damage, and longevity. Moreover, the decrease in %FRL also could be the result of a regulated response instead of a direct effect of a methionine metabolite. Because methionine is a donor of methyl groups for different cellular reactions, changes in DNA methylation also could be partially the cause of the changes in the level of expression of many genes described in CR [64]. Further research on the molecular mechanism decreasing mitochondrial ROS production and oxidative stress during caloric, protein, and methionine restriction is clearly needed.

In summary, the decrease in mtROS generation and oxidative mtDNA damage that takes place during caloric restriction is due to protein, not to lipid or carbohydrate restriction. Methionine restriction can explain all the effect of either caloric restriction or protein restriction on these parameters. In contrast, the three dietary manipulations are able to increase maximum longevity (although to a different extent) decreasing the aging rate and the incidence of degenerative and age-related diseases. Long-lived species (mammals and birds) and caloric-restricted rodents share an important characteristic, they both have low generation rates of endogenous oxidative damage. Methionine plays a key role decreasing ROS generation during CR and probably is also implicated in decreasing the susceptibility to oxidation of proteins from long-lived species. In addition, long-lived animals posses membrane lipids less sensitive to oxidative stress due to their lower degree of fatty acid unsaturation.

Acknowledgements Our results described in this review were supported by grant BFU2005-02584 from the Ministry of Science and Education and from CAM/UCM groups (910521) (to G.B.).

References

1. Harman D. The biological clock: the mitochondria. J Am Geriatr Soc 1972;20:145–147.
2. Sohal RS, Weindruch R. Oxidative stress, caloric restriction, and aging. Science 1996;273:59–63.
3. Barja G. Aging in vertebrates and the effect of caloric restriction: a mitochondrial free radical production-DNA damage mechanism? Biol Rev 2004;79:235–251.
4. Barja G. Free radicals and aging. Trends Neurosci 2004;27:595–600.
5. Barja G, Cadenas S, Rojas C, Pérez-Campo R, López-Torres M. Low mitochondrial free radical production per unit O_2 consumption can explain the simultaneous presence of high longevity and high aerobic metabolic rate in birds. Free Radic Res 1994;21:317–328.
6. Barja G, Cadenas S, Rojas C, López-Torres M, Pérez-Campo R. A decrease of free radical production near critical targets as a cause of maximum longevity. Comp Biochem Physiol 1994;108B:501–512.
7. Gredilla R, Barja G, López-Torres M. Effect of short-term caloric restriction on H_2O_2 production and oxidative DNA damage in rat liver mitochondria, and location of the free radical source. J Bioenerg Biomembr 2001;33:279–287.

8. Gredilla R, Barja G. The role of oxidative stress in relation to caloric restriction and longevity. Endocrinology 2005;146:3713–3717.

9. Barja G, Herrero A. Oxidative damage to mitochondrial DNA is inversely related to maximum life span in the heart and brain of mammals. FASEB J 2000;14:312–318.

10. Gredilla R, Sanz A, López-Torres M, Barja G. Caloric restriction decreases mitochondrial free radical generation at complex I and lowers oxidative damage to mitochondrial DNA in the rat heart. FASEB J 2001;15:1589–1591.

11. Sanz A, Caro P, Barja G. Protein restriction without strong caloric restriction decreases mitochondrial oxygen radical production and oxidative DNA damage in rat liver. J Bioenerg Biomembr 2004;36:545–552.

12. Pamplona R, Portero-Otín M, Riba D, Requena JR, Thorpe SR, López-Torres M, Barja G. Low fatty acid unsaturation: a mechanism for lowered lipoperoxidative modification of tissue proteins in mammalian species with long life spans. J Gerontol 2000;55:B286–B291.

13. Pamplona R, Barja G. Mitochondrial oxidative stress, aging and caloric restriction: the protein and methionine connection. Biochim Biophys Acta 2006;1757:496–508.

14. Pamplona R, Barja G, Portero-Otín M. Membrana fatty acid unsaturation, protection against oxidative stress, and maximum life span. A homeoviscous-longevity adaptation? Ann NY Acad Sci 2002;959:475–490.

15. Ruiz MC, Ayala V, Portero-Otín M, Requena JR, Barja G, Pamplona R. Protein methionine content and MDA-lysine adducts are inversely related to maximum life span in the heart of mammals. Mech Ageing Dev 2005;126:1106–1114.

16. Barja de Quiroga G, López-Torres M, Pérez-Campo R. Relationship between antioxidants, lipid peroxidation and aging. In: Emerit I, Chance B, eds. Free radicals and aging. Basel, Switzerland: Birkhäuser, 1992:109–123.

17. Benzi CD, Moretti A. Age- and peroxidative stress- related modifications of the cerebral enzymatic activities linked to mitochondria and the glutathione system. Free Radic Biol Med 1995;12:77–101.

18. López-Torres M, Pérez-Campo R, Rojas C, Cadenas S, Barja G. Maximum life span in vertebrates: correlation with liver antioxidant enzymes, glutathione system, ascorbate, urate, sensitivity to peroxidation, true malondialdehyde, in vivo H_2O_2, and basal and maximum aerobic capacity. Mech Ageing Dev 1993;70:177–199.

19. López-Torres M, Pérez-Campo R, Cadenas S, Rojas C, Barja G. The rate of free radical production as a determinant of the rate of aging: evidence from the comparative approach. J Comp Physiol B 1998;168:149–158.

20. Harris SB, Weindruch R, Smith GS, Mickey MR, Walford RL. Dietary restriction alone and in combination with oral ethoxyquine/2-mercaptoethylamine in mice. J Gerontol 1990; 45: B141–B147.

21. López-Torres M, Pérez-Campo R, Rojas C, Cadenas S, Barja G. Simultaneous induction of SOD, glutathione reductase, GSH and ascorbate in liver and kidney correlates with survival throughout the lifespan. Free Radic Biol Med 1993;15:133–142.

22. Jaarsma D, Haasdijk ED, Grashorn JAC, Hawkins R, Van Duijn W, Verspaget HW, London J, Holstege JC. Human Cu/Zn superoxide dismutase (SOD1) overexpression in mice causes mitochondrial vacuolization, axonal degeneration, and premature motoneuron death and accelerates motoneuron disease in mice expressing a familial amyotrophic lateral sclerosis mutant SOD1. Neurobiol Dis 2000;7:623–643.

23. Huang TT, Carlsson EJ, Gillespie AM, Shi Y, Epstein CJ. Ubiquitous expression of CuZn superoxide dismutase does not extend lifespan in mice. J Gerontol 2000;55A:B5–B9.

24. Mockett RJ, Sohal RS, Orr WC. Overexpression of glutathione reductase extends survival in transgenic *Drosophila melanogaster* under hyperoxia but not in normoxia. FASEB J 1999;13:1733–1742.

25. Schriner SE, Linford NL, Martin GM et al. Extension of murine life span by overexpression of catalase targeted to mitochondria. Science 2005;308:1909–1911.

26. Muller FL, Mele J, Van Remmen V, Richardson A. Proving the in vivo relevance of oxidative stress in aging using knockout and transgenic mice. In: Von Zglinicki T, ed. Aging at molecular level. Boston, MA: Kluwer, 2003:131–144.

27. Sanz A, Pamplona R, Barja G. Is the mitochondrial free radical theory of aging intact? Antioxid Redox Signal 2006;8:582–599.
28. Ku HH, Brunk UT, Sohal RS. Relationship between mitochondrial superoxide and hydrogen peroxide production and longevity of mammalian species. Free Radic Biol Med 1993;15:621–627.
29. Barja G, Herrero A. Localization at complex I and mechanism of the higher free radical production of brain non-synaptic mitochondria in the short-lived rat than in the longevous pigeon. J Bioenerg Biomembr 1998;30:235–243.
30. Herrero A, Barja G. Sites and mechanisms responsible for the low rate of free radical production of heart mitochondria in the long-lived pigeon. Mech Ageing Dev 1997;98:95–111.
31. Herrero A, Barja G. H_2O_2 production of heart mitochondria and aging rate are slower in canaries and parakeets than in mice: sites of free radical generation and mechanisms involved. Mech Ageing Dev 1998;103:133–146.
32. Ku HH, Sohal RS. Comparison of mitochondrial pro-oxidant generation and antioxidant defences between rat and pigeon: possible basis of variation in longevity and metabolic potential. Mech Ageing Dev 1993;72: 67–76.
33. St-Pierre J, Buckingham JA, Roebuck SJ, Brand MD. Topology of superoxide production from different sites in the respiratory chain. J Biol Chem 2002;277:44784–44790.
34. Kudin AP, Bimpong-Buta NY, Vielhaber S, Elger CE Kunz WS. Characterization of superoxide producing sites in isolated brain mitochondria. J Biol Chem 2004;279:4127–4135.
35. Hekimi S, Guarente L. Genetics and the specificity of the aging process. Science 2003;299:1351–1354.
36. Krause F, Scheckhuber CQ, Werner A, Rexroth S, Reifschneider NH, Dencher NA, Osiewacz IID. Supramolecular organization of cytochrome c oxidase- and alternative oxidase-dependent respiratory chains in the filamentous fungus *Podospora anserina*. J Biol Chem 2004;279:26453–26461.
37. Stroikin Y, Dalen H, Brunk UT, terman A. Testing the "garbage" accumulation theory of ageing: mitotic activity protects cells from death induced by inhibition of autophagy. Biogerontol 2005;6:39–47.
38. Boveris A, Cadenas E, Stoppani AOM. Role of ubiquinone in the mitochondrial generation of hydrogen peroxide. Biochem J 1976;156:435–444.
39. Takeshige K, Minakami S. NADH- and NADPH-dependent formation of superoxide anions by bovine heart submitochondrial particles and NAD-ubiquinone reductase preparation. Biochem J 1979;180:129–135.
40. Turrens JF, Boveris A. Generation of superoxide anion by the NADH dehydrogenase of bovine heart mitochondria. Biochem J 1980;191:421–427.
41. Genova ML, Ventura B, Giulano G, Bovina C, Formiggini G, Parenti Castelli G, Lenaz G. The site of production of superoxide radical in mitochondrial complex I is not a bound ubisemiquinone but presumably iron-sulphur cluster N2. FEBS Lett 2001;505:364–368.
42. Kushnareva Y, Murphy AN, Andreyev A. Complex I-mediated reactive oxygen species generation: modulation by cytochrome c and NAD(P)+ oxidation-reduction state. Biochem J 2002;368:545–553.
43. Lambert AJ, Brand MD. Inhibitors of the quinine-binding site allow rapid superoxide production from mitochondrial NADH: ubiquinone oxidoreductase (complex I). J Biol Chem 2004;279:39414–39420.
44. Ohnishi T, Johnson JE Jr, Yano T, Lobrutto R, Widger WR. Thermodynamic and EPR studies of slowly relaxing ubisemiquinone species in the isolated bovine heart complex I. FEBS Lett 2005;579:500–506.
45. Chen Q, Chen YR, Chen CL, Zhang L, Green-Church KB, Zweier JL. Superoxide production by NADH dehydrogenase induces self-inactivation with specific protein radical formation. J Biol Chem 2005;280:37339–37348.
46. Lehninger AL. Principles of biochemistry, 4th edn., New York: Freeman and Co, 2005:721–722.
47. Liu Y, Fiskum G, Schubert D. Generation of reactive oxygen species by the mitochondrial electron transport chain. J Neurochem 2002;80:780–787.

48. Herrero A, Barja G. Localization of the site of oxygen radical generation inside the Complex I of heart and nonsynaptic brain mammalian mitochondria. J Bioenerg Biomembr 2000;32:609–615.
49. Muller FL, Liu Y, Van Remmen H. Complex III releases superoxide to both sides of the inner mitochondrial membrane. J Biol Chem 2004;279:49064–49073.
50. Fraga CG, Shigenaga MK, Park JW, Degan P, Ames BN. Oxidative damage to DNA during aging: 8-hydroxy-2′-deoxyguanosine in rat organ DNA and urine. Proc Natl Acad Sci U S A 1990;87:4533–4537.
51. Mecocci P, MacGarvey U, Kaufman AE, Koontz D, Shoffner JM, Wallace DC, Beal F. Oxidative damage to mitochondrial DNA shows age-dependent increases in human brain. Ann Neurol 1993;34:609–616.
52. Asunción JG, Millan A, Pla R, Bruseghini L, Esteras A, Pallardó FV, Sastre J, Viña J. Mitochondrial glutathione oxidation correlates with age-associated oxidative damage to mitochondrial DNA. FASEB J 1996;10:333–338.
53. Herrero A, Barja G. Effect of aging on mitochondrial and nuclear DNA oxidative damage in the Herat and brain throughout the life-span of the rat. J Am Aging Assoc 2001;24:45–50.
54. Hirano T, Yamaguchi R, Asami S, Iwamoto N, Kasai H. 8-hydroxyguanine levels in nuclear DNA and its repair in rat organs associated with age. J Gerontol 1996;51A:B303–307.
55. Foksinski M, Rozalski R, Guz J, Ruszowska B, Sztukowska P, Ptwowarski M, Klungland A, Olinski R. Urinary excretion of DNA repair products correlates with metabolic rates as well as with maximum life spans of different mammalian species. Free Radic Biol Med 2004;37:1449–1454.
56. Herrero A, Barja G. 8-oxodeoxyguanosine levels in heart and brain mitochondrial and nuclear DNA of two mammals and three birds in relation to their different rates of aging. Aging Clin Exp Res 1999;11:294–300.
57. Wang E, Wong A, Cortpassi G. The rate of mitochondrial mutagenesis is faster in mice than in humans. Mutat Res 1997;377:157–166.
58. Blanchard JL, Lynch M. Organellar genes. Why do they end up in the nucleus? Trends in genetics 2000;16:315–320.
59. Bohr VA. Repair of oxidative DNA damage in nuclear and mitochondrial DNA, and some changes with aging in mammalian cells. Free Radic Biol Med 2002;32:804–812.
60. Barja G. The flux of free radical attack through mitochondrial DNA is related to aging rate. Aging Clin Exp Res 2000;12:342–355.
61. Crott JW, Choi SW, Branda RF, Mason JB. Accumulation of mitochondrial DNA deletions is age, tissue, and folate-dependent in rats. Mutat Res 2005;570:63–70.
62. Kraytsberg Y, Kudryavtseva E, McKee AC, Geula C, Kowall NW, Khrapko K. Mitochondrial DNA deletions are abundant and cause functional impairment in aged human *substantia nigra* neurons. Nature Genetics 2006;38:518–520.
63. Trifunovic A, Wredenberg A, Falkenberg M et al. Premature aging in mice expressing defective mitochondrial DNA polymerase. Nature 2004;429:417–423.
64. Weindruch R. Caloric restriction: life span extension and retardation of brain aging. Clin Neurosci Res 2003;2:279–284.
65. Gems D, Partridge L. Insulin IGF-1 signaling and ageing: seeing the bigger picture. Curr Opin Genet Dev 2001;11:287–292.
66. Sohal RS, Ku HH, Agarwal S, Forster, MJ, Lal H. Oxidative damage, mitochondrial oxidant generation and antioxidant defences during aging and in response to food restriction in the mouse. Mech Ageing Dev 1994;74:121–133.
67. López-Torres M, Gredilla R, Sanz A, Barja G. Influence of aging and long-term caloric restriction on oxygen radical generation and oxidative DNA damage in rat liver mitochondria. Free Radic Biol Med 2002;32:882–889.
68. Sanz A, Caro P, Ibáñez J, Gomez J, Gredilla R, Barja G. Dietary restriction at old age lowers mitochondrial oxygen radical production and leak at complex I and oxidative DNA damage in rat brain. J Bioenerg Biomembr 2005;37:83–90.
69. Drew B, Phaneuf S, Dirks A, Selman C, Gredilla R, Lezza A, Barja G, Leeuwenburgh C. Effect of aging and caloric restriction on mitochondrial energy production in gastrocnemius muscle and heart. Am J Physiol 2003;284:R474–R480.

70. Pamplona R, Portero-Otín M, Requena J, Gredilla R, Barja G. Oxidative, glycoxidative and lipoxidative damage to rat heart mitochondrial proteins is lower after four months of caloric restriction than in age-matched controls. Mech Ageing Dev 2002;123:1437–1446.
71. Stuart JA, Karahalil B, Hogue BA, Souza-Pinto NC, Bohr VA. Mitochondrial and nuclear DNA base excision repair are affected differently by caloric restriction. FASEB J 2004;18:595–597.
72. Nisoli E, Tonello C, Cardile A et al. Calorie restriction promotes mitochondrial biogenesis by inducing the expression of eNOS. Science 2005;310:314–317.
73. López-Lluch G, Hunt N, Jones B, Zhu M, Jamieson H, Hilmer S, Cascajo MV, Allard J, Ingram DK, Navas P, De Cabo R. Calorie restriction induces mitochondrial biogenesis and bioenergetic efficiency. Proc Natl Acad Sci U S A 2006;103:1768–1773.
74. Braeckman BP, Houthoofd K, Vanfleteren JR. Assessing metabolic activity in aging Caenorhabditis elegans: concepts and controversies. Aging Cell 2002;1:82–88.
75. Lin SJ, Kaeberlein M, Andalis AA, Sturtz LA, Defossez PA, Cullota VC, Fink GR, Guarente L. Caloric restriction extends Saccharomyces cerevisiae life span by increasing respiration. Nature 2002;418:344–348.
76. Yen K, Mastitis JW, Mobbs CV. Lifespan is not determined by metabolic rate: evidence from fishes and C. elegans. Exp Gerontol 2004;39:3947–3949.
77. Mair W, Piper MDW, Partridge L. Calories do not explain extension of life span by dietary restriction in Drosophila. PLOS Biol 2005;3:1305–1311.
78. Min K-J, Tatar M. Restriction of amino acids extends lifespan in Drosophila melanogaster. Mech Ageing Dev 2006;127:643–646.
79. Orentreich N, Matias JR, DeFelice A, Zimmerman JA. Low methionine ingestion by rats extends life span. J Nutr 1993;123:269–274.
80. Richie JP Jr, Leutzinger Y, Parthasarathy S, Malloy V, Orentreich N, Zimmerman JA. Methionine restriction increases blood glutathione and longevity in F344 rats. FASEB J 1994;8:1302–1307.
81. Miller RA, Buehner G, Chang Y, Harper JM, Sigler R. Methionine deficient diet extends mouse lifespan, slows immune and lens aging, alters glucose, T4, IGF-I and insulin levels, and increases hepatocyte MIF levels and stress resistance. Aging Cell 2005;4:119–125.
82. Ayala V, Naudí A, Sanz A, Caro P, Portero-Otín M, Barja G, Pamplona R. Dietary protein restriction decreases oxidative protein damage, peroxidizability index, and mitochondrial complex I content in rat liver. J Gerontol 2007;62A:352–360.
83. Sanz A, Caro P, Barja G. Effect of lipid restriction on mitochondrial free radical production and oxidative DNA damage. Ann New York Acad Sci 2006;1067:200–209.
84. Sanz A, Gómez J, Caro P, Barja G. Carbohydrate restriction does not change mitochondrial free radical generation and oxidative DNA damage. J Bioenerg Biomembr 2006;38:327–333.
85. Sanz A, Caro P, Ayala V, Portero-Otín M, Pamplona R, Barja G. Methionine restriction decreases mitochondrial oxygen radical generation and leak as well as oxidative damage to mitochondrial DNA and proteins. FASEB J 2006;20:1064–1073.
86. Malloy VL, Krajcik RA, Bailey SJ, Hristopoulos G, Plummer JD, Orentreich N. Methionine restriction decreases visceral fat mass and preserves insulin action in aging male Fischer 344 rats independent of energy restriction. Aging Cell 2006;5:305–314.
87. Hepple RT, Baker DJ, McConkey M, Murynka T, Norris R. Caloric restriction protects mitochondrial function with age in skeletal and cardiac muscles. Rejuv Res 2006;9:219–222.
88. Taylor ER, Hurrell F, Shannon RJ, Lin TK, Hirst J, Murphy MP. Reversible glutathionylation of complex I increases mitochondrial superoxide formation. J Biol Chem 2003;278:19603–19610.

Section III
Oxidative Stress in Human Aging and Diseases

10
Deregulation of Mitochondrial Function: A Potential Common Theme for Cardiovascular Disease Development

Scott W. Ballinger

Summary The pathogenesis of atherosclerosis has been intensively studied and described; yet, despite these advances, the underlying events that initiate atherogenesis are not yet fully understood. Consequently, cardiovascular disease remains the major cause of death and morbidity in the western world. Although studies are in general agreement that atherosclerosis is a chronic, inflammatory disease, and additionally, that increased oxidative stress plays a key factor in the early stages of disease development, the impact of such changes on the subcellular or organellar components and their functions that are relevant to cardiovascular disease inception are less appreciated. In this regard, studies are beginning to show that mitochondria are common targets of oxidative stress and that they may play significant roles in the regulation of cardiovascular cell function. A common theme among cardiovascular disease risk factors is that they can mediate mitochondrial damage and dysfunction, and moreover, that mitochondrial damage can significantly increase the risk for disease onset. Moreover, it has been recently suggested that the balance between mitochondrial energy efficiency, oxidant production, and thermogenesis that developed during human prehistory has long-term implications for human disease development, susceptibility, and evolution of disease. Hence, the emphasis of this review is discussion of the unique features of mitochondrial functional biology that potentially make it a central focal point in terms of the mechanistic basis of cardiovascular disease.

Keywords Atherosclerosis, cardiovascular disease (CVD) risk factors, mitochondria, oxidative stress, environmental oxidants, cigarette smoke.

1 Introduction

An estimated 71 million American adults (34.2% total population) have one or more types of cardiovascular disease (CVD). In fact, CVD claims more lives each year than the next five leading causes of death combined (cancer, chronic lower respiratory diseases, accidents, diabetes, influenza/pneumonia), and it was the underlying (primary) or a contributing cause of death in 37.3 or 58%, respectively, of Americans in 2003 [1].

From: *Aging Medicine: Oxidative Stress in Aging: From Model Systems to Human Diseases*   165
Edited by: S. Miwa, K.B. Beckman, and F.L. Muller © Humana Press, Totowa, NJ

Moreover, the prevalence of CVD in the United States increases with age: 11.2 and 6.2% of men and women respectively, between the ages 20–34 have CVD, whereas greater than half of men and women between the ages of 55–64 have CVD; more than three quarters of men and women (77.8 and 86.4%, respectively) over age 75 have CVD [1]. Hence, as the population ages, the incidence of CVD continues to rise. In addition to aging, numerous other factors influence CVD development, including environmental factors (e.g., air quality, tobacco smoke), hypercholesterolemia, and diabetes (section 4). The interaction of these factors with aging has led to the concept that CVD begins decades before the clinical manifestations of the disease; studies have now shown atherogenesis occurs in children, adolescents, and young adults; moreover, exposure to CVD risk factors during childhood or in utero increases the risk for adult disease development [2–11].

Atherosclerosis is the leading cause of death from heart disease, accounting for nearly three fourths of all deaths from CVD, and it is generally considered to be a form of progressive chronic inflammation resulting from interactions between modified lipoproteins, monocyte-derived macrophages, T cells, and cellular elements of the arterial wall [1, 12–15]. Inflammation and endothelial cell injury are thought to be the first steps in the development of atherosclerosis, leading to accumulation of lipids in injured areas of the artery. Because atherosclerosis is thought to be initiated by endothclial ccll injury, many studies have investigated the potential sources and causes of endothelial cell dysfunction and damage. Endothelial cells form the innermost surface of the artery wall, and they provide a permeable barrier for exchange and transport of substances into the artery wall. They provide a luminal surface that is resistant to leukocyte adhesion and thrombosis, and they are important for vascular tone (the generation of nitric oxide [NO] by the endothelium mediates smooth muscle cell relaxation) and the formation and secretion of growth regulatory molecules and cytokines. Beneath the endothelial layer lies the arterial intima and smooth muscle cells (SMCs), which are normally "quiescent" but change to a proliferative phenotype (and thus are important in the fibroproliferative component of atherosclerotic lesion formation) that express and respond to cytokines and growth factors during atherogenesis.

2   Atherogenesis

The pathobiology of atherosclerotic lesion formation has been described in detail by a number of reviews [12–15]. Briefly, it is known that CVD risk factors, such as tobacco smoke exposure, hypertension, hypercholesterolemia (elevated low-density lipoprotein [LDL]), hyperglycemia, obesity, insulin resistance, or a diet rich in high saturated fats, increase the expression of adhesion molecules on the surface of the endothelium, resulting in increased leukocyte infiltration through the endothelium. Once in the arterial intima, monocytes become macrophages and engulf modified lipoproteins (e.g., oxidized LDL) and take on a "foam cell" appearance. Foam cells are macrophage that are engorged with lipids, and they are characteristic of the early stages of atherogenesis.

Concomitantly, macrophages proliferate and secrete growth factors and cytokines that contribute to and further initiate inflammation. T lymphocytes also participate in athero-genesis by joining macrophages in early lesion development, forming the "fatty streak" and secrete cytokines and growth factors that promote migration of SMCs into the intima, where they proliferate to form an intermediate, atherosclerotic fibro-fatty lesion and ultimately a fibrous plaque. The "stability" (propensity for plaque rupture) of the fibrous cap/plaque is reliant upon interstitial collagen, which is influenced by relative efficiencies of collagen fiber formation and degradation. Smooth muscle cells are responsible for the majority of collagen formation in the arterial wall; however, infiltrating T lymphocytes secrete factors that inhibit SMC-mediated collagen formation and promote the production of factors that degrade collagen. Finally, T lymphocytes also promote thrombogenicity by expressing factors that stimulate macrophage production of tissue factor, a protein that can initiate blood coagulation when encountering factor VII. Overall, it is clear that inflammation plays an important role in the initiation, progression, and rupture of the atherosclerotic plaque (Fig. 10.1).

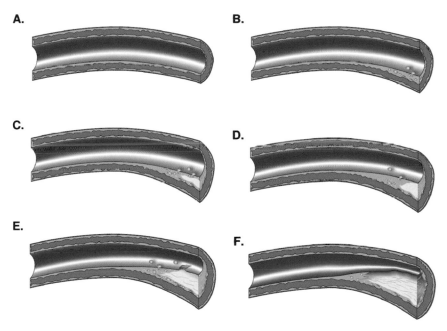

Fig. 10.1 Atherosclerotic lesion formation. (**A**) Healthy artery. (**B**) Induction of inflammation and early atherogenesis (in response to CVD risk factors); increased expression of adhesion molecules on the endothelial surface results in increased leukocyte adhesion and infiltration. (**C**) Fatty streak formation; macrophages engulf modified lipoproteins, secrete growth factors and cytokines that promote migration of smooth muscle cells into the intima. (**D**) Continued proliferation and inflammatory response result in the formation of the atherosclerotic fibro-fatty lesion with a lipid core and fibrous cap of a "vulnerable plaque" characterized by a thin-walled fibrous cap. (**E**) Inhibition of collagen formation and increased degradation of existing collagen of the fibrous cap results in plaque rupture and release of plaque contents into the lumen and thrombus formation. (**F**) Stenotic, stable plaque formation with a thick fibrous cap, which contribute to chronic ischemia

2.1   Changing Concepts of Atherogenesis

Interestingly, the majority of acute coronary events are not due to the rupture of a single flow-blocking lesion (atherosclerotic plaques that cause significant arterial narrowing). In fact, the artery can enlarge to compensate for plaque development and retain blood flow to the heart [16–20]. Serial angiogram data show that only about 15% of myocardial infarctions are associated with extreme arterial narrowing [21]. In contrast, it is now becoming evident that atherogenesis tends to progress "outward," or away from the lumen. Studies have now shown that atherosclerosis can be substantial within an artery without producing a stenotic effect [20, 21]. Presence of stenotic lesions usually indicates that that the disease has progressed in a widespread and substantial manner. Hence, individuals may seem to have angiographically "normal" coronary arteries, yet still have substantial atherosclerosis associated with outward remodeling or compensatory enlargement of the artery. Finally, it has been shown that many individuals will have multiple disrupted plaques when presented clinically, suggesting that the concept of a single, vulnerable plaque as the culprit for an acute coronary event is an oversimplification [22–27]. Infarct studies to date have shown that >20% of individuals presenting acute coronary syndrome will have multiple disrupted plaques [25–28].

2.2   Mechanisms of CVD

Whilst there has been significant progress in understanding the pathological processes that are involved in determining CVD progression and development, the continuing status of CVD as the leading cause of death and morbidity in the Western world for the past century implies that there is still a significant lack of understanding regarding the mechanisms of CVD risk and initiation. As discussed previously, it is becoming evident that atherosclerosis is a disease initiated and influenced by inflammation; however, the mechanisms that initiate this inflammatory state are not completely understood. A plethora of studies have revealed that among the potential initiating causes of CVD, free radical- or oxidative stress-mediated changes within the cardiovascular milieu are among the most popular postulated mechanisms [29–33].

Oxidative stress is caused by reactive species (RS), a collective grouping of reactive oxygen and nitrogen species (ROS and RNS, respectively) that are capable of disrupting cell function and exerting cytotoxic effects when generated in unregulated amounts. The concept that oxidative stress is important in the pathogenesis of CVD was conceived from studies that noted the cytotoxic and atherogenic properties of oxidized LDL (oxLDL) cholesterol [34–38]. Subsequently, it has become apparent that oxLDL cholesterol can be produced via several pathways involving oxidative stress [31]; moreover, increased vascular oxidant stress has several biological effects, including the peroxidation of polyunsaturated fatty acids in membrane or plasma lipoproteins, direct inhibition of mitochondrial respiratory chain enzymes, inactivation of membrane sodium channels, and DNA damage

[39–43]. Furthermore, increased vascular lesion formation and functional defects have been linked to increased levels of oxidatively modified proteins and lipoproteins, including oxidized LDL, lipid peroxidation products, nitrated and chlorinated tyrosines, and advanced glycation end products have been identified in patients with CVD [29–34, 44–47]. Hence, these findings are consistent with the notion that RS play a role in atherogenesis by modifying essential molecules such as lipids, proteins, and DNA and by altering their biological functions [48–51]. These events result in a proinflammatory environment and endothelial dysfunction, increasing the propensity for atherogenesis.

3 Mitochondrial Paradigm for CVD Development

Correspondingly, it may be no coincidence that a shared feature among common CVD risk factors such as hypercholesterolemia, diabetes, hypertension, stress, age and tobacco smoke exposure is increased levels of RS [31, 37, 38, 52–61]. However, the actual mechanistic associations of the majority of these factors (particularly environmental factors) to the initiation or development of CVD are poorly understood. For example, although it is known that increased oxidative stress originating from a variety of endogenous sources is a common theme among CVD risk factors [37, 38, 62–68], it is not yet clear whether these factors act to alter cellular function through similar or dissimilar mechanisms, or whether there are CVD risk factor-specific, or, common cellular "targets" that are relevant to CVD development. Another aspect of CVD development to be considered is that the events that lead to atherosclerosis usually begin decades before the clinical manifestations of the disease become evident.

Accordingly, the etiology of atherosclerosis may entail "subtle" changes in the endothelial/intimal environment that over time, collectively result in the initiation and progression of atherogenesis. Hence, the mechanisms involved in CVD development would likely involve cellular components that (1) play important roles in a variety of cell functions (e.g., growth, death, signaling, and bioenergetics); (2) are susceptible to oxidative damage; and (3) are capable of gradual decline or dysfunction that would coincide with disease development. In these respects, the mitochondrion meets these basic conditions. Hence, it is suggested herein that among the potential mediating mechanisms of atherogenesis and CVD, mitochondrial damage and dysfunction is a significant initiating event.

3.1 Mitochondria and Their DNA

Mitochondria are ancient bacterial symbionts with their own DNA, RNA, and protein synthesis systems. Each cell contains hundreds of mitochondria and thousands of mitochondrial DNA (mtDNA) copies, which are maternally inherited. The mtDNA is a closed, circular, double-stranded molecule attached to the matrix side of the inner mitochondrial membrane, and it is present in approximately 5–10 copies per mitochondrion. In mammals, the mtDNA is ~16 kilobases and encodes 13 polypeptides that are essential for oxidative phosphorylation (OXPHOS) plus two rRNAs (12S and 16S) and 22 tRNAs that are required for mitochondrial protein synthesis (the mtDNA has a different genetic code than the nucleus). The mtDNA encoded polypeptide genes are structural subunits for four of the five OXPHOS enzyme complexes; whereas all the subunits of complex II (succinate dehydrogenase) are encoded by the nuclear DNA (nDNA), the remaining four enzyme complexes (I, III–V) contain polypeptide subunits encoded by both the nDNA and mtDNA (Fig. 10.2). The nuclear DNA (nDNA) codes for all other mitochondrial proteins, including the mitochondrial DNA polymerase γ subunits, the mitochondrial RNA polymerase components, the mitochondrial transcription factor, the mitochondrial ribosomal proteins and elongation factors, and the mitochondrial metabolic enzymes. Numerous types of age-related diseases have been related to mtDNA mutations, mitochondrial dysfunction, or both in humans, including neurodegeneration, neuromuscular disease, deafness, blindness, diabetes, and cardiovascular disease [69]. Many forms of mitochondrial diseases can show marked clinical heterogeneity, and there are several aspects of mitochondrial genetics that contribute to such variability; however, a detailed treatise of the features of mitochondrial genetics is beyond the scope of this review, and in this regard there are several detailed reviews available [70–75]. The focus of this review is upon more global aspects of mitochondrial function and their potential roles in CVD development and susceptibility.

3.2   Mitochondrial Oxidative Phosphorylation

Mitochondria generate energy via OXPHOS, which "couples" electron transport with proton translocation for the production of ATP. Mitochondrial OXPHOS consists of five multiple subunit enzyme complexes embedded within the mitochondrial inner membrane, which coordinately function to generate mitochondrial ATP (Fig. 10.2). Electrons donated by NADH + H⁺ (derived from dietary carbohydrates and fats) to complex I (NADH dehydrogenase) or from succinate to complex II (succinate dehydrogenase) are shuttled to ubiquinone (CoQ) to yield ubisemiquinone (CoQH) and then ubiquinol (CoQH$_2$). Electrons are transferred from CoQH$_2$ to complex III (ubiquinol:cytochrome c oxidoreductase), which transfers them to cytochrome c. Electrons are transferred from cytochrome c to complex IV (cytochrome c oxidase) and ultimately to ½ O$_2$ to give water. The energy released during the movement of electrons along the electron transport chain is used to pump protons out of the inner membrane at complexes I, III, and IV, creating an electrochemical gradient across the inner membrane that provides the potential energy for ATP formation from ADP and P$_i$ at complex V (ATP synthase). Mitochondrial ATP is transported out of the mitochondrion by the nuclear encoded adenine nucleotide translocase (ANT), which is a trans-inner membrane enzyme that mediates exchange of ATP and ADP out and into the mitochondrial matrix, respectively.

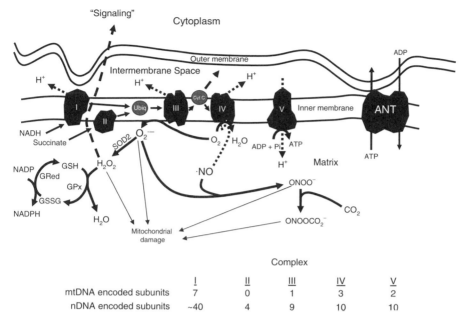

	Complex				
	I	II	III	IV	V
mtDNA encoded subunits	7	0	1	3	2
nDNA encoded subunits	~40	4	9	10	10

Fig. 10.2 Mitochondrial oxidative phosphorylation and oxidant generation. Mitochondrial oxidative phosphorylation and electron transport represent a series of multisubunit protein complexes that are encoded by both the mtDNA and nDNA. Reducing equivalents are provided by NADH to complex I (7 mtDNA encoded subunits, ~40 nDNA encoded subunits) and succinate to complex II (4 nDNA encoded subunits). Electrons are shuttled to complex III (1 mtDNA encoded subunit, 9 nDNA encoded subunits) via ubiquinol. Cytochrome c carries electrons from complex III to complex IV (three mtDNA encoded subunits, 10 nDNA encoded subunits), with the reduction of O_2 to form H_2O (a reaction inhibited by NO). As electrons move through the series of enzyme complexes, protons (H^+) are pumped across the inner membrane at complexes I, III, and IV, creating an electromotive differential across the inner membrane. Complex V (ATP synthase, two mtDNA encoded subunits, 10 nDNA encoded subunits) uses the potential energy created by this proton gradient to condense a molecule of ADP with P_i to form ATP. ATP is subsequently transported out of the mitochondrial matrix by the ANT, which is also responsible for moving cytoplasmic ADP into the mitochondrial matrix. Of the electrons entering the transport chain, it has been estimated that 2–4% escape the confines of the "chain" to form $O_2^{\cdot-}$. Inhibitors of electron transport (e.g., NO) can increase mitochondrial oxidant ($O_2^{\cdot-}$) production. NO can impair electron flow at cytochrome c oxidase (complex IV) by oxidizing the heme group of cytochrome aa_3, and it is competitive with O_2. NO also inhibits the bc1 segment of complex III, resulting in increased autooxidation of ubisemiquinone and $O_2^{\cdot-}$ production. At low NO concentrations, the dismutation of $O_2^{\cdot-}$ (via SOD2) and formation of H_2O_2 is favored, which can subsequently diffuse from the mitochondrion and act as a signaling molecule, or, is reduced by GSH to form H_2O. At higher NO concentrations, $O_2^{\cdot-}$ reacts with NO to form $ONOO^-$, which may promote cytochrome c release and apoptosis, mtDNA damage, inhibit electron transport, and inactivate mitochondrial proteins (e.g., mitochondrial damage). In the presence of CO_2, $ONOO^-$ yields nitrosoperoxycarbonate ($ONOOCO_2^-$), which has been proposed to diminish oxidation and increase nitration reactions

3.3   *Mitochondrial Oxidant Production and Regulation*

Although mitochondria have been stereotyped as "cellular powerplants," they are also the primary sources of endogenous cellular ROS (Fig. 10.2). In fact, the primary function of the mitochondrion in some cells may be as a signaling organelle [76]. During electron transport, oxygen can pick up electrons directly from the ubiquinone site in complex III and flavin mononucleotide group of complex I to generate superoxide ($O_2^{\cdot-}$) [77–81]. Mitochondrial generated $O_2^{\cdot-}$ can be converted to hydrogen peroxide (H_2O_2) by mitochondrial manganese superoxide dismutase (SOD2). Under basal conditions, mitochondrial H_2O_2 generation represents up to 2% of the oxygen consumption by this organelle, with mitochondrial H_2O_2 fluxes altered by drugs or toxins such as electron transport inhibitors, uncouplers, redox cycling molecules, or by local (endogenous) and exogenous environmental changes [78, 82–86]. H_2O_2 can be converted to H_2O by the glutathione redox system, react with $O_2^{\cdot-}$, or, in the presence of transition metals, yield highly reactive hydroxyl radicals (•OH) [87]. Mitochondrial generated H_2O_2 can act as a signaling molecule (H_2O_2 is freely diffusible), stimulating growth and survival pathways or cytochrome *c* release and caspase activation, at low or high H_2O_2 concentrations, respectively [88–92].

It also has been suggested that mitochondria are important in the regulation of many genes involved in cellular defense mechanisms, pathogen defenses, immunological responses, and expression of cytokines and cell adhesion molecules via the generation of ROS by the mitochondrial respiratory chain that serve as the intermediate second messengers for the activation of nuclear factor-κB (NFκB) by tumor necrosis factor (TNF)α and interleukin (IL)-1 [93–96]. Similarly, inhibition of mitochondrial respiratory function prevents H_2O_2-induced transactivation of the epidermal growth factor receptor and stimulation of downstream targets c-Jun NH_2-terminal kinase and Akt and H_2O_2-induced growth factor transactivation (both vascular endothelial growth factor-2 receptor and platelet-derived growth factor (PDGF)β receptor in endothelium and fibroblasts, respectively). Growth factor receptor transactivation and its downstream signaling in response to H_2O_2 is abrogated by mitochondrial targeted antioxidants, but not by nontargeted counterparts, suggesting the involvement of mitochondrial oxidants in these events [97]. Antioxidant expression and cytokines are also important factors in influencing the steady-state levels of specific mitochondrial oxidants, and thus they play a significant role in cell function and redox signaling in the mitochondrion. Cytokines such as PDGF and TNFα have been shown to either directly or secondarily alter cellular and/or mitochondrial oxidant production and SOD2 expression [98–100].

Several factors regulate mitochondrial oxidant generation, including local concentrations of both RNS and ROS, mitochondrial antioxidants, cytokines, electron transport efficiency, metabolic reducing equivalent availability (NADH and $FADH_2$), uncoupling protein (UCP) activities, and overall organelle integrity (damage to membranes, DNA, and proteins). NO concentration regulates mitochondrial electron transport by modulating complex IV activity, which influences ATP production, $O_2^{\cdot-}$ generation, and oxygen diffusion [101, 102]. Low NO concentrations can modulate mitochondrial respiration and oxygen consumption through reversible binding at complex IV, whereas higher concentrations of NO result in increased peroxynitrite ($ONOO^-$) formation (via reaction with $O_2^{\cdot-}$) and can contribute to cytotoxicity [104–107]. In the mitochondrion, it is likely that $ONOO^-$ reacts with CO_2 (the principle site of CO_2 is the mitochondrion) to form a nitrosoperoxycarbonate intermediate ($ONOOCO_2^-$). It has been shown that $ONOOCO_2^{\cdot-}$ efficiently mediates nitration reactions (thus diminishing oxidation yields) [50, 108]. In this manner, the relative levels of NO and $O_2^{\cdot-}$ are important in influencing mitochondrial respiratory regulation and the production of downstream reactive species (e.g., H_2O_2, $ONOO^-$).

The relative efficiency of the mitochondrion to convert caloric energy into ATP also influences oxidant production. Highly efficient or "tightly" coupled mitochondria are the most economical in terms of generating the maximum amount of ATP with a minimal loss of caloric energy, whereas "loosely" coupled mitochondria are less energetically efficient. A related feature to this coupling is that tightly coupled mitochondria are more prone to donate electrons to O_2 to form $O_2^{\cdot-}$ compared with loosely coupled mitochondria. Hence, although tightly coupled mitochondria generate ATP more efficiently compared with loosely coupled mitochondria, they also produce greater levels of $O_2^{\cdot-}$. The potential ramifications of these differences in energy and oxidant production is discussed in greater detail later in this review (see Section 5).

Mitochondrial UCPs are a family of mitochondrial anion carriers important in proton conductance across the inner membrane; thus, they play a role in regulating mitochondrial membrane potential, thermogenesis, and mitochondrial ROS production [109–115]. UCPs modulate proton "leak" across the inner membrane and therefore influence membrane potential and the production of mitochondrial oxidants. The greatest levels of mitochondrial $O_2^{\cdot-}$ production are associated with high membrane potentials; hence, proton leak reduces mitochondrial oxidant formation. Consequently, low UCP expression is thought to be associated with reduced proton leakage, high membrane potentials, and increased $O_2^{\cdot-}$ production. Interestingly, it has been noted that $O_2^{\cdot-}$ activates UCPs indirectly, through lipid peroxidation products, and reactive aldehydes (4-hydroxynonenal), suggesting that a feedback mechanism of oxidant stress may contribute to the regulation of UCP expression [116].

3.4 Mitochondrial Damage and Function

Numerous reports have shown that mitochondria are sensitive to both ROS- and RNS-mediated damage and alterations in function [78, 103, 117–121]. The susceptibility of the mtDNA to damage is thought to be due to several factors, including the lack of both protective histone and nonhistone proteins and a relatively limited DNA repair capability compared with the nucleus. The mtDNA is attached to the matrix side of the mitochondrial inner membrane, placing it in proximity to reactive lipophilic species and reactive lipid oxidation products (generated within the membrane) that are capable of modifying the mtDNA. In vitro studies have shown that ROS and RNS induce a variety of effects in both vascular endothelial cells and SMCs, including preferential and sustained mtDNA damage, altered mitochondrial transcript levels, and decreased mitochondrial protein synthesis [121]. Increased mtDNA damage and decreased transcriptional and translational functions would potentially decrease the efficiency of mitochondrial replication and assembly of respiratory chain proteins, further increasing mitochondrial dysfunction and susceptibility to the damaging effects of ROS/RNS. In support of this notion are studies that show high levels of NO decrease the levels of mitochondrial respiratory proteins in endothelial cells and increase the susceptibility to cell death [105]. In vivo studies have shown that cardiovascular mtDNA damage is increased in both humans and animals with CVD [122–125]. Atherosclerotic lesions in brain microvessels from Alzheimer's (AD) patients and rodent AD models also have significantly more mtDNA deletions and abnormalities, (as do the endothelium and perivascular cells), suggesting the mitochondria within the vascular wall can be central targets for oxidative stress induced damage [126]. Chronic ischemia increases both mtDNA deletions in human heart tissue [124] and cardiac mitochondrial sensitivity to inhibitors of cellular respiration [127]. Animal studies have shown that vascular mtDNA damage is increased in animal models of atherosclerosis and that this damage occurs before or coincidental with disease development [122, 123]. Moreover, deficiencies in mitochondrial antioxidants (SOD2), regulatory proteins (UCPs), or both that modulate mitochondrial oxidant production have been shown to promote the onset of CVD in vivo, consistent with the notion that mitochondrial generated oxidants can play a contributory role in atherogenesis [122, 128]. Furthermore, decreased SOD2 activity increased susceptibility to ischemia/reperfusion mediated cardiac damage and resistance to cardiac preconditioning [129]. In contrast, overexpression of mitochondrial antioxidants and/or UCPs has been shown to protect against the effects of ischemia/reperfusion and oxidative stress [130–132]. Ex vivo studies in rat heart have shown that ischemia reduces myocardial ANT activity, and reperfusion further contributes to the loss of both ANT and OXPHOS capacities [120]. Using a mouse model for myocardial infarction (MI), it was found that previous MI is associated with increased RS (left ventricle), and decreased mtDNA copy number, mitochondrial encoded gene transcripts, and related enzymatic activities (complexes I, III, and IV). However, nuclear encoded genes (complex II) and citrate synthase, are unaffected [133]. Finally, cardiotoxic RS generators increase mtDNA deletions and lipid peroxidation in the myocardial mitochondria; overexpression of mitochondrial antioxidants prevents these effects and increases cardiac tolerance to ischemia [130].

Interestingly, studies examining the role of UCPs have yielded somewhat conflicting results; one study reported that overexpression of UCP-1 (a protein typically expressed in brown adipose tissue) in aortic smooth muscle cells in vivo increases oxidant formation, hypertension, and exacerbates atherosclerotic lesion formation in a cholesterol-independent manner [134]. Although these results were somewhat unexpected, because overexpression of UCP-1 in skeletal muscle reduces blood pressure and increases insulin sensitivity in genetically obese mice and studies investigating the impact of UCP-2 (a widely expressed UCP including the cardiovasculature) have shown that it decreases ROS generation and mitochondrial overload in cardiomyocytes, and influences atherosclerotic lesion formation in vivo [128, 131, 135], the presence of multiple UCP isoforms raises the possibility of distinct functional importance of the different UCP proteins, potentially explaining the contrasting experimental results between UCP-1 and UCP-2 described above. Similarly, studies examining the effects of cytosolic SOD (SOD1) overexpression failed to reduce atherogenesis [136], whereas reduced SOD2 levels hastened the onset of atherogenesis in mice [122]. Consequently, the impact of endogenous UCP and SOD expression in the absence or overexpression of specific UCPs or SODs (e.g., UCP-2 or SOD2 levels during UCP-1 or SOD1 overexpression, respectively) may shed light on the relative importance of each in CVD development. Nonetheless, these studies do suggest that altered metabolic and antioxidant efficiency (and thus $O_2^{\cdot-}$ formation) due to changes in UCP activity, SOD activity, or both can contribute to vascular dysfunction.

4   CVD Risk Factors and Mitochondrial Damage

Many "proatherogenic" factors (i.e., hypercholesterolemia, hyperglycemia, cigarette smoke exposure) have been associated with significant alterations in mitochondrial function and damage [2, 121–123]. For example, incubation of endothelial cells with concentrations of oxLDL that are associated with cytoprotection induces mitochondrial complex I activity in a manner that seems to be dependent on the induction of oxidative stress [137]. Similarly, oxLDL treatment of human macrophage results in the alteration of several mitochondrial functions including antioxidant expression (SOD2), oxidant generation, and membrane potentials, which in turn, may contribute to further oxLDL formation [138–140].

Cholesterol administration in rabbits is associated with impaired mitochondrial function [141]. SOD2 activity and glutathione (GSH) concentration are significantly increased in the atherosclerotic intima when compared with the media of the aorta in hyperlipidemic rabbits; interestingly, SOD2 activity and GSH concentration also are found to be inversely related with age and plaque size, suggesting that they may be related to the early stages of atherosclerotic lesion formation in that they represent an early "protective" response to increased mitochondrial oxidation that occurs during the initial events of CVD risk factor exposure, atherogenesis, or both [138]. Finally, hypercholesterolemic mice sustain significantly higher levels of mtDNA damage in their aortic tissues compared to age-matched normocholesterolemic controls [122].

Hyperglycemia induces increased $O_2^{\cdot-}$ generation in endothelial cells in vitro, and studies suggest that the majority of this $O_2^{\cdot-}$ is produced by the mitochondrion [142, 143]. It has been proposed that high glucose levels increase the mitochondrial inner membrane potential as a result of overproduction of electron donors (e.g., NADH and succinate/$FADH_2$) by the citric acid cycle, which results in increased production of $O_2^{\cdot-}$ [142–144]. The finding that overexpression of the mitochondrial antioxidant SOD2 prevents the hyperglycemia associated production of $O_2^{\cdot-}$, as do mitochondrial electron complex I inhibitors or uncouplers of oxidative phosphorylation, supports this notion [142, 144]. Interestingly, these same factors also prevent glucose-induced activation of phosphokinase C and NFκB in endothelial cells and decrease IL-1b, IL-1β/TNFα/interferon-γ activation of NFκB and the induction of inducible nitric-oxide synthase (iNOS) in insulin-producing cells (RINm5F) [145, 146]. In contrast, suppression of SOD2 increases activation of NFκB and the iNOS promoter; catalase, Gpx, and SOD1 did not affect the cytokine-induced activation of NFκB and the iNOS promoter [147]. Similar to the effects of hyperglycemia alone, mitochondrial electron transport inhibitors or antioxidants completely inhibit the effects of leptin, a hormone that is increased in insulin-resistant relatives of diabetics [148–151]. Leptin is a circulating hormone that increases fatty acid oxidation, is involved in body weight control, and is secreted mainly from adipose tissues [152, 153]. A leptin receptor has been identified on endothelial cells, and leptin has been shown to promote both angiogenesis and inflammation [154, 155]. It also has been noted that leptin increases oxidative stress in endothelial cells in a dose-dependent manner, and its effects are additive with glucose [151]. Additionally, angiotensin II (AngII) stimulates vascular cell adhesion molecule-1 (VCAM-1), activates NFκB, and stimulates degradation of both inhibitor of κB (IκB)a and IκBb [156]; degradation of IκBs is not inhibited by SOD1 or catalase, whereas inhibitors of mitochondrial respiration inhibit AngII induced IκB degradation, suggesting that AngII induces intracellular oxidative stress (presumably mitochondrial) in endothelial cells that stimulates IκB degradation and NFκB activation. This activation enhances the expression of VCAM-1 and probably other genes involved in the early stages of atherosclerosis. These findings suggest that mitochondrial oxidant production plays an important role in the redox related regulation of these factors.

Of the 500,000 annual CVD related deaths in the United States each year, 20% are directly attributed to tobacco smoke exposure; 40,000 of these deaths are related to second-hand tobacco smoke exposure [1, 157]. However, despite these major cardiovascular health consequences, and the fact that tobacco smoke exposure is one of the strongest predictors of CVD development, molecular mechanisms of how tobacco smoke exposure specifically results in cardiovascular cell dysfunction and CVD development are not well understood. Additionally, it has only been recently appreciated that prenatal and childhood exposure to tobacco smoke plays a significant role in adult CVD risk, yet even these effects of prenatal and childhood tobacco smoke exposure on adult CVD development are currently in the early stages of investigation [2]. One apparent observation is that the mitochondrion seems to be a common target for components of tobacco smoke exposure and perhaps other environmental oxidants associated with disease development.

Tobacco smoke exposure has several cardiovascular and mitochondrial effects, including decreased arterial oxygen carrying capacity (via increased serum carboxyhemoglobin) associated with increased OXPHOS dysfunction in cardiac cells [52, 158–160], and increased generation of mitochondrial RS in platelets

[161]. Rat myocardial cytochrome oxidase activity is significantly decreased after exposure to secondhand smoke; activity continues to decline with prolonged exposures to tobacco smoke [158], and it seems to enhance myocardial OXPHOS dysfunction during reperfusion injury [52]. Similarly, cigarette smoke exposure increases sensitivity to heart ischemia/reperfusion injury in rats [52]. Second-hand smoke exposure in mice causes significant aortic mtDNA damage, reduced ANT activity, increased nitration and inactivation of SOD2, and mitochondrial dysfunction compared with unexposed mice, and when combined with other CVD risk factors (e.g., hypercholesterolemia), it accelerates both mitochondrial damage and atherogenesis [123]. It also has been shown that benzo-[a]-pyrene (a component of cigarette smoke) induces an atherogenic phenotype associated with up-regulated expression of mtDNA transcripts in rat smooth muscle cells; cultures made from these cells have increased growth rates and marked enhancement of proliferation to serum mitogens [162].

5 Mitochondrial Function and Genetics May Influence Disease Susceptibility

Why some individuals exposed to the same CVD risk factor with identical risk factor profiles will go on to develop CVD, whereas others will not, is simply not understood. One explanation is that CVD is a multifactorial disorder that involves both environmental and genetic factors. It has been estimated that 70–80% of CVD is attributable to modifiable nongenetic factors, which is consistent with the notion that environmental factors heavily influence the risk of CVD development [163]. A corollary to this is that an individual's response to an environmental CVD risk factor is likely to be genetically influenced. Numerous studies have looked for potential associations between polymorphic gene mutations and CVD development; however, many of the noted associations reported in small-scale studies were not found in larger scale studies, or they were not as predictive for CVD risk as plasma markers (e.g., cholesterol levels) [164–170]. Studies also have investigated potential associations between smoking-related CVD risk and gene polymorphisms in small-scale studies [165, 166]. Similarly, these associations were not observed in larger scale studies, or it was found that DNA adduct formation in vascular tissues (a feature of tobacco smoke exposure that has been reported to correlate with CVD severity) did not correlate with genotype [167, 171]. Currently, it seems that <5% of CVD results from single mutations, such as those regulating lipoprotein synthesis [172]. Hence, in the majority of instances it is likely that individual CVD development is the result of multigenetic factors that modulate biological responses to an environmental challenge. Consequently, it has been hypothesized that multiple genes involved in many aspects of vascular regulation, lipoprotein metabolism, inflammation, metabolic control and redox tone (the balance between oxidant generation and neutralization by antioxidants) and their interaction with environmental factors will influence CVD susceptibility. Recent studies on CVD susceptibility

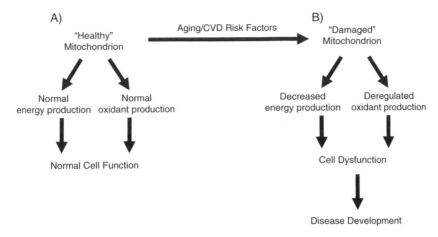

Fig. 10.3 The impact of age and CVD risk factor exposure on mitochondrial oxidative phosphorylation and oxidant generation. (**A**) Healthy mitochondria generate sufficient ATP and regulated amounts of mitochondrial oxidants to mediate proper cell growth, programmed cell death, resulting in "normal" cell function. (**B**) Aging and/or exposure to CVD risk factors results in increased mitochondrial damage, causing decreased ATP production, and/or dysregulation of mitochondrial oxidant production. Depending upon the demands of the cell (e.g., cells with high energy demands are sensitive to changes in ATP production, whereas those primarily reliant on mitochondria as signaling organelles for proper function are sensitive to changes in mitochondrial oxidant production), deficits in energy production, and/or inappropriate levels of mitochondrial oxidant production lead to cell dysfunction and ultimately, disease development

and resistance support the concept of groups of genes being involved in atherogenic susceptibility, and although important in advancing the understanding of gene "groups" that may be involved in influencing individual biological responses to an environmental challenge, they do not address the underlying issue of why these differences exist, nor how they might have developed [173, 174]. In this respect, none of the current studies has considered the potential role of the "other" eukaryotic genome, the mtDNA or the mitochondrion, and the possibility for mitochondrial–nuclear interaction in influencing individual susceptibility for CVD development.

As discussed in sections 3 and 4, the efficacy of mitochondrial function and oxidant production is dependent upon a number of factors including local concentrations of both RNS and ROS, mitochondrial antioxidants, cytokines, electron transport efficiency, metabolic reducing equivalent availability (NADH and $FADH_2$), UCP activities, and overall organelle integrity (damage to membranes, DNA, and proteins). Of these factors, the coupling efficiency of the mitochondrion has perhaps received the least attention in relation to CVD development. Tightly coupled mitochondria have (1) electron transport that is highly efficient at pumping protons out of the mitochondrial inner membrane (complexes I, III, and IV); and (2) an ATP synthase (complex V) that is highly efficient at converting proton flow through its proton channel into ATP. Consequently, tightly coupled mitochondria are the most economical in terms of generating the maximum amount of ATP with a minimal loss of caloric energy (e.g., heat, or thermogenesis). Additionally, tightly coupled mitochondria will usually have higher membrane potentials (relative to "less coupled" mitochondria) and a greater proportion of electron carriers in a reduced state (occupied by electrons); thus, they are also more prone to donate electrons to oxygen to form $O_2^{\cdot-}$. In contrast, loosely coupled mitochondria have (1) reduced proton pumping efficiency (complexes I, III, and IV) within electron transport; and/ or (2) require more protons to make ATP by the ATP synthase (complex V). Hence, greater electron flow is required for proton pumping and ATP generation, resulting in a less energetically efficient but more thermogenic mitochondrion. Additionally, by virtue of the greater electron flow in loosely coupled mitochondria, electron carriers retain fewer electrons (remain oxidized), which decreases mitochondrial ROS production. Consequently, mitochondrial coupling efficiency determines the proportion of caloric energy used by the mitochondrion to perform work (ATP generation) versus that used to maintain body temperature (thermogenesis), and its propensity to generate oxidants. Hence, the "trade-off" for being energetically efficient (tightly coupled mitochondria) is increased and decreased capacities in basal $O_2^{\cdot-}$ production and thermogenesis, respectively, whereas increased capacity for thermogenesis results in decreased capacities in basal energy and oxidant production. In this regard, it has been proposed that environmental factors influenced prehistoric human radiation patterns and survival by selecting for aspects of mitochondrial function (ATP production and thermogenesis) [71, 72, 175]. With the development of climate control and excess availability of calories seen in Western societies today, however, these variants in mitochondrial function and genetics may influence an individual's predisposition toward disease development in that individuals with more tightly coupled mitochondria have increased basal levels of endogenous oxidant stress and thus they will be susceptible to diseases related to oxidative stress, whereas those with loosely coupled mitochondria will be susceptible to diseases of energetic insufficiency.

Evidence supporting this concept has only recently been recognized. In addition to the original studies that have linked mtDNA mutations with human disease, several studies have shown that specific mtDNA mutations and haplotypes are associated with increased risk for diseases thought or known to have an environmental component in humans (e.g., deafness, blindness, Alzheimer's disease, diabetes, and cancer) [125, 175–190]. Similarly, studies have shown that the mtDNA haplotype can influence tumor growth and age-related deafness [185, 191]. Finally, it has been shown that human longevity significantly cosegregates with mtDNA haplotypes that have temperate and arctic origins [187, 188]. Consequently, it is hypothesized that certain mitochondrial haplotypes that are associated with "cold adaptation" (e.g., less coupled mitochondria) will be associated with increased life span, but increased predilection for clinical illnesses such as blindness and central nervous system defects [72]. Alternatively, mitochondrial haplotypes thought to be associated with tightly coupled mitochondria seem to be more prone to certain types of cancer and neurodegenerative diseases associated with somatic mutation or oxidative stress [72, 175, 185]. Hence, although it is widely accepted that CVD development is influenced by a combination of genetic, environmental, and behavioral factors that influence an individual's biological response to known CVD risk factors and although studies have shown such differences occur, in general, they have not provided a clear mechanistic basis or explanation for these observed differences in CVD susceptibility. An important consideration currently lacking from these analyses is the potential role for mitochondrial–nuclear interaction that evolved during human prehistory in influencing individual susceptibility to contemporary environmental CVD risk factors.

6 Conclusions

Mitochondria are commonly stereotyped as cellular powerplants; however, they also play critical roles in cell regulation, growth, thermogenesis and apoptosis by generating oxidants that are used in these processes. A common theme among CVD risk factors seems to be mitochondrial damage and dysfunction in the cardiovasculature, and moreover, this mitochondrial damage sustained during important developmental stages (e.g. gestation) may significantly increase the risk for adult disease onset (Fig. 10.3). However, perhaps more provocative, it has been suggested that the balance between energy efficiency, oxidant production, and thermogenesis in the mitochondrion that developed during human prehistory has long-term implications for human disease development, susceptibility, and evolution of disease [71, 72, 175]. Consequently, the mitochondrion is a multifunctional organelle, whose primary function is dependent upon the current requirements and environment of the cell. Hence, the primary function of a mitochondrion in an endothelial cell may be the regulated generation of oxidants for cell signaling, whereas within a cardiac myocyte it may be the generation of ATP, or a combination of functions (e.g., ATP and oxidant generation). This unique feature of mitochondrial functional biology potentially makes the mitochondrion a central focal point in terms of the mechanistic basis of many forms of age-related diseases.

Acknowledgements This work was supported by National Institutes of Health grants ES11172, HL77419 and DK79626.

References

1. American Heart Association. Heart Disease and Stroke Statistics-2006 Update. Circulation 2006; 105:1–67.
2. Yang Z, Knight CA, Mamerow M, Vickers K, Penn A, Postlethwait E et al. Prenatal environmental tobacco smoke exposure promotes adult atherogenesis and mitochondrial damage in apoE$^{-/-}$ mice fed a chow diet. Circulation 110, 3715–3720. 2004.
3. Napoli C, De Nigris F, Welch JS et al. Maternal hypercholesterolemia during pregnancy promotes early atherogenesis in LDL receptor-deficient mice and alters aortic gene expression determined by microarray. Circulation 2002; 105(11):1360–1367.

4. Ignarro LJ, Cirino G, Casini A, Napoli C. Nitric oxide as a signaling molecule in the vascular system: an overview. J Cardiovasc Pharmacol 1999; 34(6):879–886.

5. Napoli C, D'Armiento FP, Mancini FP et al. Fatty streak formation occurs in human fetal aortas and is greatly enhanced by maternal hypercholesterolemia: intimal accumulation of low density lipoprotein and its oxidation precede monocyte recruitment into early atherosclerotic lesions. J Clin Invest 1997; 100(11):2680–2690.

6. Napoli C, Glass CK, Witztum JL, Deutsch R, D'Armiento FP, Palinski W. Influence of maternal hypercholesterolaemia during pregnancy on progression of early atherosclerotic lesions in childhood: fate of early lesions in children (FELIC) study. Lancet 1999; 354(9186):1234–1241.

7. Napoli C, Palinski W. Maternal hypercholesterolemia during pregnancy influences the later development of atherosclerosis: clinical and pathogenic implications. Eur Heart J 2001; 22(1):4–9.

8. Palinski W, Napoli C. The fetal origins of atherosclerosis: maternal hypercholesterolemia, and cholesterol-lowering or antioxidant treatment during pregnancy influence in utero programming and postnatal susceptibility to atherogenesis. FASEB J 2002; 16(11):1348–1360.

9. Oral H, Sivasubramanian N, Dyke DB, Mehta RH, Grossman PM, Briesmiester K et al. Myocardial proinflammatory cytokine expression and left ventricular remodeling in patients with chronic mitral regurgitation. Circulation 2003; 107(6):831–837.

10. Tutar E, Kapadia SR, Ziada KM, Hobbs RE, L'Allier PL, Rincon G et al. Coronary atherosclerosis begins at young age: intravascular ultrasound evidence of disease in individuals <20 years old. Circulation 1999; 100(18):524.

11. Tuzcu EM, Kapadia SR, Tutar E et al. High prevalence of coronary atherosclerosis in asymptomatic teenagers and young adults: evidence from intravascular ultrasound. Circulation 2001; 103(22):2705–2710.

12. Dollery CM, Libby P. Atherosclerosis and proteinase activation. Cardiovasc Rese 2006; 69(3):625–635.

13. Libby P, Theroux P. Pathophysiology of coronary artery disease. Circulation 2005; 111(25):3481–3488.

14. Libby P. Inflammation and cardiovascular disease mechanisms. Am J Clin Nutr 2006; 83(2):456S 460S.

15. Libby P. Inflammatory components of the atherosclerotic plaque. Atheroscler Suppl 2006; 7(3):168.

16. Glagov S, Vito R, Giddens DP, Zarins CK. Microarchitecture and composition of artery walls: relationship to location, diameter and the distribution of mechanical-stress. J Hypertens 1992; 10:S101–S104.

17. Glagov S. Intimal hyperplasia, vascular modeling, and the restenosis problem. Circulation 1994; 89(6):2888–2891.

18. Glagov S, Bassiouny HS, Giddens DP, Zarins CK. Pathobiology of plaque modeling and complication. Surg Clin North Am 1995; 75(4):545–556.

19. Glagov S, Bassiouny HS, Sakaguchi Y, Goudet CA, Vito RP. Mechanical determinants of plaque modeling, remodeling and disruption. Atherosclerosis 1997; 131:S13–S14.

20. Glagov S, Weisenberg E, Zarins C, Stankunavicius R, Kolletis G. Compensatory enlargement of human atherosclerotic coronary arteries. N Engl J Med 1987; 316:371–375.

21. Hackett D, Davies G, Maseri A. Pre-existing coronary stenosis in patients with first myocardial infarction are not necessarily severe. Eur Heart J 1988; 9:1317–1323.

22. Rioufol G, Finet G, Andre-Fouet X et al. Multiple atherosclerotic plaque rupture in acute coronary syndromes. Arch Mal Coeur Vaiss 2002; 95(3):157–165.

23. Rioufol G, Finet G, Ginon I et al. Multiple atherosclerotic plaque rupture in acute coronary syndrome: a three-vessel intravascular ultrasound study. Circulation 2002; 106(7):804–808.

24. Rioufol G, Gilard M, Finet G. Upstream or downstream longitudinal patterns of ruptured coronary plaques by IVUS. Eur Heart J 2005; 26:715.

25. Schoenhagen P, Stone GW, Nissen SE et al. Coronary plaque morphology and frequency of ulceration distant from culprit lesions in patients with unstable and stable presentation. Arterioscler Thromb Vasc Biol 2003; 23(10):1895–1900.

26. Kotani J, Mintz GS, Castagna MT, Pinnow E, Berzingi CO, Bui AB et al. Intravascular ultrasound analysis of infarct-related and non-infarct-related arteries in patients who presented with an acute myocardial infarction. Circulation 2003; 107(23):2889–2893.
27. Hong MK, Mintz GS, Lee CW et al. Comparison of coronary plaque rupture between stable angina and acute myocardial infarction: a three-vessel intravascular ultrasound study in 235 patients. Circulation 2004; 110(8):928-933.
28. Tanaka A, Shimada K, Sano T et al. Multiple plaque rupture and c-reactive protein in acute myocardial infarction. J Am Coll Cardiol 2005; 45(10):1594–1599.
29. Carpenter KL, Taylor SE, van der Veen C, Mitchinson MJ. Evidence of lipid oxidation in pulmonary artery atherosclerosis. Atherosclerosis 1995; 118:169–172.
30. Halliwell B. Free radicals, reactive oxygen species and human disease: a critical evaluation with special reference to atherosclerosis. Br J Exp Pathol 1989; 70:737–757.
31. Berliner JA, Heinecke JW. The role of oxidized lipoproteins in atherogenesis. Free Radic Biol Med 1996; 20:707–727.
32. Massaeli H, Pierce GN. Involvement of lipoproteins, free radicals, and calcium in cardiovascular disease processes. Cardiovasc Res 1995; 29:597–603.
33. Freeman BA, White CR, Gutierrez H, Paler-Martinez A, Tarpey MM, Rubbo H. Oxygen radical-nitric oxide reactions in vascular diseases. Adv Pharmacol 1995; 34:45–69.
34. Steinberg D, Parthasarathy S, Carew TE, Khoo JC, Witztum JL. Beyond cholesterol. Modifications of low-density lipoprotein that increase its atherogenicity. N Engl J Med 1989; 321:1196–1197.
35. Witztum JL, Steinberg D. Role of oxidized low density lipoprotein in atherogenesis. J Clin Invest 1991; 88:1785–1792.
36. Reid V, Mitchinson MJ, Skepper J. Cytotoxicity of oxidised low density lipoprotein to mouse peritoneal macrophages: an ultrastructural study. J Pathol 1993; 171:321–328.
37. Holland JA, Ziegler LM, Meyer JW. Atherogenic levels of low-density lipoprotein increase hydrogen peroxide generation in cultured human endothelial cells: Possible mechanism of heightened endocytosis. J Cell Physiol 1996; 166:144–151.
38. Ohara Y, Peterson TE, Harrison DG. Hypercholesterolemia increases endothelial superoxide anion production. J Clin Invest 1993; 91:2546–2551.
39. Diaz MN, Frei B, Vita JA, Keaney JF. Mechanisms of disease: antioxidants and atherosclerotic heart disease. N Engl J Med 1997; 337(6):408–416.
40. Cai H, Harrison DG. Endothelial dysfunction in cardiovascular diseases: the role of oxidant stress. Circ Res 2000; 87(10):840–844.
41. Harrison D, Griendling KK, Landmesser U, Hornig B, Drexler H. Role of oxidative stress in atherosclerosis. Am J Cardiol 2003; 91(3):7A–11A.
42. Dhalla NS, Temsah RM, Netticadan T. Role of oxidative stress in cardiovascular diseases. J Hypertens 2000; 18(6):655–673.
43. Iuliano L. The oxidant stress hypothesis of atherogenesis. Lipids 2001; 36:S41–S44.
44. Keith M, Geranmayegan A, Sole MJ, Kurian R et al. Increased oxidative stress in patients with congestive heart failure. J Am Coll Cardiol 1998; 31:1352–1356.
45. Holvoet P, Perez G, Zhao Z, Brouwers E, Bernar H, Collen A. Malondialdehyde-modified low density lipoproteins in patients with atherosclerotic disease. J Clin Invest 1995; 95:2611–2619.
46. Schmidt AM, Hori O, Chen JX, Li JF, Crandall J, Zhang J et al. Advanced glycation endproducts interacting with their endothelial receptor induce expression of vascular cell adhesion molecule-1 (VCAM-1) in cultured human endothelial cells and mice. A potential mechanism for the accelerated vasculopathy of diabetes. J Clin Invest 1995; 96:1395–1403.
47. Hazen SL, Heinecke JW. 3-chlorotyrosine, a specific marker of myeloperosidase-catalyzed oxidation, is markedly elevated in low density lipoprotein isolated from human atherosclerotic intima. J Clin Invest 1997; 99:2075–2081.
48. Radi R, Rodriguez M, Castro L, Telleri R. Inhibition of mitochondrial electron transport by peroxynitrite. Arch Biochem Biophys 1994; 308:89–95.
49. Radi R, Denicola A, Freeman BA. Peroxynitrite reactions with carbon dioxide-bicarbonate. Methods Enzymol 1999; 301:353–367.

50. Ischiropoulos H. Biological tyrosine nitration: a pathophysiological function of nitric oxide and reactive oxygen species. Arch Biochem Biophys 1998; 356:1–11.
51. Minetti M, Scorza G, Pietraforte D. Peroxynitrite induces long lived tyrosyl radicals in oxyhemoglobin of red blood cells through a reaction involving CO_2 and a ferryl species. Biochemistry 1999; 38:2078–2087.
52. van Jaarsveld H, Kuyl JM, Alberts DW. Exposure of rats to low concentrations of cigarette smoke increases myocardial sensitivity to ischaemia/reperfusion. Basic Res Cardiol 1992; 87:393–399.
53. Alexander RW. Atherosclerosis as disease of redox-sensitive genes. Trans Am Clin Climatol Assoc 1998; 109:129–145.
54. Alexander RW. Hypertension and the pathogenesis of atherosclerosis: oxidative stress and the mediation of arterial inflammatory response: a new perspective. Hypertension 1995; 25:155–161.
55. Berliner JA, Navab M, Fogelman AM, Frank JS, Demer LL, Edwards PA et al. Atherosclerosis: basic mechanisms. Oxidation, inflammation, and genetics. Circulation 1995; 91:2488–2496.
56. Parthasarathy S, Rankin SM. Role of oxidized low density lipoprotein in atherogenesis. Prog Lipid Res 1992; 31:127–143.
57. Taylor AE, Johnson DC, Kazemi H. Environmental tobacco smoke and cardiovascular disease. Circulation 1992; 86:1–4.
58. Glantz S, Parmley W. Passive smoking and heart disease; epidemiology, physiology, and biochemistry. Circulation 1991; 83:1–12.
59. Glantz S, Parmley W. Passive smoking and heart disease. J Am Med Assoc 1995; 273:1047–1053.
60. Steenland K. Passive smoking and the risk of heart disease. J Am Med Assoc 1992; 267:94–99.
61. Steenland K, Thun M, Lally C, Heath C. Environmental tobacco smoke and coronary heart disease in the American Cancer Society CPS-II Cohort. Circulation 1996; 94:622–628.
62. Reilly M, Delanty N, Lawson JA, Fitzgerald GA. Modulation of oxidant stress in vivo in chronic cigarette smokers. Circulation 1996; 94:19–25.
63. Reilly M, Pratico D, Delanty N et al. Increased formation of distinct F_2 isoprostanes in hypercholesterolemia. Circulation 1998; 98:2822–2828.
64. Pratico D, Tangirala RK, Rader DJ, Rokach J, FitzGerald GA. Vitamin E suppresses isoprostane generation in vivo and reduces atherosclerosis in apoE deficient mice. Nat Med 1998; 4(10):1189–1192.
65. Pratico D, Barry OP, Lawson JA, Adiyaman M, Hwang S-W, Khanapure SP et al. IPF_{2a-I}: an index of lipid peroxidation in humans. Proc Natl Acad Sci U S A 1998; 95:3449–3454.
66. Ito H, Torii M, Suzuki T. Decreased superoxide dismutase activity and increased superoxide anion production in cardiac hypertrophy of spontaneously hypertensive rats. Clin Exp Hypertens 1995; 17:803–816.
67. Watts GF, Playford DA. Dyslipoproteinaemia and hyperoxidative stress in the pathogenesis of endothelial dysfunction in a non-insulin dependent diabetes mellitus: an hypothesis. Atherosclerosis 1998; 141:17–30.
68. Rousselot DB, Bastard JP, Jaudon MC. Consequences of the diabetic status on the oxidant/antioxidant balance. Diabetes Metab 2000; 26:163–176.
69. Wallace D. Mitochondrial diseases in man and mouse. Science 1999; 283:1482–1488.
70. Wallace DC. A mitochondrial paradigm for degenerative diseases and ageing. Novartis Found Symp 2001; 235:247–263.
71. Wallace DC. The mitochondrial genome in human adaptive radiation and disease: On the road to therapeutics and performance enhancement. Gene 2005; 354:169–180.
72. Wallace DC. A mitochondrial paradigm of metabolic and degenerative diseases, aging, and cancer: a dawn for evolutionary medicine. Annu Rev Genet 2005; 39:359–407.
73. Ballinger SW. Mitochondrial dysfunction in cardiovascular disease. Free Radic Biol Med 2005; 38(10):1278–1295.
74. Ballinger SW, Shoffner JM, Wallace DC. Mitochondrial myopathies–genetic-aspects. Curr Top Bioenerg 1994; 17:59–98.

75. Wallace DC, Shoffner JM, Trounce I et al. Mitochondrial-DNA mutations in human degenerative diseases and aging. Biochim Biophys Acta 1995; 1271(1):141–151.
76. Quintero M, Colombo SL, Godfrey A, Moncada S. Mitochondria as signaling organelles in the vascular endothelium. Proc Natl Acad Sci U S A 2006; 103(14):5379–5384.
77. Zhang Y, Marcillat O, Giulivi C, Ernster L, Davies KJ. The oxidative inactivation of mitochondrial electron transport chain components and ATPase. J Biol Chem 1990; 265:16330–16336.
78. Turrens JF. Mitochondrial formation of reactive oxygen species. J Physiol 2003; 552(2):335–344.
79. Lambert A, Brand M. Superoxide production by NADH:ubiquinone oxidoreductase (complex I) depends on the pH gradient across the mitochondrial inner membrane. Biochem J 2004.
80. Liu Y, Fiskum G, Schubert D. Generation of reactive oxygen species by the mitochondrial electron transport chain. J Neurochem 2002; 80:780–787.
81. Han D, Canali R, Rettori D, Kaplowitz N. Effect of glutathione depletion on sites and topology of superoxide and hydrogen peroxide production in mitochondria. Mol Pharmacol 2003; 64(5):1136–1144.
82. Forman HJ, Boveris A. In: Prior WA, ed. Free radicals in biology. Orlando, Florida: Academic Press, 1982: 65–90.
83. Forman HJ, Boveris A. Superoxide radical and hydrogen peroxide in mitochondria. In: Prior WA, ed. Free radicals in biology. Orlando, Florida: Academic Press, 1982: 65–90.
84. Boveris A, Turrens JF. Production of superoxide anion by the NADH dehydrogenase of mammalian mitochondria. In: Bannister J, Hill H, eds. Chemical and biochemical aspects of superoxide and superoxide dismutase. Amsterdam: Elsevier, 1980: 84–91.
85. Turrens JF, Boveris A. Generation of superoxide anion by the NADH dehydrogenase of bovine heart mitochondria. Biochem J 1980; 191:421–427.
86. Chance B, Sies H, Boveris A. Hydroperoxide metabolism in mammalian organs. Physiol Rev 1979; 59:527–605.
87. Ide T, Tsutsui H, Kinugawa S et al. Direct evidence for increased hydroxyl radicals orginating from superoxide in the failing myocardium. Circ Res 2000; 86:152–157.
88. Baas AS, Berk BC. Differential activation of mitogen-activated protein kinases by H_2O_2 and $O_2^{\cdot-}$ in vascular smooth muscle cells. Circ Res 1995; 77(1):29–36.
89. Chen K, Vita JA, Berk BC, Keaney JF Jr. c-Jun-N-terminal kinase activation by hydrogen peroxide in endothelial cells involves Src-dependent epidermal growth factor receptor trans-activation. J Biol Chem 2001; 276(19):16045 16050.
90. Guyton KZ, Liu Y, Gorospe M, Xu Q, Holbrook NJ. Activation of mitogen-activated protein kinase by H_2O_2. J Biol Chem 1996; 271:4138–4142.
91. Kim SM, Byun JS, Jung YD et al. The effects of oxygen radicals on the activity of nitric oxide synthase and guanylate cyclase. E. Exp Mol Med 1998; 30(4):221–6.
92. Sundaresan M, Yu ZX, Fererans VJ, Irani K, Finkel T. Requirement for generation of H2O2 for platelet-derived growth factor signal transduction. Science 1995; 270:296–299.
93. Schulze-Osthoff K, Los M, Baeuerle PA. Redox signaling by transcription factors NF-kB and AP-1 in lymphocytes. Biochem Pharmacol 1995; 50:735–741.
94. Devary Y, Rosette C, DiDonato JA, Karin M. NF-κB activation by ultraviolet light not dependent on a nuclear signal. Science 1993; 261:1442–1445.
95. Mohan N, Meltz MM. Induction of nuclear factor κB after low-dose ionizing radiation involves a reactive oxygen intermediate signaling pathway. Radic Res 1994; 140:97–104.
96. Schulze-Osthoff K, Beyaert R, Vandervoorde V, Haegeman G, Fiers W. Depletion of the mitochondrial electron transport abrogates the cytotoxic and gene-inductive effects of TNF. EMBO J 1993; 12:3095–3104.
97. Chen K, Thomas SR, Albano A, Murphy MP, Keaney JF. Mitochondrial function is required for hydrogen peroxide-induced growth factor receptor transactivation and downstream signaling. J Biol Chem 2004; 279:35079–35086.

98. Manna SK. Overexpression of manganese superoxide dismutase suppresses tumor necrosis factor-induced apoptosis and activation of nuclear transcription factor-kappaB and activated protein-1. J Biol Chem 1998; 273:13245–13254.

99. Schulze-Osthoff K, Bakker AC, Vanhaesebroeck B, Beyaert R, Jacob WA, Fiers W. Cytotoxic activity of tumor necrosis factor is mediated by early damage of mitochondrial functions. Evidence for the involvement of mitochondrial radical generation. J Biol Chem 1992; 267(8):5317–5323.

100. Maehara K, Oh-Hashi K, Isobe K-I. Early growth-responsive-1-dependent manganese superoxide dismutase gene transcription mediated by platelet-derived growth factor. FASEB J 2001; 15(11):2025–2026.

101. Clementi E, Brown G, Foxwell N, Moncada S. On the mechanism by which vascular endothelial cells regulate their oxygen consumption. Proc Natl Acad Sci U S A 1999; 96:1559–1562.

102. Moncada S, Erusalimsky JD. Does nitric oxide modulate mitochondrial energy generation and apoptosis? Nat Rev Mol Cell Biol 2002; 3(3):214–220.

103. Cassina A, Radi R. Differential inhibitory action of nitric oxide and peroxynitrite on mitochondrial electron transport. Arch Biochem Biophys 1996; 328:309–316.

104. Ramachandran A, Levonen AL, Brookes PS et al. Mitochondria, nitric oxide, and cardiovascular dysfunction. Free Radic Biol Med 2002; 33(11):1465–1474.

105. Ramachandran A, Moellering DR, Ceaser E, Shiva S, Xu J, Darley-Usmar VM. Inhibition of mitochondrial protein synthesis results in increased endothelial cell susceptibility to nitric oxide-induced apoptosis. Proc Natl Acad Sci U S A 2002; 99(10):6643–6648.

106. Kissner R, Nauser T, Bugnon P, Lye PG, Loppenol WH. Formation and properties of peroxynitrite as studied by laser flash photolysis, high-pressure stopped-flow technique, and pulse radiolysis. Chem Res Toxicol 1997; 10:1285–1292.

107. Ramachandran A, Ceaser E, Darley-Usmar VM. Chronic exposure to nitric oxide alters the free iron pool in endothelial cells: role of mitochondrial respiratory complexes and heat shock proteins. Proc Natl Acad Sci U S A 2004; 101:384–389.

108. Beckman JS. Oxidative damage and tyrosine nitration from peroxynitrite. Chem Res Toxicol 1996; 9:836–844.

109. Garlid KD, Jaburek M, Jezek P. Mechanism of uncoupling proteins action. Biochem Soc Trans 2001; 29:803–806.

110. Jaburek M, Varecha M, Gimeno RE et al. Transport function and regulation of mitochondrial uncoupling proteins 2 and 3. J Biol Chem 1999; 274:26003–26007.

111. Klingenberg M, Winkler E, Echtay K. Uncoupling protein, H+ transport and regulation. Biochem Soc Trans 2001; 29:806–811.

112. Nedergaard J, Golozoubova V, Matthias A, Shabalina I, Ohba K, Ohlson K et al. Life without UCP1: mitochondrial, cellular, and organismal characteristics of the UCP1-ablated mice. Biochem Soc Trans 2001; 29:756–763.

113. Stuart JA, Cadenas S, Jacobsons MB, Roussel D, Brand MD. Mitochondrial proton leak and the uncoupling protein 1 homologues. Biochim Biophys Acta 2001; 1504:144–158.

114. Dulloo AG, Samec S, Seydoux J. Uncoupling protein 3 and fatty acid metabolism. Biochem Soc Trans 2001; 29:785–791.

115. Nedergaard J, Cannon B. The "novel" "uncoupling" proteins UCP2 and UCP3: what do they really do? Pros and cons for suggested functions. Exp Physiol 2003; 88:65–84.

116. Talbot DA, Lambert AJ, Brand MD. Production of endogenous matrix superoxide from mitochondrial complex I leads to activation of uncoupling protein 3. FEBS Lett 2004; 556:111–115.

117. Yan L-J, Sohal RS. Mitochondrial adenine nucleotide translocase is modified oxidatively during aging. Proc Natl Acad Sci U S A 1998; 95:12896–12901.

118. Yakes FM, Van Houten B. Mitochondrial DNA damage is more extensive and persists longer than nuclear DNA damage in human cells following oxidative stress. Proc Natl Acad Sci USA 1997; 94:514–519.

119. MacMillan-Crow LA, Crow JP, Thompson JA. Peroxynitrite-mediated inactivation of manganese superoxide dismutase involves nitration and oxidation of critical gyrosine residues. Biochemistry 1998; 37:1613–1622.

120. Duan J, Karmazyn M. Relationship between oxidative phosphorylation and adenine nucleotide translocase activity in two populations of cardiac mitochondria and mechanical recovery of ischemic hearts following reperfusion. Can J Physiol Pharmacol 1989; 67:704–709.

121. Ballinger SW, Patterson WC, Yan C-N, Doan R, Burow DL, Young CG et al. Hydrogen peroxide and peroxynitrite induced mitochondrial DNA damage and dysfunction in vascular endothelial and smooth muscle cells. Circ Res 2000; 86:960–966.

122. Ballinger SW, Patterson C, Knight-Lozano CA, Burow DL, Conklin CA, Hu Z et al. Mitochondrial integrity and function in atherogenesis. Circulation 106, 544–549. 2002.

123. Knight-Lozano CA, Young CG, Burow DL et al. Cigarette smoke exposure and hypercholesterolemia increase mitochondrial damage in cardiovascular tissues. Circulation 105, 849–854. 2002.

124. Corral-Debrinski M, Stepien G, Shoffner JM, Lott MT, Kanter K, Wallace DC. Hypoxemia is associated with mitochondrial DNA damage and gene induction: implications for cardiac disease. J Am Med Assoc 1991; 266:1812–1816.

125. Corral-Debrinski M, Shoffner JM, Lott MT, Wallace DC. Association of mitochondrial DNA damage with aging and coronary atherosclerotic heart disease. Mutat Res 1992; 275:169–180.

126. Aliev G, Seyidova D, Neal ML et al. Atherosclerotic lesions and mitochondria DNA deletions in brain microvessels as a central target for the development of human AD and AD-like pathology in aged transgenic mice. Ann NY Acad Sci 2002; 977:45–64.

127. Brookes PS, Zhang J, Dai L et al. Increased sensitivity of mitochondrial respiration to inhibition by nitric oxide in cardiac hypertrophy. J Mol Cell Cardiol 2001; 33(1):69–82.

128. Blanc J, Alves-Guerra MC, Esposito B, Rousset S, Gourdy P, Ricquier D et al. Protective role of uncoupling protein 2 in atherosclerosis. Circulation 2003; 107:388–390.

129. Asimakis G, Lick S, Patterson W. Postischemic recovery of contractile function is impaired in SOD2$^{+/-}$ but not SOD1$^{+/-}$ mouse hearts. Circulation 2002; 105(8):981–986.

130. Chen Z, Siu B, Ho Y-S, Vincent R, Chua CC, Hamdy RC et al. Overexpression of MnSOD protects against myocardial ischemia/reperfusion injury in transgenic mice. J Mol Cell Cardiol 1998; 30:2281–2289.

131. Teshima Y, Akao M, Jones SP, Marban E. Uncoupling protein-2 overexpression inhibits mitochondrial death pathway in cardiomyocytes. Circ Res 2003; 93:192–200.

132. Bienengraeber M, Ozcan C, Terzic A. Stable transfection of UCP1 confers resistance to hypoxia/reoxygenation in a heart-derived cell line. J Mol Cell Cardiol 2003; 35:861–865.

133. Ide T, Tsutsui H, Hayashidani S et al. Mitochondrial DNA damage and dysfunction associated with oxidative stress in failing hearts after myocardial infarction. Circ Res 2001; 88:529–535.

134. Bernal-Mizrachi C, Gates AC, Weng S et al. Vascular respiratory uncoupling increases blood pressure and atherosclerosis. Nature 2005; 435(7041):502–506.

135. Bernal-Mizrachi C, Weng S, Li B et al. Respiratory uncoupling lowers blood pressure through a leptin-dependent mechanism in genetically obese mice. Arterioscler Thromb Vasc Biol 2002; 22(6):961–968.

136. Yang H, Roberts LJ, Shi MJ et al. Retardation of atherosclerosis by overexpression of catalase or both Cu/Zn-superoxide dismutase and catalase in mice lacking apolipoprotein E. Circ Res 2004; 95(11):1075–1081.

137. Ceaser E, Ramachandran A, Levonen AL, Darley-Usmar VM. Oxidized low-density lipoprotein and 15-deoxy-delta12,14-PGJ2 increase mitochondrial complex I activity in endothelial cells. Am J Physiol 2003; 285:H2298–H2308.

138. Kinscherf R, Deigner H-P, Usinger C et al. Induction of mitochondrial manganese superoxide dismutase in macrophages by oxidized LDL: its relevance in atherosclerosis of humans and heritable hyperlipidemic rabbits. FASEB J 1997; 11:1317–1328.

139. Asmis R, Begley JG. Oxidized LDL promotes peroxide mediated mitochondrial dysfunction and cell death in human macrophages. Circ Res 2003; 92:e20–e29.

140. Mabile L, Meilhac O, Escargueil-Blanc I, Troly M, Pieraggi M-T, Salvayre R et al. Mitochondrial function is involved in LDL oxidation mediated by human cultured endothelial cells. Arterioscler Thromb Vasc Biol 1997; 17:1575–1582.

141. Mikaelian NP, Khalilov EM, Ivanov AS, Fortinskaia ES, Lopukhin IM. Mitochondrial enzymes in circulating lymphocytes during hemosorption for experimental hypercholesterolemia. Biull Eksp Biol Med 1983; 96(9):35–37.

142. Nishikawa T, Edelstein D, Du XL, Yamagishi S, Matsumura T, Kaneda Y et al. Normalizing mitochondrial superoxide production blocks three pathways of hyperglycaemic damage. Nature 2000; 404(6779):787–790.

143. Nishikawa T, Edelstein D, Brownlee M. The missing link: a single unifying mechanism for diabetic complications. Kidney Int 2000; 58:26–30.

144. Vega-Lopez S, Devaraj S, Jialal I. Oxidative stress and antioxidant supplementation in the management of diabetic cardiovascular disease. J Invest Med 2004; 52(1):24–32.

145. Venugopal SK, Devaraj S, Yang T, Jialal I. alpha-Tocopherol decreases superoxide anion release in human monocytes under hyperglycemic conditions via inhibition of protein kinase C-alpha. Diabetes 2002; 51(10):3049–3054.

146. Venugopal SK, Devaraj S, Yang TTC. Alpha tocopherol decreases superoxide anion release in THP-1 cells under hyperglycemic conditions through inhibition of PKC-a. Diabetes 2002; 51:3049–3054.

147. Azevedo-Martins AK, Lortz S, Lenzen S, Curi R, Eizirik DL, Tiedge M. Improvement of the mitochondrial antioxidant defense status prevents cytokine-induced nuclear factor-kB activation in insulin-producing cells. Diabetes 2003; 52:93–101.

148. Nyholm B, Fisker S, Lund S, Moller N, Schmitz O. Increased circulating leptin concentrations in insulin-resistant first-degree relatives of patients with non-insulin-dependent diabetes mellitus: relationship to body composition and insulin sensitivity but not to family history of non-insulin-dependent diabetes mellitus. Eur J Endocrinol 1997; 136(2):173–179.

149. Rudberg S, Persson B. Serum leptin levels in young females with insulin-dependent diabetes and the relationship to hyperandrogenicity and microalbuminuria. Horm Res 1998; 50(6):297–+.

150. Maffei M, Halaas J, Ravussin E et al. Leptin levels in human and rodent: measurement of plasma leptin and Ob RNA in obese and weight reduced subjects. Nat Med 1995; 1(11):1155–1161.

151. Yamagishi S, Edelstein D, Du XL, Kaneda Y, Guzman M, Brownlee M. Leptin induces mitochondrial superoxide production and monocyte chemoattractant protein-1 expression in aortic endothelial cells by increasing fatty acid oxidation via protein kinase A. J Biol Chem 2001; 276(27):25096–25100.

152. Halaas JL, Gajiwala KS, Maffei M, Cohen SL, Chait BT, Rabinowitz D et al. Weight-reducing effects of the plasma-protein encoded by the Obese gene. Science 1995; 269(5223):543–546.

153. Pelleymounter MA, Cullen MJ, Baker MB et al. Effects of the Obese gene-product on bodyweight regulation in Ob/Ob mice. Science 1995; 269(5223):540–543.

154. Sierra-Honigmann MR, Nath AK, Murakami C, Garcia-Cardena G, Papapetropoulos A, Sessa WC et al. Biological action of leptin as an angiogenic factor. Science 1998; 281(5383):1683–1686.

155. Bouloumie A, Marumo T, Lafontan M, Busse R. Leptin induces oxidative stress in human endothelial cells. FASEB J 1999; 13(10):1231–1238.

156. Pueyo ME, Gonzales W, Nicoletti A, Savoie F, Arnal J-F, Michel J-B. Angiotensin II stimulates endothelial vascular cell adhesion molecule-1 via nuclear transcription factor-kB activation induced by intracellular oxidative stress. Arterioscler Thromb Vasc Biol 2000; 20:645–651.

157. American Heart Association. Heart disease and stroke statistics-2005 Update. 1a-60. 2005. Dallas, TX, American Heart Association.

158. Gvozdjak J, Gvozdjakova A, Kucharska J, Bada V. The effect of smoking on myocardial metabolism. Czech Med 1987; 10:47–53.

159. Gvozdjakova A, Kucharska J, Gvozdjak J. Effect of smoking on the oxidative processes of cardiomyocytes. Cardiology 1992; 1992:81–84.
160. van Jaarsveld H, Kuyl JM, Alberts DW. Antioxidant vitamin supplementation of smoke exposed rats partially protects against myocardial ischemic/reperfusion injury. Free Radic Res Commun 1992; 17:263–269.
161. Davis J, Shelton L, Watnabe I, Arnold J. Passive smoking affects endothelium and platelets. Arch Intern Med 1989; 149:386–389.
162. Lu KP, Alejandro NF, Taylor KM, Joyce MM, Spencer TE, Ramos KS. Differential expression of ribosomal L31, Zis, gas-5 and mitochondrial mRNAs following oxidant induction of proliferative vascular smooth muscle cell phenotypes. Atherosclerosis 2002; 160:273–280.
163. Willett WC. Balancing life-style and genomics research for disease prevention. Science 2002; 296(5568):695–698.
164. Connelly JJ, Wang T, Cox JE et al. GATA2 is associated with familial early-onset coronary artery disease. PloS Genet 2006; 2(8):1265–1273.
165. Humphries SE, Talmud PJ, Hawe E, Bolla M, Day INM, Miller GJ. Apolipoprotein E4 and coronary heart disease in middle-aged men who smoke: a prospective study. Lancet 2001; 358(9276):115–119.
166. Talmud PJ, Humphries SE. Gene: environment interaction in lipid metabolism and effect on coronary heart disease risk. Curr Opin Lipidol 2002; 13(2):149–154.
167. Keavney B, Parish S, Palmer A et al. Large-scale evidence that the cardiotoxicity of smoking is not significantly modified by the apolipoprotein E epsilon 2/epsilon 3/epsilon 4 genotype. Lancet 2003; 361(9355):396–398.
168. Wheeler JG, Keavney BD, Watkins H, Collins R, Danesh J. Four paraoxonase gene polymorphisms in 11,212 cases of coronary heart disease and 12,786 controls: meta-analysis of 43 studies. Lancet 2004; 363(9410):689–695.
169. Ishida BY, Blanche PJ, Nichols AV, Yashar M, Paigen B. Effects of atherogenic diet consumption on lipoproteins in mouse strains C57Bl/6 and C3H. J Lipid Res 1991; 32(4):559–568.
170. Wang XS, Ria M, Kelmenson PM et al. Positional identification of TNFSF4, encoding OX40 ligand, as a gene that influences atherosclerosis susceptibility. Nat Genet 2005; 37(4):365–372.
171. van Schooten FJ, Hirvonen A, Maas LM et al. Putative susceptibility markers of coronary artery disease: association between VDR genotype, smoking, and aromatic DNA adduct levels in human right atrial tissue. FASEB J 1998; 12(13):1409–1417.
172. Stephens JW, Humphries SE. The molecular genetics of cardiovascular disease: clinical implications. J Int Med 2003; 253(2):120–127.
173. Tabibiazar R, Wagner RA, Spin JM, Ashley EA, Narasimhan B, Rubin EM et al. Mouse strain-specific differences in vascular wall gene expression and their relationship to vascular disease. Arterioscler Thromb Vasc Biol 2005; 25(2):302–308.
174. Tabibiazar R, Wagner RA, Ashley EA et al. Signature patterns of gene expression in mouse atherosclerosis and their correlation to human coronary disease. Physiol Genomics 2005; 22(2):213–226.
175. Ruiz-Pesini E, Mishmar D, Brandon M, Procaccio V, Wallace DC. Effects of purifying and adaptive selection on regional variation in human mtDNA. Science 2004; 303:223–226.
176. Ballinger SW, Shoffner JM, Hedaya EV et al. Maternally transmitted diabetes and deafness associated with a 10.4 kb mitochondrial DNA deletion. Nat Genet 1992; 1:11–15.
177. Corral-Debrinski M, Horton T, Lott MT, Shoffner JM, Beal MF, Wallace DC. Mitochondrial DNA deletions in human brain: regional variability and increase with advanced age. Nat Genet 1992; 2(4):324–329.
178. Shoffner JM, Lott MT, Lezza AM, Seibel P, Ballinger SW, Wallace DC. Myoclonic epilepsy and ragged-red fiber disease (MERRF) is associated with a mitochondrial DNA tRNA(Lys) mutation. Cell 1990; 61(6):931–937.
179. Wallace DC, Singh G, Lott MT et al. Mitochondrial DNA mutation associated with Leber's hereditary optic neuropathy. Science 1988; 242(4884):1427–1430.

180. Wallace DC, Zheng XX, Lott MT et al. Familial mitochondrial encephalomyopathy (MERRF): genetic, pathophysiological, and biochemical characterization of a mitochondrial DNA disease. Cell 1988; 55(4):601–610.
181. Brown MD, Sun FZ, Wallace DC. Clustering of Caucasian Leber hereditary optic neuropathy patients containing the 11778 or 14484 mutations on an mtDNA lineage. Am J Hum Genet 1997; 60(2):381–387.
182. Brown MD, Starikovskaya E, Derbeneva O et al. The role of mtDNA background in disease expression: a new primary LHON mutation associated with Western Eurasian haplogroup J. Hum Genet 2002; 110(2):130–138.
183. Brown MD, Zhadanov S, Allen JC et al. Novel mtDNA mutations and oxidative phosphorylation dysfunction in Russian LHON families. Hum Genet 2001; 109(1):33–39.
184. Brown MD, Derbeneva OA, Starikovskaya Y, Allen J, Sukernik RI, Wallace DC. The influence of mtDNA background on the disease process: a new primary Leber's Hereditary Optic Neuropathy mtDNA mutation requires European haplogroup J for expression. American Journal of Human Genetics 2001; 69(4):578.
185. Petros JA, Baumann AK, Ruiz-Pesini E et al. mtDNA mutations increase tumorigenicity in prostate cancer. Proc Natl Acad Sci U S A 2005; 102(3):719–724.
186. Chagnon P, Gee M, Filion M, Robitaille Y, Belouchi M, Gauvreau D. Phylogenetic analysis of the mitochondrial genome indicates significant differences between patients with Alzheimer disease and controls in a French-Canadian founder population. Am J Med Genet 1999; 85(1):20–30.
187. De Benedictis G, Rose G, Carrieri G, De Luca M, Falcone E, Passarino G et al. Mitochondrial DNA inherited variants are associated with successful aging and longevity in humans. FASEB J 1999; 13(12):1532–1536.
188. Rose G, Passarino G, Carrieri G et al. Paradoxes in longevity: sequence analysis of mtDNA haplogroup J in centenarians. Eur J Hum Genet 2001; 9(9):701–707.
189. van der Walt JM, Nicodemus KK, Martin ER et al. Mitochondrial polymorphisms significantly reduce the risk of Parkinson disease. Am J Hum Genet 2003; 72(4):804–811.
190. van der Walt JM, Dementieva YA, Martin ER et al. Analysis of European mitochondrial haplogroups with Alzheimer disease risk. Neurosci Lett 2004; 365(1):28–32.
191. Johnson KR, Zheng QY, Bykhovskaya Y, Spirina O, Fischel-Ghodsian N. A nuclear-mitochondrial DNA interaction affecting hearing impairment in mice. Nat Genet 2001; 27(2):191–194.

11
Oxidative Stress in Type 2 Diabetes Mellitus

Muhammad A. Abdul-Ghani and Ralph A. DeFronzo

Summary Type 2 diabetes (T2DM) is associated with significant morbidity and mortality, which primarily results from vascular damage. Two major defects contribute to the pathogenesis of T2DM: (1) impaired insulin secretion in response to glucose and other stimuli, i.e., β cell failure; and (2) impaired insulin action in the liver and peripheral (muscle and adipose) tissues, i.e., insulin resistance. In this review, we summarize the molecular mechanisms that contribute to insulin resistance, β cell failure, and vascular damage in T2DM, with emphasis on the contribution of oxidative stress to these mechanisms.

Keywords Type 2 diabetes, β cell, insulin resistance, diabetic complications, oxidative stress.

1 Introduction

The prevalence of type 2 diabetes mellitus has increased progressively to epidemic proportions over the past 20 years. More than 150 million individuals worldwide were affected by the disease in the year 2000, and this number is expected to rise to more than 300 million by the year 2025 [1]. The prevalence of T2DM increases with age, and it is estimated that ~30 million individuals over the age of 65 in developed countries are affected by the disease [1]. Type 2 diabetes is closely associated with obesity, and the epidemic of obesity that we are witnessing has resulted not only in a marked increase in the incidence of diabetes but also in a decrease in the age of the disease onset. Type 2 diabetes, once a rare disease in youth, now is frequently encountered among adolescents and children [2].

Subjects with T2DM have increased morbidity and mortality, the majority of which results from vascular damage, including both microvascular (retinopathy, nephropathy, and neuropathy) and macrovascular (heart attack, stroke, and amputation) complications [3]. Diabetes is the leading cause of blindness and end stage renal disease in westernized countries [4], T2DM individuals have two- to threefold increased risk for cardiovascular disease, e.g., heart attack and stroke. Clinically

From: *Aging Medicine: Oxidative Stress in Aging: From Model Systems to Human Diseases*   191
Edited by: S. Miwa, K.B. Beckman, and F.L. Muller © Humana Press, Totowa, NJ

significant morbidity often develops before the diagnosis of diabetes is made [5]. Between one third to one half of all people with diabetes have evidence for organ or tissue damage [6]. Although not everyone with diabetes is destined to develop complications, a recent epidemiological study [6] reported that two or more complications are apparent in almost one fifth of people with diabetes.

The increased prevalence of T2DM, together with its associated morbidity and mortality, have placed a heavy burden on the health care system. In the United States, for example, the estimated annual cost for the treatment of diabetes exceeds $100 billion [7]. Therefore, a clear understanding of the pathophysiology of T2DM and the development of strategies to prevent the disease and its complications have important public health and economic implications.

2 Normal Glucose Tolerance

In normal glucose tolerant (NGT) subjects, the plasma glucose concentration during the postabsorptive state (e.g., after an overnight fasting) is maintained below 100 mg/dl. The majority (~70–75%) of the total body glucose uptake (~2 mg/kg min) takes place in insulin-insensitive tissues, mainly the brain, erythrocytes, and splanchnic (gut plus liver) tissues, whereas the remaining 20–25% occurs in insulin-sensitive tissues, primarily muscle [8]. The rate of glucose uptake by all tissues in the body during the postabsorptive state is precisely matched by the rate of endogenous glucose production, primarily by the liver [8] and to a smaller extent by the kidney [9].

After glucose ingestion, the increase in plasma glucose concentration stimulates insulin secretion. The combination of hyperglycemia plus hyperinsulinemia combines to suppress hepatic glucose production and to stimulate glucose uptake by splanchnic and peripheral tissues to dispose of the ingested glucose and restore normoglycemia [10–12]. The rise in plasma glucose concentration after glucose ingestion is influenced both by the ability of hyperinsulinemia and hyperglycemia to suppress endogenous glucose production and to augment the disposal of the ingested glucose by peripheral tissues and the liver. Thus, the maintenance of normal glucose tolerance is dependent on adequate secretion of insulin from the β cells in response to nutrient stimuli and on the normal action of insulin in liver and peripheral tissues.

3 Pathophysiology of T2DM

T2DM subjects manifest two major defects [13]: (1) impaired insulin secretion in response to glucose and other stimuli, i.e., β cell failure; and (2) impaired insulin action in the liver and peripheral (muscle and adipose) tissues, i.e., insulin resistance. Both insulin resistance and β cell failure are present long before the onset of overt diabetes. The earliest detectable abnormality in glucose metabolism is an increase in insulin resistance in liver, muscle, and other insulin target tissues. Both genetic and environmental factors, e.g., obesity and sedentary lifestyle, contribute to the development of insulin resistance [14] (see Subsection 3.1). In response to insulin resistance, the β cell increases its secretion of insulin under fasting conditions and in response to nutrient stimuli, and this results in hyperinsulinemia. Longitudinal and cross-sectional studies have demonstrated that increased insulin resistance is accompanied by hyperinsulinemia, which initially is sufficient to offset the insulin resistance and maintain normal glucose tolerance [15–17]. Most overweight and obese individuals are insulin resistant, but the majority, ~70%, maintain normal glucose tolerance throughout life due to "healthy" β cells, which are capable of fully compensating for the insulin resistance. However, when the β cells fail to adequately compensate for the insulin resistance, glucose homeostasis deteriorates. Initially, this is manifest as an impaired glucose tolerance and later as overt diabetes [15].

3.1 Insulin Resistance

Insulin is the principal anabolic hormone in the body. It acts on multiple organs to enhance the storage and/or oxidation of substrates [13, 14, 18]. In the liver insulin inhibits hepatic glucose production and stimulates glycogen synthesis [19]. In skeletal muscle, insulin stimulates glucose uptake by causing the translocation of the GLUT4 glucose transporter from the cytosol to the cell surface membrane and by enhancing glucose oxidation and glycogen synthesis [14, 20]. In adipocytes, insulin stimulates glucose uptake and retards lipolysis by inhibiting hormone-sensitive lipase, thereby restraining the release of free fatty acids (FFA) from adipocytes [21]. The dose-response curve relating plasma insulin concentration and its physiological actions has an ED_{50} of 15–20 µU/ml for suppression of lipolysis in adipocytes, 30–40 µU/ml for suppression of hepatic glucose production, and 70–80 µU/ml for stimulation of glucose uptake in skeletal muscle [21]. During states of insulin resistance, there is an ~2-fold increase in ED_{50} [21]. Initially, the hyperinsulinemia that accompanies the insulin resistance is sufficient to offset the increase in ED_{50} and maintain near normal glucose homeostasis. However, as β cell failure progresses and "relative hypoinsulinemia" ensues, insulin's multiple actions are disturbed. In adipocytes, this results in impaired inhibition of lipolysis and an increase in plasma FFA concentration. In liver, the suppression of hepatic glucose production by insulin is impaired, whereas in muscle glycogen synthesis and glucose oxidation are reduced, leading to diminished glucose uptake and hyperglycemia. Thus, subjects with type 2 diabetes manifest insulin resistance in multiple organs, e.g., liver, skeletal muscle, and adipocytes, which results in hyperglycemia and elevated plasma FFA levels. Elevated plasma FFA levels cause insulin resistance in muscle and liver and impair insulin secretion.

4   The Insulin Receptor

For insulin to exert its biological effects on glucose metabolism, it must first bind
to specific receptors that are present on the cell surface of all insulin target tissues [22].
The insulin receptor is a glycoprotein consisting of two α subunits and two β subunits
linked by disulfide bonds [22, 23]. The α subunit of the insulin receptor is entirely
extracellular and contains the insulin-binding domain. The β subunit has an extracellular domain, a transcellular domain, and an intracellular domain that expresses
insulin-stimulated kinase activity directed toward its own tyrosine residues [23].
Insulin receptor phosphorylation of the β subunit, with subsequent activation of
insulin receptor tyrosine kinase, represents the first step in the action of insulin on
glucose metabolism [22].

5   Insulin Receptor Signal Transduction

After activation, insulin receptor tyrosine kinase phosphorylates specific intracellular proteins, of which at least nine have been identified [22]. Four of these belong
to the family of insulin receptor substrate proteins: IRS-1, IRS-2, IRS-3, and IRS-4
(the others include Shc, Cbl, Gab-1, p60[dok], and APS). In muscle, IRS-1 serves as
the major docking protein that interacts with the insulin receptor tyrosine kinase
and undergoes tyrosine phosphorylation in regions containing amino acid
sequence motifs (YXXM or YMXM) that, when phosphorylated, serve as recognition sites for proteins containing src-homology 2 (SH2) domains [22–24].
Mutation of these specific tyrosine residues severely impairs the ability of insulin
to stimulate glycogen and deoxyribonucleic acid synthesis, establishing the
important role of IRS-1 in insulin signal transduction [25]. In liver, IRS-2 serves
as the primary docking protein that undergoes tyrosine phosphorylation and
mediates the effect of insulin on hepatic glucose production, gluconeogenesis,
and glycogen formation [26].

   In muscle, the phosphorylated tyrosine residues on IRS-1 mediate an association
with the SH2 domains of the p85-kd regulatory subunit of phosphatidylinositol (PI)-
3 kinase, leading to activation of the enzyme [27] PI-3 kinase is a heterodimeric
enzyme composed of a p85-kd regulatory subunit and a p110-kd catalytic subunit.
The latter catalyzes the 3~ phosphorylation of PI, PI-4 phosphate, and PI-4,5 diphos-
phate, resulting in the stimulation of glucose transport [28]. Activation of PI-3
kinase by phosphorylated IRS-1 also leads to activation of glycogen synthase [22]
via a process that involves activation of protein kinase B (PKB)/Akt and subsequent inhibition of kinases such as glycogen synthase kinase 3 [29] and activation of
protein phosphatase 1 [30]. Inhibitors of PI-3 kinase impair glucose transport [31]
by interfering with the translocation of GLUT4 transporters from their intracellular location [23] and block the activation of glycogen synthase [32] and hexokinase II
expression [33].

6 Molecular Mechanism of Insulin Resistance

In insulin resistance states, there is a major defect in the insulin signaling pathway [13]. Reduced insulin-stimulated glucose transport in skeletal muscle is due to a defect in the ability of insulin to activate the insulin receptor and initiate the signal transduction cascade that eventually leads to the mobilization of GLUT4 to the muscle surface membrane [13]. Type 2 diabetic and most obese nondiabetic subjects are insulin resistant. Insulin resistance also can be induced in lean healthy insulin-sensitive individuals by elevating the plasma FFA concentration with lipid infusion. All insulin-resistant states are characterized by a defect in the ability of insulin to increase IRS-1–associated PI 3-kinase activity, and this defect is due to impaired tyrosine phosphorylation of IRS-1 by the insulin receptor [34]. An important mechanism responsible for this insulin-signaling defect is excessive serine phosphorylation of IRS-1. There are > 70 potential serine phosphorylation sites on IRS-1 [35–37], and, in general, serine phosphorylation negatively regulates IRS signaling. Recent studies have demonstrated increased serine phosphorylation of IRS-1 on Ser302, Ser307, Ser612, and Ser632 in insulin-resistant rodent models [37–40], and as in lean insulin-resistant offspring of type 2 diabetic parents [41]. Furthermore, high fat diet-induced insulin resistance can be abrogated in rodent models when specific Ser/Thr kinases (c-Jun NH_2-terminal kinase [JNK], inhibitor of nuclear factor κB kinase β subunit [IKKβ], S6 kinase 1, and protein kinase C-θ) were either knocked down or pharmacologically inhibited [37, 40–42]. Further evidence for the role of IRS serine phosphorylation in the pathogenesis of insulin resistance comes from studies in which mutation of three key serine residues to alanine in mouse skeletal muscle provided protection against high fat diet-induced insulin resistance in vivo [43]. Collectively, these data indicate that serine phosphorylation on key residues of IRS-1 plays an important role in the pathogenesis of insulin resistance in skeletal muscle.

The intracellular mechanisms responsible for activation of serine kinases is still unclear, but it seems to involve activation of protein kinase C (PKC) [37]. Recent studies have demonstrated that insulin resistance in skeletal muscle caused by lipid infusion is associated with activation of novel PKC isoforms, mainly PKCβ, θ, and δ. Activation of PKC results from increased intracellular levels of "toxic" lipid metabolites, e.g., long-chain fatty acyl CoAs and diacylglycerol [44]. The intramyocellular concentrations of both of these metabolites increase during lipid infusion and fat feeding, and they are known activators of PKC [31]. Serine phosphorylation of IRS also can result from the activation of stress kinases, e.g., JNK and NF-κB. These kinases also are activated by increased intracellular concentrations of lipid metabolites. The important role of these stress kinases in the pathophysiology of insulin resistance is demonstrated by the observation that knocking down these kinases abrogates high fat diet-induced insulin resistance in mice [40, 42].

7 Reactive Oxygen Species (ROS) and Insulin Action

For >30 years, it has been known that hydrogen peroxide and other oxidants exert insulin-like effects. Czech and colleagues reported that generation of hydrogen peroxide from thiols and Cu^{2+} stimulates glucose use and inhibits lipolysis in adipocytes [45]. Latter studies demonstrated that hydrogen peroxide stimulates glucose transport, glycogen synthesis, lipid synthesis, and amino acid transport (reviewed in [46]). In their efforts to elucidate the mechanism of hydrogen peroxide action, Hayes and Lockwood [47] reported that H_2O_2 enhances insulin signaling by increasing insulin receptor tyrosine phosphorylation. This effect of hydrogen peroxide is due to inhibition of tyrosine phosphatase (PTP-1B). During insulin action a burst of hydrogen peroxide, generated via activation of NADPH oxidase, inhibits PTP-1B [48]. The physiological importance of hydrogen peroxide in normal insulin signaling was demonstrated using catalase and inhibitors of NADPH oxidase [49]. Both resulted in reduced insulin-stimulated tyrosine phosphorylation of the insulin receptor and IRS-1.

7.1 Possible Role of Oxidative Stress in Pathogenesis of Insulin Resistance

Although a low steady-state level of ROS is required for normal insulin action, chronic unphysiologic increases in ROS interfere with insulin action and result in insulin resistance [46]. Cross-sectional studies have demonstrated a strong correlation between markers of increased oxidative stress and the severity of insulin resistance in subjects with type 2 diabetes [50, 51]. The correlation between elevated markers of oxidative stress and insulin resistance is not only restricted to diabetes but also is evident in insulin-resistant obese nondiabetic individuals [52] and in subjects with the metabolic syndrome [53, 54]. These observations suggest that increased oxidative stress is an early event that contributes to the pathogenesis of insulin resistance, rather than a consequence of the diabetic environment, e.g., chronic hyperglycemia or elevated plasma FFA levels.

If increased oxidative stress is casually linked to insulin resistance, treatment with antioxidants should improve the defect in insulin action. In vivo studies in experimental animals and small clinical trials in humans have provided support for this concept. Treatment of diabetic rats with α-lipoic acid, a physiological antioxidant, significantly improved insulin sensitivity [55], and small clinical trials have demonstrated that α-lipoic acid ameliorates insulin resistance by ~25% in subjects with T2DM [56, 57]. Other antioxidants, e.g., vitamin C and E and N-acetylcysteine, also have been shown to improve insulin sensitivity in diabetic patients [58, 59].

Further support for a possible role of increased oxidative stress in the pathogenesis of insulin resistance derives from animal and cell culture studies. In cultured hepatoma cells [60], 50 μM hydrogen peroxide results in impaired insulin signaling, due to serine phosphorylation of IRS-1. The insulin resistance caused by hydrogen peroxide is associated with increased JNK and NF-κB activity in hepatoma cells, indicating an important role for these stress kinases in the pathogenesis of insulin resistance. Similarly, hydrogen peroxide has been shown to inhibit insulin action in cultured adipocytes and in skeletal and vascular smooth muscle cells. This effect of hydrogen peroxide is associated with increased NF-κB activity [61–63].

A possible role for endogenous ROS production in the pathogenesis of insulin resistance also is suggested by studies in which the endogenous ROS level was altered by impairing or enhancing the antioxidant defense mechanisms. Pharmacological inhibition of glutathione synthase in rats doubled the plasma concentration of lipid peroxidation products and reduced total glucose disposal in skeletal muscle, measured with the euglycemic insulin clamp [64]. Conversely, antioxidants that alleviate oxidative stress by scavenging ROS have been reported to reduce insulin resistance. α-Lipoic acid, a potent antioxidant, protects muscle cells from the insulin resistance caused by hydrogen peroxide [65], and it augments insulin-stimulated glucose uptake in cultured skeletal muscle cells from diabetic animals and in rat muscle in vitro. In vivo, α-lipoic acid increases total glucose disposal in insulin-resistant Zucker fatty rats. In humans, administration of α-lipoic acid improves glucose metabolism and insulin sensitivity in subjects with T2DM. Treatment with N-acetylcysteine, an antioxidant, prevents hyperglycemia-induced insulin resistance in rats [66]. Collectively, these studies suggest an important role for increased oxidative stress in the development of insulin resistance.

7.2 Molecular Mechanism of Oxidative Stress-Induced Insulin Resistance

As discussed above, impaired tyrosine phosphorylation of IRS-1 consistently has been demonstrated in insulin-resistant individuals. A large body of evidence supports an important role for serine phosphorylation in the impaired tyrosine phosphorylation of insulin signaling molecules. A chronic increase in oxidative stress is a strong activator of stress kinases, e.g., JNK and NF-κB. Increased activity of theses kinases has been implicated in development of insulin resistance in multiple animal models [67]. In cultured skeletal muscle cells, H_2O_2 increases JNK and NF-κB activity and causes insulin resistance. These effects of hydrogen peroxide can be blocked with α-lipoic acid [63], suggesting an important role of these stress kinases in ROS-related insulin resistance.

A second molecular mechanism by which ROS has been shown to cause insulin resistance is related to the production of peroxynitrate. Most of the superoxide produced in cells is converted to hydrogen peroxide by the enzyme superoxide dismutase (SOD). Under conditions of increased cellular superoxide concentration (overproduction or decreased degradation), superoxide interacts with NO to produce peroxynitrate ($ONOO^-$), which can modify proteins by nitrosation of tyrosine residues. Nitrosation of tyrosine residues on insulin signaling molecules is another

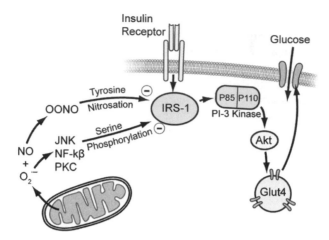

Fig. 11.1 Schematic representation of the insulin signaling pathway. Overproduction of ROS by the mitochondria activates stress kinases that serine phosphorylate IRS-1 and increase cellular hypoxynitrate, which nitrosates the IRS-1. Both nitrosation and serine phosphorylation of the IRS-1 impair insulin signaling and causes insulin resistance

mechanism by which increased oxidative stress can cause insulin resistance [68] (Fig. 11.1). In fibroblasts expressing the human insulin receptor, SIN-1, a constitutive producer of peroxynitrate, inhibits insulin signaling and reduces insulin-stimulated glucose transport. The defect in insulin signaling is due to nitrosation of four tyrosine residues in IRS-1, including Tyr 939, which is critical for association of IRS-1 with the p85 subunit of PI-3 kinase [69]. Nitrosation of IRS-1 also has been demonstrated in mice fed a high fat diet and in ob/ob diabetic mice [70]. The results indicate that, at least in experimental animals, nitrosation of IRS-1 leads to the development of insulin resistance.

7.3 β Cell Failure

In response to intravenous glucose injection, insulin is secreted from the β cell in a biphasic pattern, with an early burst of insulin release within the first 10 min (first phase insulin secretion) followed by a progressively increasing phase (second phase) of insulin secretion that persist as long as the hyperglycemic stimulus is present [71]. Loss of first phase insulin secretion is a characteristic and early abnormality in patients destined to develop T2DM [17, 72]. The decline in β cell function starts long before diabetes become evident, and the continuing decline in β cell function in subjects with type 2 diabetes contributes to the progressive nature of the disease [73]. Impaired first phase insulin secretion becomes evident when the fasting plasma glucose concentration exceeds 90–95 mg/dl [74]. Similarly, decreased insulin secretion after glucose ingestion, e.g., oral glucose tolerance test, has been demonstrated when the 2-h plasma glucose concentration exceeds 100 mg/dl [75], a value well within the range considered to represent normal glucose tolerance.

There is considerable debate regarding the etiology of β cell failure: is it secondary to impaired function of a normal number of β cells or to a decrease in the number of β cells. Both functional failure and loss of β cell mass have been demonstrated in subjects with T2DM [75, 76]. Quantitation of β cell mass in T2DM individuals at autopsy has demonstrated a 40–60% reduction compared with age- and body mass index-matched nondiabetic subjects [76]. Furthermore, subjects with T2DM had a three- to 10-fold increase in markers of apoptosis in the remaining beta cells, suggesting that the reduced beta cell mass in diabetic subjects most likely results from increased β cell death [76]. Although these studies clearly demonstrate that accelerated β cell death occurs in T2DM subjects and results in decreased β cell mass, these abnormalities cannot account for the total impairment in insulin secretion observed in T2DM subjects. Functional studies have demonstrated that by the time diabetes is diagnosed ~80% of β cell function has been lost [75]. Therefore, it is likely that functional defects and loss of β cell mass contribute to the progressive β cell failure in T2DM.

Many factors have been shown to contribute to β cell failure [77]. Studies in first degree relatives of type 2 diabetes subjects [78, 79] and in identical twins [80] have provided strong evidence for a genetic basis of β cell failure. Normal-glucose-tolerant offspring of T2DM parents and the NGT twin of an identical diabetic twin pair have a significant reduction in both first and second phase insulin secretion [78–80], indicating that the impairment in insulin secretion begins at the state of normal glucose tolerance.

Acquired factors also contribute to the impairment in β cell function. Glucotoxicity [81] and lipotoxicity [82, 83] are amongst the acquired factors that can lead to impaired insulin secretion.

Although glucose is the principal physiological stimulus for insulin secretion, a chronic increase in the plasma glucose concentration can lead to impaired β cell function. An important role for glucotoxicity is supported by the observation that improved glycemic control, however achieved, leads to enhanced insulin secretion [84, 85]. More direct support for the glucotoxic action of increased plasma glucose concentration on insulin secretion comes from studies in humans and experimental animal in which small increases in the plasma glucose concentration have been shown to impair insulin secretion. In normal healthy subjects, a 50 mg/dl increase in plasma glucose concentration for as short as 90 min obliterates first phase insulin secretion [86]. Similarly, in partially pancreatectomized NGT rats, a 16-mg/dl increment in the mean day-long plasma glucose concentration impairs first phase insulin secretion [87]. Furthermore, correction of chronic hyperglycemia with phlorizin, an inhibitor of renal tubular glucose transport, in partially pancreatectomized diabetic rats restores both the first and second phases of insulin secretion, indicating that in this animal model the reduction in insulin secretion results from chronic hyperglycemia and is reversible upon restoration of normoglycemia [87].

Chronically elevated plasma FFA levels also have been shown to impair β cell function, and this has been referred to as lipotoxicity. Studies in experimental animals and in humans have demonstrated that short-term, physiologic elevation of the plasma FFA level augments insulin secretion [88]. However, more prolonged exposure of β cell to elevated plasma FFA levels impairs insulin secretion both in vitro and in vivo. When the pancreatic β cell line MIN6 is cultured for 3 days with increasing concentrations of palmitic acid, a dose-dependent decrease in insulin content and glucose-stimulated insulin secretion is observed [89]. Similar deleterious effects of FFA have been demonstrated in pancreatic islets in vitro. Pancreatic islets cultured in the presence of elevated FFA levels manifest reduced insulin gene expression [90, 91] and an increased rate of β cell death [83, 92, 93]. Studies in humans have demonstrated that physiological elevation of the plasma FFA concentration in lean healthy individuals to levels similar to those observed in T2DM subjects (~700–800 μM) impairs both first and second phases insulin secretion, as measured with the hyperglycemic clamp [94]. Two points are worth noting: (1) the physiologic elevation in plasma FFA impaired insulin secretion only in subjects with a strong positive family history of T2DM [94], and (2) the lipotoxic effect on β cell function was markedly enhanced in the presence of hyperglycemia [91, 95]. These observations suggest that multiple factors (genetic and environmental) synergistically interact to promote β cell failure and demonstrate the complexity of mechanisms that lead to β cell demise.

7.4   Cellular Mechanisms of β Cell Failure and Role of ROS

Chronic exposure of the β cell to hyperglycemia has been shown to increase ROS production in islets, mainly via enhancing metabolic flux through the tricarboxylic acid (TCA) cycle and increasing oxidative phosphorylation. Extramitochondrial mechanisms to explain increased ROS generation in the β cell also have been described. These include glyceraldehyde autoxidation, increased flux through the hexosamine pathway, and activation of protein kinase C [96]. The intrinsic level of antioxidant enzymes, e.g., superoxide dismutase and catalase, in the β cell is relatively low compared with other tissues [97]. Thus, conditions which increase ROS generation place the islet at greater risk of oxidative damage by ROS (Fig. 11.2).

Glucose is the principal physiological stimulus for insulin secretion. Glucose must be metabolized by the β cell to increase the ATP concentration, which is the signal that triggers insulin secretion. Chronic hyperglycemia and elevated FFA levels can interfere with normal β cell function via several mechanisms: (1) inhibition of glucose-stimulated insulin secretion (GSIS); (2) inhibition of insulin synthesis; and (3) stimulation of apoptotic pathways, which lead to β cell death. As discussed above, all three of these mechanisms have been reported to contribute to the progressive β cell failure that is characteristic of subjects with T2DM.

In vitro and in vivo studies have demonstrated that elevated glucose and FFA concentrations inhibit insulin gene expression and GSIS and accelerate β cell death. In cultured β cell lines and in islets isolated from diabetic animals [98, 99], chronic hyperglycemia inhibits insulin gene expression, leading to decreased insulin

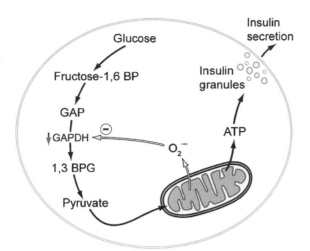

Fig. 11.2 Role of increased beta cell ROS production in beta cell failure. Normal mitochondrial ATP production is essential for insulin secretion by the β cell. Increased cellular ROS levels inhibit GAPDH and reduce the flux of metabolites into the mitochondria. Decreased substrate availability impairs ATP generation, which in turn inhibits insulin secretion

content and impaired insulin secretion. The decrease in insulin synthesis caused by chronic hyperglycemia is due to the loss of two critical proteins, PDX-1 and MafA, that activate the insulin gene promoter. This deleterious effect of hyperglycemia is mediated by generation of ROS in the β cell [100], can be prevented with antioxidants [101], and involves activation of JNK kinase. Dominant-negative JNK overexpression protects against the inhibitory effect of oxidative stress on insulin gene expression [102].

Although glucose is the principal physiological stimulus for insulin secretion, chronic exposure of pancreatic islets to high glucose concentrations impairs GSIS. Because glucose is the major fuel source for ATP production, one would expect elevation of the glucose concentration to increase β cell ATP content and potentiate GSIS. However, measurement of β cell ATP content has revealed reduced ATP levels in response to chronic (as opposed to acute) hyperglycemia. This paradoxical effect of chronic hyperglycemia on beta cell ATP production results from overexpression of UCP2 in the β cell [103]. In response to chronic hyperglycemia, increased flux through the TCA cycle increases the mitochondrial redox potential and maximally hyperpolarizes the inner mitochondrial membrane, a situation that markedly increases mitochondrial ROS generation. To protect the mitochondria from increased ROS levels, expression of UCP2 is enhanced and this uncouples the inner mitochondrial membrane, reduces the voltage gradient across the inner mitochondrial membrane, and ameliorates superoxide production. Because the inner mitochondrial membrane potential provides the driving force for ATP synthesis, the decreased in voltage gradient across the inner mitochondrial membrane results in reduced mitochondrial ATP synthesis and impaired GSIS. The important role of UCP2 overexpression in mediating the impaired insulin secretory response after exposure to chronic hyperglycemia has been demonstrated in UCP2 knockout animals. In β cells isolated from UCP2 knockout mice, hyperglycemia caused a marked increase in superoxide production but failed to inhibit GSIS [103]. These results indicate that overproduction of ROS plays a pivotal role in mediating the deleterious action of chronic hyperglycemia on insulin secretion.

Additional support for the important role of accelerated ROS production in mediating the deleterious effect of chronic hyperglycemia on insulin secretion derives from studies using antioxidants that scavenge ROS and ameliorate the oxidative stress in the β cell [104, 105]. Thus, *N*-acetyl cysteine and astaxanthin prevent the deleterious effects of chronic hyperglycemia on GSIS and insulin synthesis in cultured β cells. In humans, antioxidants improve glycemic control and augment insulin secretion in subjects with T2DM [106]. These observations suggest that ROS overproduction plays an important role in mediating the deleterious effect of chronic hyperglycemia on insulin secretion.

Islets cultured in medium containing high glucose exhibit increased apoptosis, accompanied by decreased expression of the antiapoptotic gene Bcl-xL, overexpression of the proapoptotic genes Bad, Bid, and Bik [107], and overproduction of ROS in the β cell has been implicated in hyperglycemia-induced β cell apoptosis [108, 109].

7.5 Molecular Etiology of Type 2 Diabetic Complications

Vascular endothelium, in large and small vessels, is injured by chronic exposure to hyperglycemia. Because the contribution of oxidative stress to the pathogenesis of atherosclerosis is discussed elsewhere in this book, we limit our discussion to the role of oxidative stress in microvascular complications.

Two landmark studies, the DCCT [110] and the UKPDS [111], unequivocally have documented that control of the plasma glucose concentration in subjects with T2DM markedly reduces the risk of microvascular complications, indicating that chronic exposure of tissues to hyperglycemia triggers pathological processes that lead to organ damage. All cells in the body of T2DM subjects are exposed to chronic hyperglycemia. However, only cells which are unable to efficiently control glucose influx through their plasma membrane, e.g., vascular endothelium of the retina, kidney, and the peripheral nerves are deleteriously affected by hyperglycemia. Intracellular hyperglycemia increases the flux through glycolysis and TCA cycle, resulting in increased mitochondrial NADH levels and increased production of superoxide. Overproduction of ROS in the cell activates a number of metabolic pathways that cause vascular damage and lead to microvascular complications. Four metabolic pathways have been shown to be activated by chronic hyperglycemia, and their activation contributes to vascular damage [112]. (1) Polyol pathway: Aldose reductase is an intracellular enzyme that reduces toxic aldehydes to inactive alcohols. When exposed to increased intracellular glucose concentrations, aldose reductase converts glucose to sorbitol. This process consumes NADH, leading to a reduction in the intracellular concentration of critical antioxidants, e.g., glutathione, rendering the cell vulnerable to oxidative damage [113]. (2) Intracellular advanced glycated end product (AGE) precursors: AGEs are glycated proteins which originally were thought to originate from the non-enzymatic reaction between extracellular proteins and glucose. However, the rate of AGE generation from intracellular glucose metabolites, e.g., dicarbonyls, is orders of magnitude faster than with glucose itself, and it is now thought that intracellular (and not extracellular) hyperglycemia is the primary initiating event in the formation of both intracellular and extracellular AGEs [114]. AGEs cause cell damage by modification of intracellular proteins, e.g., transcription factors [115], and extracellular and circulating proteins, e.g., matrix molecules and albumin [116]. (3) PKC activation: Intracellular hyperglycemia increases the production of diacylglycerol (DAG) in cultured microvascular cells from the retina and kidney of diabetic animals. The increase in intracellular DAG level results from accelerated synthesis from increased levels of the glycolytic precursor glycerol-3 phosphate. DAG is a strong activator of nine of the 11 isoforms of PKC. Hyperglycemia activates the β and δ subunits of PKC in glomeruli and retinal vascular cells from diabetic animals [117, 118]. Activation of PKC in vascular endothelium impairs endothelial function by decreasing NO synthase, and increasing endothelin 1 and plasminogen activator-1, which lead to endothelial damage [119–123]. (4) Hexosamine pathway activity: Under conditions of increased glycolytic flux, e.g., chronic hyperglycemia, some of the fructose-6 phosphate is converted by the enzyme glutamine:fructose-6 phosphate amidotransferase to glucoseamine-6 phosphate, which then is converted to N-acetyl glucoseamine (GlcNAc). Increased GlcNAc levels modify the structure and expression of cellular proteins, e.g., the transcription factor Sp1, which regulates the expression of PAI-1 [124]. In aortic endothelial cells, hyperglycemia causes a 2.4-fold increase in hexosamine pathway activity, resulting in increased Sp1 O-linked GlcNAc. In cardiomyocytes, hyperglycemia increases nuclear O-GlcNAacylaion, leading to impaired calcium cycling [125].

Several lines of evidence suggest that overproduction of mitochondrial ROS is the process that links hyperglycemia to the activation of the four metabolic pathways involved in diabetic complications: (1) Agents that inhibit mitochondrial metabolism, e.g., carbonyl cyanide p-trifluoromethoxyphenylhydrazone, block the production of ROS during hyperglycemic conditions and prevent the activation of PKC, and they inhibit the production of sorbitol, and AGEs in bovine aortic endothelial cells [126]. (2) Overexpression of either UCP2 or Mn-SOD in aortic endothelial cells inhibits hyperglycemia-stimulated production of ROS, prevents the activation of PKC, and inhibits the production of sorbitol and AGEs [126]. (3) In cultured endothelial cells, removal of the electron transport chain by depleting mitochondrial DNA blocks ROS production in response to hyperglycemia and inhibits the ability of hyperglycemia to augment polyol pathway flux, PKC activity, AGE formation, and hexosamine pathway flux [112]. These results suggest that conditions that block the formation of ROS in response to hyperglycemia prevent activation of the metabolic pathways that cause vascular injury.

The mechanism by which overproduction of superoxide activates these metabolic pathways is likely mediated through the inhibition of glyceraldehyde phosphate dehydrogenase (GAPDH [127]). GAPDH is a key glycolytic enzyme that catalyzes the conversion of glyceraldehyde 3 phosphate to 1,3-diphosphoglycerate (Fig. 11.3). GAPDH activity is markedly reduced in diabetic subjects and animals and in cells grown under hyperglycemic conditions. GAPDH activity is highly sensitive to superoxide, and overproduction of superoxide during hyperglycemic conditions inhibits GAPDH activity, leading to the accumulation of upstream glycolytic metabolites (Fig. 11.3), which serve as precursors for the metabolic pathways that initiate the cell damage. Overexpression of UCP2 or Mn-SOD, which blocks hyperglycemia-induced production of ROS, prevents GAPDH inhibition in endothelial cells and prevents the activation of the metabolic pathways involved in cell damage.

Because overproduction of ROS is a key step in hyperglycemia-induced cell injury, one would predict that antioxidants, which inhibit the rise in intracellular ROS levels, would prevent the development of diabetic complications. Indeed, many of the hyperglycemia-induced biological changes, including excess sorbitol production, urinary albumin excretion, and elevated HbA_{1c} levels, can be reduced with antioxidant therapy [128]. Furthermore, antioxidant therapy with α-lipoic acid and vitamins C and E have been favorable effects on blood flow and nerve conduction velocity [129]. However, a large prospective clinical trial with vitamin E failed to show any benefit in preventing vascular complications in diabetic patients [130].

Although the negative results of clinical trials with antioxidants is disappointing, further studies are needed. In these clinical trials, markers of oxidative stress were not examined, and it remains unknown whether the antioxidant treatment actually reduced oxidative stress. Interestingly, reanalysis of the data from one of these

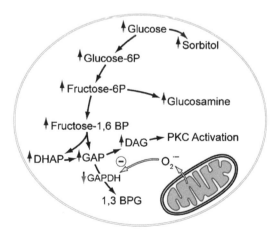

Fig. 11.3 Role of ROS in diabetic vascular complications. Inhibition of GAPDH by excessive ROS production leads to an increase in the level of upstream metabolites which augment substrate flux into alternative pathways, e.g., aldose reductase and hexosamine, which lead to vascular damage

negative studies, HOPE, demonstrated that in the subgroup with haptoglobin 2-2, treatment with vitamin E produced a 55% reduction in the rate of myocardial infarction [131]. Thus, the role of antioxidant therapy in preventing diabetic vascular complications needs further evaluation.

8 Summary and Conclusions

A large amount of data suggests an important role for ROS in the pathogenesis of type 2 diabetes and associated vascular complications. Increased ROS levels impair insulin signaling, leading to insulin resistance, and they inhibit mitochondrial ATP production in β cells, resulting in impaired impairs insulin secretion. Despite the multitude of in vitro studies and in vivo experiments in animal models of diabetes, convincing evidence for an important role of ROS in the pathogenesis of diabetes mellitus still is lacking, and no study convincing has demonstrated an actual increase in ROS production or levels in type 2 diabetic subjects compared with normal healthy individuals. Although the evidence supporting a role of ROS in the development of diabetic vascular complications is more impressive, the results of large clinical trials with antioxidant therapy have been disappointing and raise concern about the role of ROS in the pathogenesis of diabetes and its complications. Several key questions need to be addressed before any conclusions concerning the role of ROS in the pathogenesis of T2DM and diabetic complications can be established: (1) in humans with T2DM, are ROS elevated, and, if so, in which tissues are they elevated; (2) what is the source of the excess ROS production, mitochondrial versus extramitochondrial; thus, physiological ROS production by NADH oxidase seems to be important for normal insulin signaling, whereas overproduction of ROS by mitochondria is toxic and impairs insulin signaling; (3) based upon the prior considerations, more specific antioxidant therapies may be required to achieve positive therapeutic results; (4) levels of ROS and ROS production rates must be monitored in clinical trials assessing the efficacy of antioxidant therapies to ensure that excess ROS production truly has been inhibited; moreover, although prevention of excess production of ROS may be beneficial, complete inhibition of ROS production may interfere with normal cell signaling and physiologic function.

References

1. King H, Aubert RE, Herman WH, Global burden of diabetes, 1995–2025:prevalence, numerical estimates, and projections Diabetes Care 1998;21:1414.
2. Kahn SE, Hull RL, Utzschneider KM. Mechanisms linking obesity to insulin resistance and type 2 diabetes. Nature 2006;444:840–6.
3. Schalkwijk CG, Stehouwer CD. Vascular complications in diabetes mellitus: the role of endothelial dysfunction. Clin Sci (Lond) 2005;109:143–59.

4. He Z, King GL. Microvascular complications of diabetes. Endocrinol Metab Clin North Am 2004;33:215–38.
5. Harris MI, Klein R, Welborn TA, Knutman MW. Onset of NIDDM occurs at least 4–7 yr before clinical diagnosis. Diabetes Care 1992;15:815–9.
6. Morgan CL, Currie CJ, Stott NCH, Smithers M, Butler CC, Peters JR. The prevalence of multiple diabetes-related complications. Diabet Med 2000; 17:146–51.
7. Huse DM, Oster G, Killen AR, Lacey MJ, Colditz GA. The economic costs of non-insulin-dependent diabetes mellitus. JAMA 1989;262:2708.
8. DeFronzo RA. Pathogenesis of type 2 diabetes mellitus: metabolic and molecular implications for identifying diabetes genes. Diabetes Rev 1997;5:117–269.
9. Gerich JE, Meyer C, Woerle HJ, Stumvoll M. Renal gluconeogenesis. Its importance in human glucose homeostasis. Diabetes Care 2001;24:382–391.
10. DeFronzo RA, Ferrannini E. Regulation of hepatic glucose metabolism in humans. Diabetes Metab Rev 1987;3:415–59.
11. Katz LD, Glickman MG, Rapoport S, Ferrannini E, DeFronzo RA. Splanchnic and peripheral disposal of oral glucose in man. Diabetes 1983;32:675–679.
12. Mari A, Wahren J, DeFronzo RA, Ferrannini E. Glucose absorption and production following oral glucose: comparison of compartmental and arteriovenous-difference methods. Metabolism 1994;43:1419–25.
13. DeFronzo RA. Pathogenesis of type 2 diabetes mellitus. Med Clin North Am 2004;8:787–835.
14. DeFronzo RA. Pathogenesis of type 2 diabetes mellitus: metabolic and molecular implications for identifying diabetes genes. Diabetes Rev 1997;5:117–269.
15. DeFronzo RA. Lilly Lecture 1987. The triumvirate: beta-cell, muscle, liver. A collusion responsible for NIDDM. Diabetes 1988;37:667–87.
16. Gastaldelli A, Ferrannini E, Miyazaki Y, Matsuda M, DeFronzo RA; San Antonio metabolism study. beta-Cell dysfunction and glucose intolerance: results from the San Antonio metabolism (SAM) study. Diabetologia 2004;47:31–9.
17. Kahn SE. The relative contributions of insulin resistance and beta-cell dysfunction to the pathophysiology of type 2 diabetes. Diabetologia 2004;46:3–19.
18. DeFronzo RA, Gunnarsson R, Bjorkman O, Olsson M, Wahren J. Effects of insulin on peripheral and splanchnic glucose metabolism in noninsulin-dependent (type II) diabetes mellitus. J Clin Invest 1985;76:149–55.
19. DeFronzo RA and Ferrannini E Regulation of hepatic glucose metabolism in humans. Diabetes Metab Rev 1987;3:415–59.
20. Ferrannini E, Bjorkman O, Reichard GA Jr et al. The disposal of an oral glucose load in healthy subjects. A quantitative study. Diabetes 1985;34:580–8.
21. Groop LC, Bonadonna RC, DelPrato S et al. Glucose and free fatty acid metabolism in non-insulin-dependent diabetes mellitus. Evidence for multiple sites of insulin resistance. J Clin Invest 1989;84:205–13.
22. Saltiel AR, Kahn CR. Insulin signaling and the regulation of glucose and lipid metabolism. Nature 2001;414:799–806.
23. Shepherd PR, Kahn BB. Glucose transporters and insulin action. Implications for insulin resistance and diabetes mellitus. N Engl J Med 1999;341:248–257.
24. Virkamaki A, Ueki K, Kahn CR. Protein-protein interaction in insulin signaling and the molecular mechanisms of insulin resistance. J Clin Invest 1999;103:931–943.
25. White MF, Livingston JN, Backer JM et al. Mutation of the insulin receptor at tyrosine 960 inhibits signal transmission but does not affect its tyrosine kinase activity. Cell 1992;54:641–649.
26. Kerouz NJ, Horsch D, Pons S, Kahn CR. Differential regulation of insulin receptor substrates-1 and -2 (IRS-1 and IRS-2) and phosphatidylinositol 3-kinase isoforms in liver and muscle of the obese diabetic (ob/ob) mouse. J Clin Invest 1997;100:3164–3172.
27. Sun XJ, Miralpeix M, Myers MG Jr et al. The expression and function of IRS-1 in insulin signal transmission. J Biol Chem 1993;267:22662–22672.

28. Ruderman N, Kapeller R, White MF, and Cantley LC. Activation of phosphatidylinositol-3-kinase by insulin. Proc Natl Acad Sci U S A 1990;87:1411–1415.

29. Cross DA, Alessi DR, Cohen P, Andjelkovich M, and Hemmings BA. Inhibition of glycogen synthase kinase-3 by insulin mediated by protein kinase B. Nature 1995;378:785–789.

30. Brady MJ, Nairn AC, Saltiel AR. The regulation of glycogen synthase by protein phosphatase 1 in 3T3-L1 adipocytes. Evidence for a potential role for DARPP-32 in insulin action. J Biol Chem 1997;272:29698–29703.

31. Okada T, Sakuma L, Fukui Y, Hazeki O, and Ui M. Essential role of phosphatidylinositol 3-kinase in insulin-induced glucose transport and antilipolysis in rat adipocytes. J Biol Chem 1994;269:3568–3573.

32. Cross D, Alessi D, Vandenheed J et al. The inhibition of glycogen synthase kinase-3 by insulin or insulin-like growth factor 1 in the rat skeletal muscle cell line L6 is blocked by wortmannin but not rapamycin. Biochem J 1994;303:21–26.

33. Osawa H, Sutherland C, Robey R, Printz R, Granner D, Analysis of the signaling pathway involved in the regulation of hexokinase II gene transcription by insulin. J Biol Chem 1996;271:16690–16694.

34. Kashyap SR, Defronzo RA. The insulin resistance syndrome: physiological considerations. Diab Vasc Dis Res 2007;4:13–9.

35. Mothe I, Van Obberghen E. Phosphorylation of insulin receptor substrate-1 on multiple serine residues, 612, 632, 662, and 731, modulates insulin action. J Biol Chem 1996;271:11222–11227.

36. Petersen KF, Shulman GI. Etiology of insulin resistance. Am J Med 2006;119(5 Suppl 1): S10–6.

37. Kim JK, Fillmore JJ, Sunshine MJ et al. PKC-theta knockout mice are protected from fat-induced insulin resistance. J Clin Invest 2004;114:823–827.

38. Um SH, Frigerio F, Watanabe M et al. Absence of S6K1 protects against age- and diet-induced obesity while enhancing insulin sensitivity. Nature 2004;431:200–205.

39. Furukawa N, Ongusaha P, Jahng WJ et al. Role of Rho-kinase in regulation of insulin action and glucose homeostasis. Cell Metab 2005;2:119–129.

40. Hirosumi J, Tuncman G, Chang LF et al. A central role for JNK in obesity and insulin resistance. Nature 2002;420:333–336.

41. Morino K, Petersen KF, Dufour S et al. Reduced mitochondrial density and increased IRS-1 serine phosphorylation in muscle of insulin-resistant offspring of type 2 diabetic parents. J Clin Invest 2005;115:3587–3593.

42. Yuan MS, Konstantopoulos N, Lee JS et al. Reversal of obesity- and diet-induced insulin resistance with salicylates or targeted disruption of IKK beta. Science 2001;293:1673–1677.

43. Morino K, Petersen KF, Shulman GI. Molecular mechanisms of insulin resistance in humans and their potential links with mitochondrial dysfunction. Diabetes 2006;55 Suppl 2:S9–S15.

44. Yu CL, Chen Y, Cline GW et al. Mechanism by which fatty acids inhibit insulin activation of insulin receptor substrate-1 (IRS-1)-associated phosphatidylinositol 3-kinase activity in muscle. J Biol Chem 2002;277:50230–50236.

45. Czech MP, Fain JN. Cu++-dependent thiol stimulation of glucose metabolism in white fat cells. J Biol Chem 1972;247:6218–23.

46. Evans JL, Maddux BA, Goldfine ID. The molecular basis for oxidative stress-induced insulin resistance. Antioxid Redox Signal 2005;7:1040–52.

47. Hayes GR, Lockwood DH. Role of insulin receptor phosphorylation in the insulinomimetic effects of hydrogen peroxide. Proc Natl Acad Sci U S A 1987;84:8115–9.

48. Mahadev K, Zilbering A, Zhu L, Goldstein BJ. Insulin-stimulated hydrogen peroxide reversibly inhibits protein-tyrosine phosphatase 1b in vivo and enhances the early insulin action cascade. J Biol Chem 2001;276:21938–42.

49. Mahadev K, Wu X, Zilbering A, Zhu L, Lawrence JT, Goldstein BJ. Hydrogen peroxide generated during cellular insulin stimulation is integral to activation of the distal insulin signaling cascade in 3T3-L1 adipocytes. J Biol Chem 2001;276:48662–9.

50. Paolisso G, D'Amore A, Volpe C et al. Evidence for a relationship between oxidative stress and insulin action in non-insulin-dependent (type II) diabetic patients. Metabolism 1994;43:1426–9.
51. Wittmann I, Nagy J. Are insulin resistance and atherosclerosis the consequences of oxidative stress? Diabetologia 1996;39:1002–3.
52. Urakawa H, Katsuki A, Sumida Y et al. Oxidative stress is associated with adiposity and insulin resistance in men. J Clin Endocrinol Metab 2003;88:4673–6.
53. Lee KU. Oxidative stress markers in Korean subjects with insulin resistance syndrome. Diabetes Res Clin Pract 2001;54 Suppl 2:S29–33.
54. Hitsumoto T, Iizuka T, Takahashi M et al. Relationship between insulin resistance and oxidative stress in vivo. J Cardiol 2003;42:119–27.
55. Jacob S, Streeper RS, Fogt DL et al. Henriksen, the antioxidant α-lipoic acid enhances insulin stimulated glucose metabolism in insulin-resistant rat skeletal muscle, Diabetes 1996;45:1024–1029.
56. Jacob S, Henriksen EJ, Schiemann AL et al. Enhancement of glucose disposal in patients with type 2 diabetes by alpha-lipoic acid. Drug Res 1995;45:872–874.
57. Jacob S, Ruus P, Hermann R et al. Oral administration of rac-α-lipoic acid modulates insulin sensitivity in patients with type 2 diabetes mellitus-a placebo controlled pilot trial. Free Radic Biol Med 1999;27:309–314.
58. Hirashima O, Kawano H, Motoyama T et al. Improvement of endothelial function and insulin sensitivity with vitamin C in patients with coronary spastic angina: possible role of reactive oxygen species. J Am Coll Cardiol 2000;35:1860–6.
59. Paolisso G, Di Maro G, Pizza G et al. Plasma GSH/GSSG affects glucose homeostasis in healthy subjects and non-insulin-dependent diabetics. Am J Physiol 1992;263(3 Pt 1):E435–40.
60. Bloch-Damti A, Potashnik R, Gual P et al. Differential effects of IRS1 phosphorylated on Ser307 or Ser632 in the induction of insulin resistance by oxidative stress. Diabetologia 2006;49(10):2463–73.
61. Rudich A, Tirosh A, Potashnik R, Hemi R, Kanety H, Bashan N. Prolonged oxidative stress impairs insulin-induced GLUT4 translocation in 3T3-L1 adipocytes. Diabetes 1998;47:1562–9.
62. Gardner CD, Eguchi S, Reynolds CM, Eguchi K, Frank GD, Motley ED. Hydrogen peroxide inhibits insulin signaling in vascular smooth muscle cells. Exp Biol Med (Maywood) 2003;228:836–42.
63. Maddux BA, See W, Lawrence JC Jr, Goldfine AL, Goldfine ID, Evans JL. Protection against oxidative stress-induced insulin resistance in rat L6 muscle cells by mircomolar concentrations of alpha-lipoic acid. Diabetes 2001;50:404–10.
64. Rudich A, Tirosh A, Potashnik R, Khamaisi M, Bashan N. Lipoic acid protects against oxidative stress induced impairment in insulin stimulation of protein kinase B and glucose transport in 3T3-L1 adipocytes. Diabetologia 1999;42:949–57.
65. Ogihara T, Asano T, Katagiri H et al. Oxidative stress induces insulin resistance by activating the nuclear factor-kappa B pathway and disrupting normal subcellular distribution of phosphatidylinositol 3-kinase. Diabetologia 2004;47:794–805.
66. Haber CA, Lam TK, Yu Z et al. N-Acetylcysteine and taurine prevent hyperglycemia-induced insulin resistance in vivo: possible role of oxidative stress. Am J Physiol 2003;285: E744–53.
67. Bloch-Damti A, Bashan N. Proposed mechanisms for the induction of insulin resistance by oxidative stress. Antioxid Redox Signal 2005;7:1553–67.
68. Kaneki M, Shimizu N, Yamada D, Chang K. Nitrosative stress and pathogenesis of insulin resistance. Antioxid Redox Signal 2007;9:319–29.
69. Nomiyama T, Igarashi Y, Taka H et al. Reduction of insulin-stimulated glucose uptake by peroxynitrite is concurrent with tyrosine nitration of insulin receptor substrate-1. Biochem Biophys Res Commun 2004;320:639–47.
70. Carvalho-Filho MA, Ueno M, Hirabara SM et al. S-Nitrosation of the insulin receptor, insulin receptor substrate 1, and protein kinase B/Akt: a novel mechanism of insulin resistance. Diabetes 2005;54:959–67.

71. DeFronzo RA, Tobin JD, Andres R. Glucose clamp technique: a method for quantifying insulin secretion and resistance. Am J Physiol 1979;237:E214–23.
72. Bergman RN, Finegood DT, Kahn SE. The evolution of β-cell dysfunction and insulin resistance in type 2 diabetes. Eur J Clin Invest 2002;32:35–45.
73. Kahn SE. Clinical review 135. The importance of β-cell failure in the development and progression of type 2 diabetes. J Clin Endocrinol Metab 2001;86:4047–4058.
74. Godsland IF, Jeffs JA, Johnston DG. Loss of beta cell function as fasting glucose increases in the non-diabetic range. Diabetologia 2004;47:1157–66.
75. Gastaldelli A, Ferrannini E, Miyazaki Y, Matsuda M, DeFronzo RA; San Antonio metabolism study. Beta-cell dysfunction and glucose intolerance: results from the San Antonio metabolism (SAM) study. Diabetologia 2004;47:31–9.
76. Butler AE, Janson J, Bonner-Weir S, Ritzel R, Rizza RA, Butler PC. Beta-cell deficit and increased beta-cell apoptosis in humans with type 2 diabetes. Diabetes 2003;52:102–10.
77. Prentki M, Nolan CJ. Islet beta cell failure in type 2 diabetes. J Clin Invest 2006; 116:18020–1812.
78. Gautier J-F, Wilson C, Weyer C et al. Low acute insulin secretory responses in adult offspring of people with early onset type 2 diabetes. Diabetes 2001;50:1828–1833.
79. Vauhkonen N, Niskanen L, Vanninen E, Kainulainen S, Uusitupa M, Laakso M. Defects in insulin secretion and insulin action in non-insulin-dependent diabetes mellitus are inherited. Metabolic studies on offspring of diabetic probands. J Clin Invest 1997;100:86–96.
80. Vaag A, Henriksen JE, Madsbad S, Holm N, and Beck-Nielsen H. Insulin secretion, insulin action, and hepatic glucose production in identical twins discordant for non-insulin-dependent diabetes mellitus. J Clin Invest 1995;95:690–698.
81. Rossetti L, Giaccari A, DeFronzo RA. Glucose toxicity. Diabetes Care 1990;13:610–630.
82. McGarry JD. Banting Lecture 2001: dysregulation of fatty acid metabolism in the etiology of type 2 diabetes. Diabetes 2002;51:7–18.
83. Shimabukuro M, Zhou Y-T, Levi M, Unger RH. Fatty acid induced β cell apoptosis: a link between obesity and diabetes. Proc Natl Acad Sci U S A 1998;95:2498–2502.
84. Vague P, Moulin J-P. The defective glucose sensitivity of the B cell in insulin dependent diabetes. Improvement after twenty hours of normoglycaemia. Metabolism 1982;31:139–142.
85. Kosaka K, Kuzuya T, Akanuma Y, Hagura R. Increase in insulin response after treatment of overt maturity onset diabetes mellitus is independent of the mode of treatment. Diabetologia 1980;18:23–28.
86. Toschi E, Camastra S, Sironi AM et al. Effect of acute hyperglycemia on insulin secretion in humans. Diabetes 2002;51 Suppl 1:S130–3.
87. Rossetti L, Shulman GI, Zawalich W, DeFronzo RA. Effect of chronic hyperglycemia on in vivo insulin secretion in partially pancreatectomized rats. J Clin Invest 1987;80:1037–1044.
88. McGarry JD, Dobbins RL. Fatty acids, lipotoxicity and insulin secretion. Diabetologia 1999;42:128–38.
89. Iizuka K, Nakajima H, Namba M, Miyagawa J, Miyazaki J, Hanafusa T, Matsuzawa Y. Metabolic consequence of long-term exposure of pancreatic beta cells to free fatty acid with special reference to glucose insensitivity. Biochim Biophys Acta 2002;1586:23–31.
90. Gremlich S, Bonny C, Waeber G, Thorens B. Fatty acids decrease IDX-1 expression in rat pancreatic islets and reduce GLUT2, glucokinase, insulin, and somatostatin levels. J Biol Chem 1997;272:30261–30269.
91. Jacqueminet S, Briaud I, Rouault C, Reach G, Poitout V. Inhibition of insulin gene expression by long-term exposure of pancreatic β-cells to palmitate is dependent upon the presence of a stimulatory glucose concentration. Metabolism 2000;49:532–536.
92. Maedler K, Spinas GA, Dyntar D, Moritz W, Kaiser N, Donath MY. Distinct effects of saturated and monounsaturated fatty acids on ß-cell turnover and function. Diabetes 2001;50:69–76.
93. Cnop M, Hannaert JC, Hoorens A, Eizirik DL, Pipeleers DG. Inverse relationship between cytotoxicity of free fatty acids in pancreatic islet cells and cellular triglyceride accumulation. Diabetes 2001;50:1771–1777.

94. Kashyap S, Belfort R, Gastaldelli A et al. A sustained increase in plasma free fatty acids impairs insulin secretion in nondiabetic subjects genetically predisposed to develop type 2 diabetes. Diabetes 2003;52:2461–74.
95. Okuyama R, Fujiwara T, Ohsumi J. High glucose potentiates palmitate-induced NO-mediated cytotoxicity through generation of superoxide in clonal beta-cell HIT-T15. FEBS Lett 2003;545:219–23.
96. Robertson RP. Chronic oxidative stress as a central mechanism for glucose toxicity in pancreatic islet beta cells in diabetes. J Biol Chem 2004;279:42351–4.
97. Tiedge M, Lortz S, Drinkgern J, Lenzen S. Relation between antioxidant enzyme gene expression and antioxidative defense status of insulin-producing cells Diabetes 1997;46:1733–1742.
98. Olson LK, Redmon JB, Towle HC, Robertson RP. Chronic exposure of HIT cells to high glucose concentrations paradoxically decreases insulin gene transcription and alters binding of insulin gene regulatory protein. J Clin Invest 1993;92:514–9.
99. Sharma A, Olson LK, Robertson RP, Stein R. The reduction of insulin gene transcription in HIT-T15 beta cells chronically exposed to high glucose concentration is associated with the loss of RIPE3b1 and STF-1 transcription factor expression. Mol Endocrinol 1995;9:1127–34.
100. Matsuoka TA, Kajimoto Y, Watada H et al. Glycation-dependent, reactive oxygen species mediated suppression of the insulin gene promoter activity in HIT cells. J Clin Invest 1997;99:144–150.
101. Tanaka Y, Gleason CE, Tran POT, Harmon JS, Robertson RP Prevention of glucose toxicity in HIT-T15 cells and Zucker diabetic fatty rats by antioxidants. Proc Natl Acad Sci USA 1999;96:10857–10862.
102. Kaneto H, Xu G, Fujii N, Kim S, Bonner-Weir S, Weir GC. Involvement of c-Jun N-terminal kinase in oxidative stress-mediated suppression of insulin gene expression. J Biol Chem 2002;277:30010–8.
103. Krauss S, Zhang CY, Scorrano L et al. Superoxide-mediated activation of uncoupling protein 2 causes pancreatic beta cell dysfunction. J Clin Invest 2003;112:1831–42.
104. Zraika S, Aston-Mourney K, Laybutt DR et al. The influence of genetic background on the induction of oxidative stress and impaired insulin secretion in mouse islets. Diabetologia 2006;49:1254–63.
105. Uchiyama K, Naito Y, Hasegawa G, Nakamura N, Takahashi J, Yoshikawa T. Astaxanthin protects beta-cells against glucose toxicity in diabetic db/db mice. Redox Rep 2002;7:290–3.
106. Paolisso G, Giugliano D, Pizza G et al. Glutathione infusion potentiates glucose-induced insulin secretion in aged patients with impaired glucose tolerance. Diabetes Care 1992;15:1–7.
107. Federici M, Hribal M, Perego L et al. High glucose causes apoptosis in cultured human pancreatic islets of Langerhans: a potential role for regulation of specific Bcl family genes toward an apoptotic cell death program. Diabetes 2001;50:1290–301.
108. Ortega-Camarillo C, Guzman-Grenfell AM, Garcia-Macedo R et al. Hyperglycemia induces apoptosis and p53 mobilization to mitochondria in RINm5F cells. Mol Cell Biochem 2006;281:163–71.
109. Martens GA, Van de Casteele M. Glycemic control of apoptosis in the pancreatic beta cell: danger of extremes? Antioxid Redox Signal 2007;9:309–17.
110. The Diabetes Control and Complications Trial Research Group: The effect of intensive treatment of diabetes on the development and progression of long-term complications in insulin-dependent diabetes mellitus. N Engl J Med 1993;329:977–986.
111. UK Prospective Diabetes Study (UKPDS) Group: intensive blood-glucose control with sulphonylureas or insulin compared with conventional treatment and risk of complications in patients with type 2 diabetes (UKPDS 33). Lancet 1998;352:837–853.
112. Brownlee M. The pathobiology of diabetic complications: a unifying mechanism. Diabetes 2005;54:1615–25.
113. Lee AY, Chung SS. Contributions of polyol pathway to oxidative stress in diabetic cataract. FASEB J 1999;13:23–30.
114. Degenhardt TP, Thorpe SR, Baynes JW. Chemical modification of proteins by methylglyoxal. Cell Mol Biol 1998;44, 1139–1145.

115. Giardino I, Edelstein D, Brownlee M. Nonenzymatic glycosylation in vitro and in bovine endothelial cells alters basic fibroblast growth factor activity: a model for intracellular glycosylation in diabetes. J Clin Invest 1994;94:110–117.

116. Shinohara M, Thornalley PJ, Giardino I et al. Overexpression of glyoxalase-I in bovine endothelial cells inhibits intracellular advanced glycation endproduct formation and prevents hyperglycemia-induced increases in macromolecular endocytosis. J Clin Invest 1998;101:1142–1147.

117. Koya D, King GL. Protein kinase C activation and the development of diabetic complications. Diabetes 1998;47:859–866.

118. Xia P, Inoguchi T, Kern TS, Engerman RL, Oates PJ, King GL: Characterization of the mechanism for the chronic activation of diacylglycerol-protein kinase C pathway in diabetes and hypergalactosemia. Diabetes 1994;43:1122–1129.

119. Koya D, Jirousek MR, Lin YW, Ishii H, Kuboki K, King GL. Characterization of protein kinase C beta isoform activation on the gene expression of transforming growth factor-beta, extracellular matrix components, and prostanoids in the glomeruli of diabetic rats. J Clin Invest 1997;100:115–126.

120. Ishii H, Jirousek MR, Koya D et al. Amelioration of vascular dysfunctions in diabetic rats by an oral PKC beta inhibitor. Science 1996;272:728–731.

121. Kuboki K, Jiang ZY, Takahara N et al. Regulation of endothelial constitutive nitric oxide synthase gene expression in endothelial cells and in vivo: a specific vascular action of insulin. Circulation 2000;101:676–681.

122. Studer RK, Craven PA, Derubertis FR. Role for protein kinase C in the mediation of increased fibronectin accumulation by mesangial cells grown in high-glucose medium. Diabetes 1993;42:118–126.

123. Feener EP, Xia P, Inoguchi T, Shiba T, Kunisaki M, King GL. Role of protein kinase C in glucose- and angiotensin II-induced plasminogen activator inhibitor expression. Contrib Nephrol 1996;118:180–187.

124. Du XL, Edelstein D, Rossetti L et al. Hyperglycemia-induced mitochondrial superoxide overproduction activates the hexosamine pathway and induces plasminogen activator inhibitor-1 expression by increasing Sp1 glycosylation. Proc Natl Acad Sci USA 2000;97:12222–12226.

125. Du XL, Edelstein D, Rossetti L et al. Hyperglycemia-induced mitochondrial superoxide overproduction activates the hexosamine pathway and induces plasminogen activator inhibitor-1 expression by increasing Sp1 glycosylation. Proc Natl Acad Sci USA 2000;97, 12222–12226.

126. Nishikawa T, Edelstein D, Du XL et al. Normalizing mitochondrial superoxide production blocks three pathways of hyperglycaemic damage. Nature 2000;404:787–790.

127. Du X, Matsumura T, Edelstein D et al. Inhibition of GAPDH activity by poly(ADP-ribose) polymerase activates three major pathways of hyperglycemic damage in endothelial cells. J Clin Invest 2003;112:1049–1057.

128. Triggiani V, Resta F, Guastamacchia E et al. Role of antioxidants, essential fatty acids, carnitine, vitamins, phytochemicals and trace elements in the treatment of diabetes mellitus and its chronic complications. Endocr Metab Immune Disord Drug Targets 2006;6:77–93.

129. Ziegler D, Hanefeld M, Ruhnau KJ et al. Treatment of symptomatic diabetic polyneuropathy with the antioxidant alpha-lipoic acid: a 7-month multicenter randomized controlled trial (ALADIN III Study). ALADIN III Study Group. alpha-Lipoic acid in diabetic neuropathy. Diabetes Care 1999;22:1296–301.

130. Yusuf S, Dagenais G, Pogue J, Bosch J, Sleight P. Vitamin E supplementation and cardiovascular events in high-risk patients. The Heart Outcomes Prevention Evaluation Study Investigators. N Engl J Med 2000;342:154–60.

131. Levy AP, Gerstein HC, Miller-Lotan R et al. The effect of vitamin E supplementation on cardiovascular risk in diabetic individuals with different haptoglobin phenotypes. Diabetes Care 2004;27:2767.

12
DNA Oxidative Damage and Cancer

Jeffrey A. Stuart and Melissa M. Page

Summary Cancer cells have undergone several distinct transformational events leading to alterations of the normal growth and proliferation regulatory pathways. The probabilities of these events are increased by agents that damage and mutate DNA. It is now well established that oxygen is one such mutagen. Within the cell, the metabolic conversion of oxygen to reactive forms, particularly the hydroxyl radical, is an initial step in this process. Hydroxyl radical-mediated oxidative damage to DNA results in a variety of mutagenic lesions. However, this is a normal occurrence in every cell, and a host of proteins is involved in surveillance of the genome and removal of the damage. The link between DNA oxidative damage and cancer is evident from animal models lacking these DNA repair and antioxidant proteins. Mice deficient in repair of DNA oxidative damage or reactive oxygen species (ROS) detoxification typically are susceptible to cancer. Mitochondria, as a major source of intracellular ROS, and organelles essential for the maintenance of metabolic homeostasis, can also play critical roles in the initiation and promotion of cancer. Important mechanistic details of mitochondrial participation in tumorigenesis have recently been uncovered.

Keywords Reactive oxygen species (ROS), DNA repair, base excision repair (BER), 8-oxodG, oxidative damage, mitochondria, cancer, mutation, mitochondrial DNA (mtDNA).

1 Introduction

Ageing and cancer are closely related biological phenomena, both linked to oxygen free radicals and the progressive loss of genomic fidelity (Fig. 12.1; [1]). Most human premature ageing disorders (progeria) involve defects in the repair of DNA lesions, and many of these are characterized also by increased incidence of spontaneous tumors [1, 2]. By contrast, it is generally thought that longer lived animals have more robust DNA repair activities that allow them to avoid carcinogenesis for longer. Thus, DNA is considered one of the more important targets of intracellular

From: *Aging Medicine*: *Oxidative Stress in Aging: From Model Systems to Human Diseases* 213
Edited by: S. Miwa, K.B. Beckman, and F.L. Muller © Humana Press, Totowa, NJ

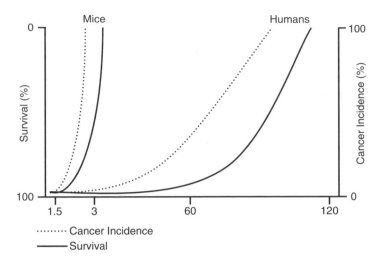

Fig. 12.1 Ageing and cancer. In both mouse and human ageing, mortality and the incidence of cancer are highly correlated (modified from [1])

reactive oxygen species (ROS), with cellular transformation and tumorigenesis being possible outcomes of DNA damage. Here, we outline the evidence that endogenously produced ROS is sufficient to elicit carcinogenesis in mammals. We briefly describe the chemistry and biology of some common oxidative DNA modifications and their repair. We then discuss experimental results implicating deficient DNA repair and aberrant mitochondrial respiration in cancer.

2   Oxygen, Reactive Oxygen, and Cancer

The participation of oxygen in DNA damage, mutation, genomic instability, cellular senescence, and transformation has been demonstrated in mouse embryonic fibroblasts (MEFs) [3]. Generally, fibroblasts growing under standard culture conditions (atmospheric oxygen levels) are actually hyperoxic relative to their environment in vivo (~170 mmHg, versus approximately 30 mmHg in vivo; [4]). By mass action, this abundance of oxygen increases the rate of metabolic superoxide production [5]. MEFs grown under these conditions accumulate DNA mutations and gross chromosomal rearrangements [6]. Within 13 population doublings, the MEFs senesce, followed by the random transformation of some cells, which then multiply indefinitely and take over the culture. A single, simple manipulation, decreasing the oxygen concentration at which the cells are grown to more physiologically relevant levels, is sufficient to reduce mutation accumulation, prevent transformation, and prolong the replicative life span of MEFs indefinitely. This highlights the central role of oxygen in cellular transformation. That this phenomenon is mediated directly by ROS is indicated by experiments such as those of Suh et al. [7], in which the catalytic subunit of the superoxide-generating enzyme NADPH oxidase (*Mox1*) was overexpressed in NIH3T3 cells. This manipulation increased intracellular superoxide production, which of itself was sufficient to produce a transformed phenotype. When transformed cells were implanted into immunodeficient nude mice, they developed into tumors. The above-mentioned results together support the connection between oxygen, ROS, and cancer. However, in both instances, the levels of oxygen that cells were exposed to in culture were well above physiologically normal levels. It is important to consider whether animal cells produce enough ROS endogenously to elicit similar effects.

This question has been addressed using superoxide dismutase (SOD) gene knockout mice, in which oxygen levels are normal, but the ability to detoxify metabolic superoxide is compromised. Mice null for the mitochondrial SOD (manganese SOD; MnSOD) gene are either not viable [8] or they live for <1 month [9], at which time death results from massive oxidative damage in highly metabolic tissues, particularly heart and brain. However, mice containing a single copy of the MnSOD gene, surprisingly, have normal life spans [10]. MnSOD activity in the major organs (liver, heart, brain, and kidney) of the heterozygotes is reduced to approximately 50% of wild type. These mice therefore provide a useful model to investigate the relationship between mitochondrial ROS metabolism and carcinogenesis. The heterozygotes accumulate higher levels of DNA oxidative damage in both mitochondrial and nuclear DNA of most tissues by 26 months, and they have an increased incidence of tumor formation, particularly lymphomas. This evidence from MnSOD$^{+/-}$ mice supports the idea that normal levels of mitochondrial superoxide production are sufficient to induce carcinogenesis.

Evidence from mice deficient in cytosolic SOD acitivity (CuZnSOD$^{-/-}$) also indicates that endogenous ROS production is sufficient to induce tumor formation [11]. Although disruption of CuZnSOD is less acutely lethal than MnSOD, CuZnSOD$^{-/-}$ mice have significantly reduced life spans (25 months maximum life span *vs* 36 months for wild type) and higher levels of cellular, including DNA, oxidative damage. Increased incidence of cancer is observed in these mice, but, interestingly, only hepatocarcenomas. This is coincident with elevated levels of protein and lipid oxidative damage and near-complete inactivation of the cytosolic isoform of aconitase, indicating massive cytosolic oxidative stress. The absence of cancers in other tissues of the CuZnSOD$^{-/-}$ mouse is perhaps surprising. However, this could be related to the fact that CuZnSOD activity in wild-type mice is 10-fold higher in liver than in other highly oxidative tissues, which may reflect its particular importance in hepatocytes, which are the sites of extensive xenobiotic metabolism.

Thus, the evidence from the SOD knockout mouse models indicates that endogenously produced superoxide is sufficient to promote tumorigenesis in vivo. However, a complication in interpreting results from these mice is that altered ROS metabolism results in indiscriminate oxidative damage to all cellular macromolecules, including lipid, protein, and DNA. The outcome of this cumulative damage includes dramatic and broad reductions in enzymatic and mitochondrial function [12]. Therefore, although DNA oxidative damage certainly occurs in these experimental models, it may not be the proximal cause of carcinogenesis. Rather, establishing a direct link among endogenous ROS production, DNA oxidative damage, and carcinogenesis requires an experimental approach that specifically targets genes whose products repair ROS-mediated damage. In these experimental models, endogenous ROS production is not itself manipulated, so that levels of protein and lipid oxidation remain unaltered, but the ability to repair DNA oxidative damage is removed. This approach has allowed for a very specific examination of how the accumulation of DNA oxidative damage influences cancer susceptibility.

3   ROS-Mediated DNA Damage

The various reactions of ROS with DNA have been comprehensively reviewed [13]. A variety of base and sugar backbone modifications has been described, although by far most experimental attention has focused on 8-hydroxy-7,8-deoxyguanine (8-oxodG). Many of the resultant DNA adducts, including 8-oxodG, are highly mutagenic, and some also interfere with replication and/or transcription.

Hydroxyl radical (HO·), generated by Fe^{2+} catalyzed reactions with superoxide and hydrogen peroxide (Fenton reactions), is the strongest oxidant amongst the endogenous ROS, and it is this species that directly mediates most DNA damage [13]. Attack by HO· of the C8-position of guanine leads to formation of 8-hydroxyguanine. Similarly, 8-hydroxyadenine can be formed from HO· attack of adenine. Hydroxyl radical also can add to the C5- and C6-positions of thymine, initiating a series of oxidative reactions that can result in formation of thymine glycol. Furthermore, reaction of HO· with the sugar moiety of DNA can lead to formation of abasic sites and strand breaks. All of these base modifications, and abasic sites, are substrates for DNA repair pathways, and dozens of nuclear proteins participate in surveillance and repair of genomic DNA (reviewed in [14]).

Nucleotide pools are vulnerable to oxidative reactions, and HO· also can modify unincorporated nucleotide triphosphates prior to their incorporation into DNA (reviewed in [15]). Thus, 8-hydroxy-7,8-deoxyguanine triphosphate (8-oxodGTP) can represent a significant proportion of the overall dGTP pool under highly oxidative conditions, and its incorporation into DNA in place of dGTP is mutagenic, leading to G:C to T:A transversions.

4   8-OxodG and Cancer

8-oxodG is by far the most studied form of oxidative damage in DNA, thought to represent 5% of all oxidative lesions, making it perhaps the most common oxidative base modification [13]. 8-oxodG is a premutagenic lesion, and it is frequently mispaired with adenine during DNA replication [16]. Several enzymes act to minimize the mutagenic potential of 8-oxodG. First, 8-oxodGTP formed in nucleotide pools is eliminated by the enzyme mut T homologue (MTH1; named from the bacterial mut T), which hydrolyzes 8-oxodGTP to 8-oxodGMP, thus preventing incorporation of the former into DNA. Second, 8-oxodG formed opposite cytosine within DNA is excised by oxoguanine DNA glycosylase (OGG1), which initiates strand repair via the base excision repair (BER) pathway (Fig. 12.2). Third, unrepaired 8-oxodG can promote mispairing with adenine during replication, resulting in 8-oxodG:A mispairs and subsequent G:C to T:A transversions [16]. The enzyme mut Y homologue (MYH) catalyzes removal of adenine from these mispairs; thus, it works to counteract the mutagenicity of 8-oxodG.

These three enzymes, MTH1, OGG1, and MYH, have all been studied by targeted gene disruption in mice, allowing their association with cancer to be investigated (Table 12.1). Disruption of the Mth1 gene increases the spontaneous mutation rate in embryonic stem cells approximately 2-fold [17]. In Mth1$^{-/-}$ mice, an increased incidence of multiple tumors is found in lung, liver, and stomach (Table 12.1), in the absence of exogenous stressors. These results demonstrate that the ability of MTH1 to prevent mutagenesis from endogenous 8-oxodG formation suppresses tumor formation, a result that directly links 8-oxodG to cancer susceptibility in mice.

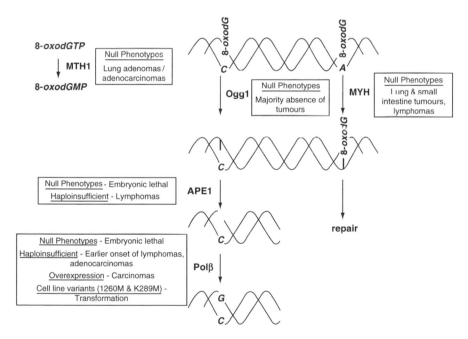

Fig. 12.2 Oxidative damage, base excision repair and carcinogenesis. 8-oxodG is the most common oxidative base lesion. Cells possess various repair mechanisms that counter the mutagenicity of this lesion. MTH1 acts to sanitize the dNTP pool by hydrolyzing 8-oxodGTP to 8-oxodGMP, thereby reducing misincorporation of modified deoxynucleotide triphosphates into the DNA strand. 8-oxodG formed within DNA is excised by Ogg1. 8-oxodG-adenine mispairs are recognized by MYH, which removes the adenines, followed by repair. The removal of a base, such as 8-oxodG, results in formation of an AP site, which is processed by APE1, an enzyme in the BER pathway. polβ then incorporates the correct nucleotide and the nicked backbone is sealed with ligase. Shown in text boxes are the effects of gene deletion, mutation or overexpression

Table 12.1 Genetic mouse models of DNA oxidative damage repair deficiency: effects on cancer

Enzyme	Treatment	Target	Phenotype	Mouse age	Localization	Reference
8-oxodG						
Mth1-/-	None	Lung	2-fold increase in mutation rate	82 weeks	Nuc/Mito	[17]
			Lung tumors	82 weeks		
Ogg1-/-	None		1.7-fold increase in 8-oxodG (liver nDNA)	13–15weeks	Nuc/Mito	[18]
			No tumors	32–44 weeks		
Ogg1-/-	None		4.2-fold increase in 8-oxodG (liver DNA)	9–14 weeks	Nuc/Mito	[19]
			No tumors	50 weeks		
Ogg1-/-	KBrO$_3$	Kidney	250-fold increase in 8-oxodG (kidney DNA)	52 weeks	Nuc/Mito	[20]
			No tumors	52 weeks		
Ogg1-/-	UVB light	Skin	Squamous cell carcinomas, sarcomas	30–44 weeks	Nuc/Mito	[23]
Myh-/-	None	ES cell	2-fold increase in mutation rate		Nuc/Mito	[21]
Myh/Ogg-/-	None	Lung	3–4 fold increase in 8-oxodG (lung, small intestine)	16–64 weeks	Nuc/Mito	[76]
			Lymphomas, tumors in lung, small intestine	52–72 weeks		[25]
BER						
Ape-/-	None	Blastocytes	Embryonic lethal	E7.5	Nuc/Mito	[28]
	IR		Increased cell loss	E3.5		
Ape-/+	None	Neurons/MEFs	Lymphomas		Nuc/Mito	[30]
	Paraquat/menadione		Decreased cell survival			
Polβ-/-	None		Embryonic lethal	E18.5	Nuc	[29]
Polβ-/+	None		Higher incidence and earlier onset of lymphomas and adenocarcinomas	104–112 weeks	Nuc	[41]
Polβ overexpressor	Inoculation of clones	Subcutaneous	Carcinoma at site of injection		Nuc	[40]
PolβI260M	None	C127 MEFs	Induce focus formation and anchorage independent growth (transformation)		Nuc	[38]
PolβK289M	None	LN12 MEFs	2.5-fold increase in mutation frequency		Nuc	[37]
PolβK289M Fen1-/-	None		Embryonic lethal	E3.5–E4.5	Nuc	[77]
Fen1-/+	IR	Blastocytes	Increased apoptosis	E3.5	Nuc	[77]
Fen1-/+	None		Non-hodgkins lymphoma B cell type	52 weeks	Nuc	[78]
PolδD407A	None		Lymphomas and carcinomas	8 weeks	Nuc	[79]
PolδD400A	None		Lymphomas and carcinomas	76 weeks	Nuc	[80]

Ogg1, oxo-8dG DNA glycosylase; BER, base excision repair; Mth1, Mut T homolog; Myh, Mut Y homologue; APE, apurinic/apyrimidinic endonuclease; FEN1, flap endonuclease; IR, ionizing radiation; MEFs, mouse embryonic fibroblasts; ES cell, embryonic stem cell; Nuc, nuclear; Mito, mitochondrial.

Surprisingly, in the majority of Ogg1$^{-/-}$ mice produced, susceptibility to cancer has typically not been associated with the absence of OGG1 activity in mice [18–20], although in a single study Hirano et al. [21] provide evidence for elevated incidence of lung tumors (Table 12.1). The elevation of 8-oxodG levels in nuclear DNA of Ogg1$^{-/-}$ mice is approximately 2-fold, and the mutation rate is increased by a similar amount. Even chronic treatment of Ogg1$^{-/-}$ mice with the exogenous oxidative stressor KBrO$_3$, which in one experiment increased 8-oxodG levels in kidney DNA by 75-fold, did not induce tumor development [20, 22]. Conversely, Kunisada et al. [23] reported an increase in 8-oxodG and skin cancer in Ogg1$^{-/-}$ mice exposed to UVB radiation, which induces ROS. Also, Ogg1 mutations have been identified as risk factors for developing lung cancer in humans [24]. Thus, there may be important differences between OGG1 function in mice and humans, and perhaps an interaction with genetic background.

One possible explanation for the absence of phenotype in most of the Ogg1$^{-/-}$ mice is that MYH activity provides an efficient backup, removing most adenines from 8-oxodG:A mispairs formed in DNA (Fig. 12.1). To investigate this possibility, Ogg1$^{-/-}$Myh$^{-/-}$ mice were developed. Although Myh gene disruption alone has no effect on tumorigenesis in mice [25], simultaneous disruption of both Ogg1 and Myh significantly increases levels of 8-oxodG in liver, lung, and small intestine (although not in brain, kidney, or spleen), and increased the incidence of tumors, including lung adenomas, lymphomas, and cancers of the reproductive system (Table 12.1). Ogg1$^{-/-}$Myh$^{-/-}$ mice had significantly reduced life spans. Perhaps most interesting, however, was that G-to-T transversions in Ogg1$^{-/-}$Myh$^{-/-}$ mice activated the *K-ras* oncogene, which is also often mutated in human lung tumors [25]. This suggests a direct pathway whereby 8-oxodG-induced mutation of an oncogene promotes carcinogenesis. In humans, mutations in Myh alone are sufficient to promote cancer. Heritable Myh mutations are associated with colorectal adenomatous polyposis. As in the Ogg1$^{-/-}$Myh$^{-/-}$ mouse, an activating G-to-T transversion is commonly observed in the *K-ras* oncogene of these tumors [26].

5 DNA Base Excision Repair and Cancer

In addition to 8-oxodG, numerous other base modifications occur in DNA due to the persistent attack of endogenously produced ROS. Of these, a majority (including 8-oxodG) are repaired by the BER pathway (Fig. 12.2), which is itself considered a tumor suppressor mechanism [27]. In its simplest form, BER includes five sequential steps: (1) recognition and removal of damaged bases, (2) incision of the resultant abasic sites, (3) processing the intermediate DNA ends, (4) DNA synthesis, and (5) ligation. In addition to repair of oxidatively damaged bases, the BER pathway participates in repair of abasic sites and single-strand breaks [14]. BER is divided into two subpathways, short-patch and long-patch, based on the number of nucleotides excised and replaced. Common to both BER subpathways are a number of DNA glycosylases, of which OGG1 is one. Glycosylase activity generates an apurinic/apyrimidinic (AP) site, which is subsequently processed by AP endonuclease (APE1), after which the dedicated DNA polymerase β (polβ) adds nucleotides as required. In the long-patch pathway (not shown in Fig. 12.2), additional DNA polymerases and other enzymes further extend the repair patch before ligation, whereas in short-patch BER polβ repair synthesis is followed immediately by ligation to restore the DNA strand.

The relationship of BER to cancer susceptibility has been studied in a number of mouse models in which individual BER genes have been targeted for disruption. Deletion of many BER genes is incompatible with mammalian life, and this is true for both Ape1 and polβ (Table 12.1) [28, 29]. However, two research groups have characterized Ape1[+/-] mice [30, 31]. These mice show evidence of oxidative stress, and they have increased spontaneous mutation rates. Tumor susceptibility in these mice has not been extensively documented, although Meira et al. [30] suggest increased incidence of several tumors in a small-scale study (Table 12.1). Also, interpretation of the role of APE1 deficiency in cancer is complicated by the fact that this protein has functions other than in BER, including its role as a redox-regulated transcription factor [32]. In addition, an APE1-independent pathway of BER involving Nei-like (NEIL) proteins 1 and 2 has been described (reviewed in [33]). This alternative pathway nonetheless requires polβ, an enzyme that plays a critical role in oxidative damage repair.

polβ is a particularly good experimental target for understanding the significance of oxidative DNA lesions in cancer susceptibility. This enzyme has no role in DNA replication, but rather it is a dedicated repair polymerase. It plays a critical role in nuclear BER, with perhaps all BER-catalyzed repair of oxidative DNA damage requiring its participation. Oxidative lesions requiring polβ-mediated repair have been estimated to be up to 100,000 per cell per day [34]. Also, in addition to its nucleotide incorporation activity, polβ catalyzes the deoxyribose phosphate (dRP) lyase reaction that is thought to be the "rate-limiting" step in BER [35]. Approximately 30% of all human tumors express mutant forms of DNA polβ [36], suggesting that somatic mutation of this gene is linked to cancer. This idea is supported by experiments demonstrating that expression of common mutants of polβ in mammalian cell lines induces mutagenesis and cellular transformation ([37, 38]; Table 12.1). For example, the K289M mutation produces a highly mutagenic polβ, in which both affinity for gapped DNA and base-pairing fidelity are compromised. Expression of the K289M polβ variant in mouse LN12 cells (twofold relative to wild type) results in a 2.5-fold increase in mutation frequency [37]. In a separate study, expression of either the I260M or K289M polβ mutants (1.5-fold relative to wild type) in mouse C127 cells induced transformation, although this was not observed when wild-type polβ was overexpressed at a similar level [38]. These results suggest that aberrant activities of mutant polβ cause mutagenesis and cellular transformation. However, it should be noted that other authors have found the BER pathway highly sensitive to overall polβ activity. In an in vitro assay measuring nucleotide incorporation, overexpression of wild-type polβ activity (sixfold) induced frameshift mutations [39]. Similarly, Bergoglio et al. [40] showed that overexpression of wild-type polβ (between 2.4- and 4.5-fold) in Chinese hamster ovary cells promoted tumorigenesis when these cells were injected into nude mice. Underexpression of polβ, as in haploinsufficient polβ mice, is also associated with increased cancer incidence, including carcinomas and lymphomas [41]. Together, these results indicate how any perturbation of native polβ activity promotes mutagenesis and cancer, suggesting that this activity is particularly critical. By implication, this also directly links the inability to efficiently repair DNA oxidative lesions arising from endogenous ROS to cancer.

Long patch BER, which requires the participation of additional enzymes, including flap endonuclease 1 (FEN1) and polymerase δ, also is implicated in oxidative damage repair and therefore tumor suppression ([14]; Table 12.1). Similarly, deficient repair of both single and double-strand breaks is carcinogenic. However, because the proteins involved in these repair pathways participate in numerous other DNA transactions, including replication, directly relating their deficiencies to ROS-mediated damage repair is less straightforward.

6 Mitochondria, ROS, and Cancer

Given the link between DNA oxidative damage and cancer, it seems intuitive that mitochondria, as the major source of ROS in most animal cells, should play a role in carcinogenesis. At least three distinct mechanisms for mitochondrial participation in carcinogenesis can be identified: (1) the mutagenicity of respiratory ROS, (2) the mitogenic potential of respiratory ROS, and (3) the interaction of mitochondrial dysfunction with intracellular signaling pathways (including apoptosis). Indeed, there is evidence for all three modes of action.

The general link between dysregulated mitochondrial ROS metabolism and carcinogenesis has been made in MnSOD$^{+/-}$ mice [42]. Similarly, mitochondrial dysfunction leading to genomic instability can be reversed with an antioxidant (N-acetyl cysteine) in mouse one-cell zygotes, suggesting that aberrant mitochondrial ROS metabolism is responsible for this phenotype [43]. These results indicate a causal role for mitochondrial ROS in cancer. Because ROS are both mutagens and mitogens, the participation of mitochondria in cancer might include contribution to both initiation and promotion events. Consistent with the latter role, overexpression of MnSOD in cancerous cells (which often underexpress MnSOD) significantly inhibits their growth [44], suggesting that mitochondrial superoxide plays an important mitogenic role. Similarly, a T8993G mutation of the ATP6 subunit of mitochondrial ATP synthase inhibits oxidative phosphorylation, increases intracellular oxidative stress, and alters apoptotic signaling pathways. Cybrid cells containing these mutant mitochondria produce highly aggressive tumors upon injection into nude mice [45, 46]. These results demonstrate the ability of aberrant mitochondrial function to promote dysregulated cell growth.

Several inherited respiratory disorders shed light on some of the specific mechanisms by which mitochondrial dysfunction can promote carcinogenesis. For example, heterozygous mutations in three of the four nuclear genes (SDHB, SDHC, SDHD) encoding subunits of respiratory complex II (succinate dehydrogenase) have been linked with development of paragangliomas and phaeochromocytomas [47–49]. These mutations prevent assembly of a functional complex II, and they increase mitochondrial ROS production [50]. Although these results are consistent with a straightforward ROS-mediated mechanism of cancer initiation, Selak et al. [51] have shown how a second mechanism is likely to contribute to the progression of these cancers. The metabolic alterations resulting from absence of competent respiratory complex II induce a stabilization of hypoxia inducible factor-1α (HIF-1α) in the SDHC mutant cells, via succinate inhibition of prolyl-hydroxylase (PHD) (Fig. 12.3).

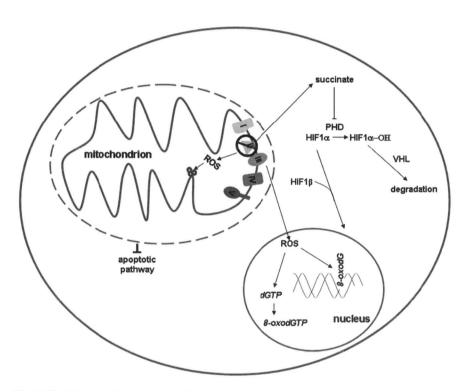

Fig. 12.3 Mitochondria and cancer. Shown are three mechanisms whereby mitochondria seem to contribute to carcinogenesis. (1) Mitochondria are the main source of ROS in many cells, and increased ROS production and associated oxidative lesions have been associated with tumorigenesis. Mitochondrial ROS damage both mitochondrial (represented as small circles) and nuclear DNA. (2) Defects in the nuclear-encoded complex II (shown as circled and struck) of the electron transport chain are associated with cancer and the Warburg effect. The accumulation of succinate inhibits the activity of PHD by product inhibition. This blocks the PHD-catalyzed hydroxylation of HIF-1α and subsequent VHL factor-mediated degradation. Therefore, even under normoxic conditions, HIF-1α accumulates and forms an active transcriptional complex with HIF-1β, increasing the transcription of enzymes associated with glycolysis and angiogenesis. This contributes to the Warburg effect, i.e., the increased reliance of cancer cells on aerobic glycolysis. (3) Mitochondria from cancer cells can be resistant to the induction of outer membrane permeabilization and thus release of apoptotic factors

HIF-1α, normally stabilized only under hypoxic conditions, is thus persistent under normoxic conditions. A similarly stabilization of HIF-1α under normoxia also is seen in other cancers (e.g., leiomyoma, leiomyosarcoma). Normally, Von Hippel Lindau (VHL) protein is involved in the rapid degradation of HIF-1α under normoxia [52]. In some cancers, however, such as clear cell renal carcinomas, VHL is dysfunctional and therefore unable to direct the degradation of HIF-1α. Stabilized HIF-1α binds HIF-1β to form a heterodimer that promotes transcription of approximately 100 genes, many of which are involved in glycolysis and angiogenesis, which are permissive of rapid tumor growth. These observations may offer an explanation for the "Warburg effect," i.e., the reliance of cancerous cells on incomplete oxidation of glucose to lactate even under normoxic conditions [53, 54]. Indeed, HIF-1α also is reported to be stabilized by mitochondrial ROS [55], suggesting multiple mechanisms by which aberrant mitochondrial metabolism could alter a putatively hypoxia-related gene transcriptional program. Interestingly, a second nuclear-encoded mitochondrial enzyme, fumarase, has been linked to tumorigenesis by a similar mechanism [56], suggesting that HIF-1α stabilization via dysfunctional oxidative phosphorylation might be a pathway common to many different cancers.

An important question in understanding a potential mitochondrial connection in cancer concerns whether random mitochondrial DNA (mtDNA) mutations in somatic cells can similarly promote cellular transformation. A high proportion of human cancers contain mtDNA homoplasmic for mutations that are absent in both germline cells and surrounding tissue cells [57]. Although some of these mutations are apparently innocuous or have even been erroneously identified [58], it remains the case that (1) a high proportion of cancers harbor homoplasmic mtDNA mutations, and (2) there is an apparent selection of mutant mtDNAs during the transformation process. However, the development of homoplasmy could occur through essentially random events [59], particularly in cells that are no longer as reliant on oxidative phosphorylation, in which case there may be little selective pressure to maintain wild-type mtDNA. Two experimental models provide the means to test whether random mtDNA mutations in somatic cells can cause cancer: mtDNA mutator mice [60, 61] and Ogg1$^{-/-}$ mice. In the former model, two groups have engineered mice with a defective polγ proofreading domain that accumulate mtDNA mutations in somatic cells at a greatly accelerated rate. Although these mice do evidence mitochondrial respiratory dysfunction and have greatly reduced life spans, there is no documented increase in the incidence of either oxidative stress or cancer [62]. Similarly, in the second experimental model, Ogg1$^{-/-}$ mice accumulate 8-oxodG in liver mtDNA to levels 20-fold higher than in wild-type mice [63], but they show no evidence of mitochondrial dysfunction or oxidative stress [64] and they are not cancer-prone.

mtDNA sustains high levels of oxidative damage in all mammalian cells, but despite frequent assertions to the contrary, repair of mtDNA oxidative damage is relatively efficient, if of limited scope (reviewed in [65]). Mammalian mitochondria possess a short-patch BER pathway. Several of the short-patch nuclear BER enzymes that function in the nucleus also localize to mitochondria, including OGG1 and APE1. MTH1 and MYH also are found in mitochondria [66, 67], which could explain the absence of effect of Ogg1 gene disruption on mitochondrial function.

It will therefore be interesting to investigate mitochondrial function/dysfunction in the Ogg1$^{-/-}$Myh$^{-/-}$ mice, which do develop cancer. Thus, there is strong evidence for mitochondrial involvement in both the initiation and promotion stages of cancer, and some of the specific molecular mechanisms connecting aberrant function and tumorigenesis have been elucidated (Fig. 12.3).

7   Conclusions

Genetically modified mice have proved invaluable tools for relating the intracellular production of oxygen radicals to carcinogenesis in mammals. Endogenous ROS production, via mitochondrial oxidative metabolism or cytosolic enzymes, provides both a source of mutagen and a potential mitogen in mammalian cells. Direct oxidative damage of DNA, via the hydroxyl radical in particular, is one means by which endogenous ROS exert their mutagenic effects. The detrimental effects of endogenous ROS are normally efficiently countered by DNA repair enzymes that remove mutagenic oxidative lesions or antioxidant enzymes that prevent them from forming. However, perturbations that compromise repair or increase mitochondrial ROS production will upset this balance and thus create an intracellular milieu that is permissive of carcinogenesis.

This general scheme is supported by a number of complementary studies done with calorie-restricted (CR) mammals. CR is a well characterized experimental model in which reduced energy intake confers enhanced longevity and resistance to cancer in most mammals (reviewed in [68]). It seems to affect many of the same molecular targets identified in genetically modified mice. In mice and rats, CR reduces mitochondrial ROS production [69–71], enhances endogenous antioxidant defenses [72] and stimulates nuclear BER activity [73] via induction of polβ [74]. These adjustments are accompanied by a significant decrease in the level of 8-oxodG in both nuclear and mitochondrial DNA of CR mice [75]. Based on the findings outlined above, it can be concluded that these responses must contribute directly to the cancer resistance of CR mice.

The findings with respect to ROS and cancer incidence are readily integrated into Harman's free radical theory of ageing. Cancer-related mortality is one determinant of life span in mammals. Indeed, the cancer susceptibility of antioxidant and DNA repair-deficient mice can be viewed as a case study within Harman's theory, illustrating both the stochastic nature of the process of free radical-mediated cellular dysfunction and the increased probability of such dysfunction that is associated with aberrant ROS metabolism.

References

1. Campisi J. Cancer and ageing: rival demons? Nat Rev Cancer 2003; 339–349.
2. Shiloh Y, Lehmann AR. Maintaining integrity. Nat Cell Biol 2004; 6:923–928.

3. Parrinello S, Samper E, Krtolica A, Goldstein J, Melov S, Campisi J. Oxygen sensitivity severely limits the replicative lifespan of murine fibroblasts. Nat Cell Biol 2003; 5:741–7.
4. Brown MF, Gratton TP, Stuart JA. Metabolic rate does not scale with body mass in cultured mammalian cells. Am J Physiol 2007; 292:R2115–R2121.
5. Turrens JF. Mitochondrial formation of reactive oxygen species. J Physiol 2003; 552:335–344.
6. Busuttil RA, Rubio M, Dolle MET, Campisi J, Vijg J. Oxygen accelerates the accumulation of mutations during the senescence and immortalization of murine cells in culture. Aging Cell 2003; 2:287–294.
7. Suh Y-A, Arnold RS, Lassegue B, Shi J, Xu X, Sorescu D, Chung AB, Griendling KK, Lambeth JD. Cell transformation by the superoxide-generating oxidase Mox1. Nature 1999; 401:79–82.
8. Li Y, Huang TT, Carlson EJ et al. Dilated cardiomyopathy and neonatal lethality in mutant mice lacking manganese superoxide dismutase. Nat Genet 1995; 11:376–381.
9. Lebowitz RM, Zhang H, Vogel H et al. Neurodegeneration, myocardial injury, and perinatal death in mitochondrial superoxide dismutase-deficient mice. Proc Natl Acad Sci U S A 1996; 93:9782–9787.
10. Van Remmen H, Ikeno Y, Hamilton M et al. Life-long reduction in MnSOD activity results in increased DNA damage and higher incidence of cancer but does not accelerate aging. Physiol Genomics 2003; 16:29–37.
11. Elchuri S, Oberley TD, Qi W et al. CuZnSOD deficiency leads to persistent and widespread oxidative damage and hepatocarcinogenesis later in life. Oncogene 2005; 24:367–380.
12. Williams MD, Van Remmen H, Conrad CC, Huang TT, Epstein CJ, Richardson A. Increased oxidative damage is correlated to altered mitochondrial function in heterozygous manganese superoxide dismutase knockout mice. J Biol Chem 1998; 273:28510–28515.
13. Dizdaroglu M, Jaruga P, Birincioglu M, Rodriguez H. Free radical-induced damage to DNA: mechanisms and measurement. Free Radic Biol Med 2002; 32:1102–1115.
14. Fortini P, Dogliotti E. Base damage and single-strand break repair: mechanisms and functional significance of short- and long-patch repair subpathways. DNA Repair 2007; 6:398–409.
15. Nakabeppu Y, Kajitani K, Sakamoto K, Yamaguchi H, Tsuchimoto D. MTH1, an oxidized purine nucleoside triphosphatase, prevents the cytotoxicity and neurotoxicity of oxidized purine nucleotides. DNA Repair 2006; 5:761–772.
16. Russo MT, De Luca G, Degan P, Bignami M. Different DNA repair strategies to combat the threat from 8-oxoguanine. Mutat. Res. 2007; 614:69–76.
17. Tsuzuki T, Egashira A, Igarashi H et al. Spontaneous tumorigenesis in mice defective in the MTH1 gene encoding 8-oxodGTPase. Proc Natl Acad Sci U S A 2001; 98:11456–11461.
18. Klungland A, Rosewell I, Hollenbach S et al. Accumulation of premutagenic DNA lesions in mice defective in removal of oxidative base damage. Proc Natl Acad Sci U S A 1999; 96:11300–13305.
19. Minowa O, Arai T, Hirano M et al. *Mmh/Ogg1* gene inactivation results in accumulation of 8-hydroxyguanine in mice. Proc Natl Acad Sci U S A 2000; 97:4156–4161.
20. Arai T, Kelly VP, Minowa O, Noda T, Nishimura S. The study using wild-type and Ogg1 knockout mice exposed to potassium bromate shows no tumor induction despite an extensive accumulation of 8-hydroxyguanine in kidney DNA. Toxicology 2006; 221:179–186.
21. Hirano S, Tominaga Y, Ichinoe A et al. Mutator phenotype of MUTYH-null mouse embryonic stem cells. J Biol Chem 2003; 278:38121–38124.
22. Arai T, Kelly VP, Minowa O, Noda T, Nishimura S. High accumulation of oxidative DNA damage, 8-hydroxyguanine, in *Mmh/Ogg1* deficient mice by chronic oxidative stress. Carcinogenesis 2002; 23:2005–2010.
23. Kunisada M, Sakumi K, Tominaga Y et al. 8-oxoguanine formation induced by chronic UVB exposure makes *Ogg1* knockout mice susceptible to skin carcinogenesis. Cancer Res 2005; 65:6006–6010.
24. Paz-Elizur T, Krupsky M, Blumenstein S, Elinger D, Schechtman E, Livneh Zvi. DNA repair activity for oxidative damage and risk of lung cancer. J Natl Cancer Inst 2003; 95:1312–1319.

25. Xie Y, Hanjing H, Cunanan C et al. Deficiencies in mouse *Myh* and *Ogg1* result in tumor predisposition and G to T mutations in Codon 12 of the *K-Ras* oncogene in lung tumors. Cancer Res 2004; 64:3096–3102.
26. Sampson JR, Jones S, Dolwani S, Cheadle JP. MutYH (MYH) and colorectal cancer. Biochem Soc Trans 2005; 33:679–683.
27. Sweasy JB, Lang T, DiMaio, D. Is base excision repair a tumor suppressor mechanism? Cell Cycle 2006; 5:250–259.
28. Ludwig DL, MacInnes MA, Takiguchi Y et al. A murine AP-endonuclease gene-targeted deficiency with post-implantation embryonic progression and ionizing radiation sensitivity. Mutat Res 1998; 409:17–29.
29. Sugo N, Aratani Y, Nagashima Y, Kubota Y, Koyama H. Neonatal lethality with abnormal neurogenesis in mice deficient in DNA polymerase β. EMBO J 2000; 19:1397–1404.
30. Meira LB, Devaraj S, Kisby GE et al. Heterozygosity for the mouse *Apex* gene results in phenotypes associated with oxidative stress. Cancer Res 2001; 61:5552–5557.
31. Huamani J, McMahan CA, Herbert DC et al. Spontaneous mutagenesis is enhanced in *Apex* heterozygous mice. Mol Cell Biol 2004; 24:8145–8153.
32. Evans AR, Limp-Foster M, Kelley MR. Going APE over ref-1. Mutat Res 2000; 461:83–108.
33. Hazra TK, Das A, Das S, Choudhury S, Kow YW, Roy R. Oxidative DNA damage repair in mammalian cells: a new perspective. DNA Repair 2007; 6:470–480.
34. Klaunig JE, Kamendulis LM. The role of oxidative stress in carcinogenesis. Annu Rev Pharmacol Toxicol 2004; 44:239–267.
35. Srivastava DK, Ber BJ, Prasad R et al. Mammalian abasic site base excision repair. Identification of the reaction sequence and rate-determining steps. J Biol Chem 1998; 273:21203–21209.
36. Starcevic D, Dalal S, Sweasy JB. Is there a link between DNA polymerase β and cancer? Cell Cycle 2004; 3:998–1001.
37. Lang T, Maitra M, Starcevic D, Li S-X, Sweasy JB. A DNA polymerase β mutant from colon cancer cells induces mutations. Proc Natl Acad Sci U S A 2004; 101:6074–6079.
38. Sweasy JB, Lang T, Starcevic D et al. Expression of DNA polymerase β cancer-associated variants in mouse cells results in cellular transformation. Proc Natl Acad Sci U S A 2005; 102: 14350–14355.
39. Chan K, Houlbrook S, Zhang Q-M, Harrison M, Hickson ID, Dianov GL. Overexpression of DNA polymerase β results in an increased rate of frameshift mutations during base excision repair. Mutagenesis 2007; 22:183–188.
40. Bergoglio V, Pillaire M-J, Lacroix-Triki M et al. Deregulated DNA polymerase β induces chromosome instability and tumorigenesis. Cancer Res. 2002; 62:3511–3514.
41. Cabelof DC, Ikeno Y, Nyska A et al. Haploinsufficiency in DNA polymerase β increases cancer risk with age and alters mortality rate. Cancer Res. 2006; 66:7460–7465.
42. Van Remmen H, Williams MD, Guo Z et al. Knockout mice heterozygous for *Sod2* show alterations in cardiac mitochondrial function and apoptosis. Am J Physiol 2001; 281:H1422–H1432.
43. Liu L, Trimarchi JR, Smith PJS, Keefe DL. Mitochondrial dysfunction leads to telomere attrition and genomic instability. Aging Cell 2002; 1:40–46.
44. Oberley LW. Mechanism of the tumor suppressive effect of MnSOD overexpression. Biomed Pharmacother 2005; 59:143–148.
45. Petros JA, Baumann AK, Ruiz-Pesini E et al. mtDNA mutations increase tumorgenicity in prostate cancer. Proc Natl Acad Sci U S A 2005; 102:719–724.
46. Shidara Y, Yamagata K, Kanamori T et al. Positive contribution of pathogenic mutations in the mitochondrial genome to the promotion of cancer by prevention from apoptosis. Cancer Res 2005; 65:1655–1663.
47. Baysal BE, Ferrell RE, Willett-Brozick JE et al. Mutations in SDHD, a mitochondrial complex II gene, in hereditary paraganglioma. Science 2000; 287:848–851.
48. Eng C, Kiuru M, Fernandez MJ, Aaltonen LA. A role for mitochondrial enzymes in inherited neoplasia and beyond. Nat. Rev. Cancer 2003; 3:193–202.

49. Gottlieb E, Tomlinson IPM. Mitochondrial tumour suppressors: a genetic and biochemical update. Nat Rev Cancer 2005; 5:857–866.
50. Slane BG, Aykin-Burns N, Smith BJ et al. Mutation of succinate dehydrogenase subunit C results in increased $O_2^{\cdot-}$, oxidative stress, and genomic instability. Cancer Res 2006; 66:7615–7620.
51. Selak MA, Armour SM, MacKenzie ED et al. Succinate links TCA cycle dysfunction to onco-genesis by inhibiting HIF-α prolyl hydroxylase. Cancer Cell 2005; 7:77–85.
52. Semenza GL. VHL and p53: tumor suppressors team up to prevent cancer. Mol Cell 2006; 22:437–439.
53. Kim J-W, Dang CV. Cancer's molecular sweet tooth and the Warburg effect. Cancer Res 2006; 66:8927–8930.
54. Semenza GL. HIF-1 mediates the Warburg effect in clear cell renal carcinoma. J Bioenerg Biomemb 2007; 39:231–234.
55. Sanjuan-Pla A, Cervera AM, Apostolova N et al. A targeted antioxidant reveals the impor-tance of mitochondrial reactive oxygen species in the hypoxic signalling of HIF-1alpha. FEBS Lett 2005; 579:2669–2674.
56. Isaacs JS, Jung YJ, Mole DR et al. HIF overexpression correlates with biallelic loss of fumu-rate hydratase in renal cancer: novel role of fumarate in regulation of HIF stability. Cancer Cell 2005; 8:143–153.
57. Zanssen S, Schon EA. Mitochondrial DNA mutations in cancer. PLoS Med 2005; 2:1082–1084.
58. Salas A, Yao Y-G, Macaulay V, Vega A, Carracedo A, Bandelt H-J. A critical reassessment of the role of mitochondria in tumorigenesis. PLOS Med 2005; 2:1158–1166.
59. Coller HA, Khrapko K, Bodyak ND, Nekhaeva E, Herrero-Jiminez P, Thilly WG. High frequency of homoplasmic mitochondrial DNA mutations in human tumors can be explained without selection. Nat Genet 2001; 28:147–150.
60. Trifunovic A, Wredenberg A, Falkenberg M et al. Premature ageing in mice expressing defec-tive mitochondrial DNA polymerase. Nature 2004; 429:417–423.
61. Kujoth GC, Hiona A, Pugh TD et al. Mitochondrial DNA mutations, oxidative stress and apoptosis in mammalian aging. Science 2005; 309:481–484.
62. Trifunovic A, Hansson, A, Wredenber A et al. Somatic mtDNA mutations cause aging pheno-types without affecting reactive oxygen species production. Proc Natl Acad Sci U S A 2005; 102:17993–17998.
63. de Souza-Pinto NC, Hogue BA, Bohr VA. DNA repair and aging in mouse liver: 8-oxodG glycosylase activity increase in mitochondrial but not in nuclear extracts. Free Radic Biol Med 2001; 30:916–923.
64. Stuart JA, Bourque BM, de Souza-Pinto NC, Bohr VA. No evidence of mitochondrial respira-tory dysfunction in OGG1-null mice deficient in removal of 8-oxodeoxyguanine from mito-chondrial DNA. Free Radic Biol Med 2005; 38:737–745.
65. Stuart JA, Brown MF. Mitochondrial DNA maintenance and bioenergetics. Biochim Biophys Acta 2006; 1757:79–89.
66. Ohtsubo T, Nishioka K, Imaiso Y et al. Identification of human MutY homolog (hMYH) as a repair enzyme for 2-hydroxyadenine in DNA and detection of multiple forms of hNYH located in nuclei and mitochondria. Nucleic Acids Res 2000; 28:1355–1364.
67. Nakabeppu Y. Regulation of intracellular localization of human MTH1, OGG1, and MYH proteins for repair of oxidative DNA damage. Prog Nucleic Acid Res Mol Biol 2001; 68:75–94.
68. Hursting SD, Lavigne JA, Berrigan D, Perkins SN, Barrett JC. Calorie restriction, aging, and cancer prevention: mechanisms of action and applicability to humans. Annu Rev Med 2003; 54:131–152.
69. Hagopian K, Harper ME, Ram JJ, Humble SJ, Weindruch R, Ramsey JJ. Long-term calorie restriction reduces proton leak and hydrogen peroxide production in liver mitochondria. Am J Physiol 2005; 288:E674–E684.

70. Sanz A, Caro P, Ibanez J, Gomez J, Gredilla R, Barja G. Dietary restriction at old age lowers mitochondrial oxygen radical production and leak at complex I and oxidative DNA damage in rat brain. J Bioenerg Biomemb 2005; 37:83–90.
71. Bevilacqua L, Ramsey JJ, Hagopian K, Weindruch R, Harper ME. Long-term caloric restriction increases UCP3 content but decreases proton leak and reactive oxygen species production in rat skeletal muscle mitochondria. Am J Physiol 2005; 289:E429–E438.
72. Wu A, Sun X, Wan F, Liu Y. Modulations by dietary restriction on antioxidant enzymes and lipid peroxidation in developing mice. J Appl Physiol 2003; 94:947–952.
73. Stuart JA, Karahalil B, Hogue BA, de Souza-Pinto NC, Bohr VA. Mitochondrial and nuclear DNA base excision repair are affected differently by caloric restriction. FASEB J 2004; 18:595–597.
74. Cabelof DC, Yanamadala S, Raffoul JJ, Guo Z, Soofi A, Heydari AR. Caloric restriction promotes genomic stability by induction of base excision repair and reversal of its age-related decline. DNA Repair 2003; 2:295–307.
75. Hamilton ML, Van Remmen H, Drake JA et al. Does oxidative damage to DNA increase with age? Proc Natl Acad Sci U S A 2001; 98:10469–10474.
76. Russo MT, De Luca G, Degan P et al. Accumulation of the oxidative base lesion 8-hydroxy-guanine in DNA of tumor-prone mice defective in both the *Myh* and *Ogg1* DNA glycosylases. Cancer Res 2004; 64:4411–4414.
77. Larsen E, Gran C, Sæther BE, Seeberg E, Klungland A. Proliferation failure and gamma radiation sensitivity of *Fen1* null mutant mice at the blastocyte stage. Mol Cell Biol 2003; 23:5346–5353.
78. Kucherlapati M, Yang K, Kuraguchi M et al. Haploinsufficiency of Flap endonuclease (Fen1) leads to rapid tumor progression. Proc Natl Acad Sci U S A 2002; 99:9924–9929.
79. Goldsby RE, Lawrence NA, Hays LE et al. Defective DNA polymerase-δ proofreading causes cancer susceptibility in mice. Nat Med 2001; 7:638–639.
80. Goldsby RE, Hays LE, Chen X et al. High incidence of epithelial cancers in mice deficient for DNA polymerase delta proofreading. Proc Natl Acad Sci U S A 2002; 99:15560–15565.

13
Oxidative Stress in Hypertension

Sean P. Didion, Sophocles Chrissobolis, and Frank M. Faraci

Summary Oxidative stress refers to increases in reactive metabolites of molecular oxygen that occur as a result of increases in formation and/or reductions in scavenging or degradation. Oxidative stress plays an important role in the pathogenesis of hypertension. In this chapter, we examine the relationship between oxidative stress in the development and maintenance of hypertension and changes in vascular structure and function. We focus primarily on enzymatic systems that contribute to oxidative stress, such as NAD(P)H oxidase, and those that limit oxidative stress, including superoxide dismutase, glutathione peroxidase, and catalase.

Keywords NAD(P)H oxidase, superoxide dismutase, glutathione peroxidase, catalase.

1 Introduction

Since the discovery of free radicals in biological tissue more than a half a century ago [1], much effort has focused on understanding the role of such radicals in vascular biology and disease, including hypertension. Free radicals, as defined by their chemical nature, refer to molecules with an unpaired electron, thus making them highly reactive. These radicals include superoxide anion and hydroxyl radical, and they have been implicated in the pathogenesis of cardiovascular disease [2]. Reactive oxygen species (ROS) are metabolites of univalent or divalent reduction of molecular oxygen, specifically superoxide and hydrogen peroxide, respectively. Reactive nitrogen species (RNS) include the free radical nitric oxide (NO) and peroxynitrite, the reaction product of NO and superoxide. ROS and RNS serve dual roles both as signaling molecules at low concentrations, and, at higher concentrations, as mediators of cellular damage [2, 3]. Steady-state levels of ROS and RNS are normally maintained by active processes and reflect their rate of production and removal by enzymatic and nonenzymatic systems [3, 4]. The term oxidative stress refers to increases in ROS that occur as a result of increases in formation and/or reductions in scavenging or degradation.

From: *Aging Medicine: Oxidative Stress in Aging: From Model Systems to Human Diseases*   229
Edited by: S. Miwa, K.B. Beckman, and F.L. Muller © Humana Press, Totowa, NJ

In this chapter, we examine the relationship between oxidative stress in the development and maintenance of hypertension as well as changes in vascular structure and function. We focus on enzymatic systems that contribute to oxidative stress, such as NAD(P)H oxidase, and those that limit oxidative stress, including superoxide dismutase (SOD), glutathione peroxidase (GPx), and catalase.

2 Linking Oxidative Stress with Hypertension

It is clear from the evidence in the literature that oxidative stress plays an important role in the pathogenesis of experimental and human hypertension. Some of the first data to implicate ROS in hypertension came from a study by Wei et al. [5], which showed that in response to acute hypertension, superoxide accumulated in the extracellular space and contributed to the impairment of cerebral vascular responses by inactivation of endothelium-derived relaxing factor, now known as NO. Moreover, the impaired vascular responses could be restored toward normal with SOD and catalase, providing strong evidence that acute hypertension produced oxidative stress and impaired vascular responses [5]. Indeed, genetic models of hypertension exhibit reductions in endothelium-dependent relaxation and increased vascular sensitivity to ROS [6–11]. Collectively, these findings underscored the impact of changes in NO bioavailability and that increases in ROS reduce the bioavailability of NO resulting in increases in vascular tone. Thus, alterations in ROS would be predicted to have a profound influence on blood pressure.

To our knowledge, some of the first evidence to implicate a role for superoxide/ROS in chronic hypertension came from two independent studies in which administration of antioxidants lowered blood pressure in spontaneously hypertensive rats (SHR) and in hypertensive individuals [12, 13]. In a study by Nakazono et al. [13], intravenous administration of a fusion protein consisting of human CuZnSOD and a peptide with high affinity for heparin sulfate on endothelial cells produced a reduction in blood pressure in SHR but not in normotensive controls. These studies suggested that superoxide near endothelium contributes to the pathogenesis of hypertension, presumably due to reductions in bioavailable NO. Based on their findings, the authors proposed that hypertension is a "free radical disease." Consistent with this concept, endogenous antioxidant levels and activity are reduced with hypertension [14–18].

Since these initial findings, oxidative stress has been shown to contribute to vascular complications such as endothelial dysfunction and vascular hypertrophy in diverse models of hypertension and in human hypertension (Table 13.1). For example, vascular superoxide levels are increased in animals made hypertensive by infusion of angiotensin II, and manipulations aimed at reducing superoxide (such as viral transfection with genes encoding SOD or treatment with pharmacological scavengers of superoxide) reduce blood pressure and improve vascular function [23, 28–30, 35, 56, 57].

Table 13.1 Examples of vascular oxidative stress and hypertrophy in experimental models of hypertension and hypertension in humans

Experimental model [reference]	Human hypertension [reference]
Adrenocorticoid [20]	Primary (essential) hypertension [44–51]
Angiotensin II [20–26]	Secondary hypertension
Dahl salt-sensitive [27,28]	(e.g., renovascular) [52–55]
Genetic:	
SHR [28–30]	
SHR-stroke prone [31]	
Blood pressure high (BPH) mice [32]	
Human Nox2 (smooth muscle specific) [33]	
Human p22phox (smooth muscle specific) [34]	
Human renin and human angiotensinogen [35,36]	
Human renin [37]	
Human Rac1 [38]	
Mineralocorticoid [39–41]	
Renovascular:	
2-Kidney 1-clip [42]	
1-Kidney 1-clip [43]	

The link between oxidative stress and angiotensin II in hypertension was propelled by the identification that angiotensin II-induced ROS formation in vascular muscle is mediated, at least in part, by vascular NAD(P)H oxidases distinct from that expressed in neutrophils [58]. Administration of angiotensin II in vivo results in increased expression of several components of the vascular NAD(P)H oxidase [20, 24, 59, 60, 61]. Increases in blood pressure and ROS produced by angiotensin II for relatively short periods of time (1–2 weeks) seem to be selective, because hypertension produced by norepinephrine was not affected by antioxidants [20]. The role of ROS in mediating more chronic effects of angiotensin II and norepinephrine are unclear, as there is evidence to suggest that norepinephrine can increase vascular ROS [62].

A major role for angiotensin II in the generation of ROS in hypertension is supported by studies in human hypertension [44–55]. For example, beneficial effects of blood pressure lowering and reduction in markers of oxidative stress occur in hypertensive patients treated with angiotensin-converting enzyme (ACE) inhibitors and angiotensin receptor blockers [63–66]. Together, these studies provide strong evidence for a role for angiotensin II-derived ROS in hypertension.

There are numerous potential enzymatic and nonenzymatic sources of ROS in blood vessels (Table 13.2). However, ROS derived from NAD(P)H oxidase may be the major source of ROS in hypertension. Increases in ROS are typically kept at relatively low levels by the activity of endogenous antioxidant enzymes, including SOD, glutathione peroxidases, and catalase (Table 13.2).

Antioxidant treatment in systems other than the vasculature, such as the kidney and brain, also have beneficial blood pressure-lowering effects in hypertension [67–71], suggesting that renal and central mechanisms that regulate blood pressure

Table 13.2 Major prooxidant and antioxidant enzymes that promote or inhibit vascular oxidative stress

Prooxidant enzyme	Antioxidant enzyme
NAD(P)H oxidase	SOD
Nitric oxide synthase (uncoupled)	Glutathione peroxidase
Mitochondrial electron transport chain (complex I and III)	Catalase
Xanthine oxidase	Nitric oxide synthase (coupled)
Cyclooxygenase	Thioreductase
Cytochrome P450 monooxygenase	
Lipooxygenase	
Myloperoxidase	

also are impacted by oxidative stress. For example, local treatment with the superoxide scavenger Tempol in the kidney and viral-mediated gene transfer of SOD in the brain limit the rise in blood pressure produced by systemic administration of angiotensin II [67, 69]. Interestingly, evidence supports an important role for hydrogen peroxide as plasma hydrogen peroxide levels are increased in hypertensive humans [44, 45]. Consistent with this concept, renal infusion of hydrogen peroxide increases blood pressure in rats [68]. Taken together, these findings implicate a potential role for superoxide and hydrogen peroxide in hypertension.

In addition to ROS, hypertension is associated with increases in peroxynitrite, a marker of oxidative/nitrosative stress and a mediator of cellular dysfunction in blood vessels [3]. Peroxynitrite produces tyrosine nitration of selective proteins, including MnSOD, and activation of poly(ADP-ribose) polymerase, another potentially important mediator of endothelial dysfunction in hypertension [4]. Moreover, peroxynitrite formation has been implicated as a mechanism of endothelial NO synthase (eNOS) uncoupling, due to oxidation of tetrahydrobiopterin and/or the zinc-thiolate complex of eNOS [4].

Clearly, oxidative stress plays a major role in the pathological consequences of hypertension, and although multiple sources of ROS may contribute to oxidative stress, we limit our discussion to NAD(P)H oxidase, which has been identified as a major source of ROS in this disease [72–74]. We also limit our discussion to the role of NAD(P)H oxidase in the vasculature.

Although much attention has focused on the role of enzymes that increase superoxide formation, much less is known regarding alterations in antioxidant enzyme expression and/or activity in relation to blood pressure. This is important because levels of antioxidant enzymes influence the overall level of oxidative stress, and genetic manipulations of these enzyme systems have a major influence on blood pressure and on vascular function and structure [4, 75]. Although nonenzymatic processes and molecules (such as thioredoxin and vitamin C) also affect levels of ROS, we limit our discussion to some of the enzymatic systems that reduce oxidative stress. For each, and where data are available, we discuss the effects of genetic alterations on blood pressure, oxidative stress, and blood vessels.

3   NAD(P)H Oxidase: A Major Prooxidant Enzyme

The vascular NAD(P)H oxidase is a multisubunit complex consisting of the membrane-bound subunits of the Nox isoforms (a gp91^{phox} homolog of the neutrophil oxidase) and p22^{phox}, and the cytosolic subunits p47^{phox}, p67^{phox}, and Rac1 (a G protein) [73]. The Nox isoforms and p22^{phox} are integral components of the oxidase, and collectively they form the membrane component necessary for the transfer of electrons from NAD(P)H to oxygen [73]. Expression of the various components of the oxidase varies within the vasculature and within cells types [76, 77]. Hypertension produced by angiotensin II infusion produces increased expression of gp91^{phox}, p22^{phox}, and p67^{phox} and increased NAD(P)H oxidase activity in the vasculature [22, 25, 60, 61]. Pharmacologic or molecular inhibition of NAD(P)H oxidase lowers blood pressure, reduces oxidative stress, and improves vascular responses in several models of hypertension [74, 78]. Most of the NAD(P)H oxidase subunits have been genetically manipulated in mice, and they have been found to have significant effects on blood pressure in models of hypertension (discussed below). In addition, ACE inhibitors and angiotensin receptor blockers lower blood pressure and decrease ROS (and reduce NAD(P)H oxidase expression) in SHR and stroke-prone SHR [79, 80]. Together, these findings provide strong evidence for NAD(P)H oxidase in the regulation of blood pressure and vascular function during angiotensin II-induced hypertension.

3.1   Nox Isoforms

Although there are several isoforms of Nox proteins, including Nox1–5 and Duox1-2, we discuss those isoforms primarily restricted to the vasculature. Blood vessels have been found to express Nox1, Nox2 and Nox4 [81–83]. Nox1 is expressed in both endothelium and smooth muscle, Nox2 is expressed in endothelium and adventitia fibroblasts and in smooth muscle of resistance vessels (but not in large blood vessels), whereas Nox4 is expressed in all vascular cell types [76]. Because NAD(P)H oxidase is a major source of superoxide and all Nox isoforms function as catalytic subunits of the NAD(P)H oxidase, alterations in Nox isoforms would be predicted to have profound effects on vascular function and structure and on blood pressure.

3.1.1   Nox1

Under baseline conditions, blood pressure is not altered in mice deficient in expression in Nox1 [84, 85]. The initial pressor response to angiotensin II administration is not affected in homozygous Nox1-deficient mice; however, the maintained response to chronic angiotensin infusion is markedly reduced [84, 85]. These findings suggest that a Nox1-containing NAD(P)H oxidase plays a critical role in the maintenance but not in the development of angiotensin II-induced hypertension. Nox1 deficiency was also associated with a partial reduction in angiotensin II-induced increases in ROS [85]. Somewhat surprisingly, Nox1 deficiency did not blunt the hypertrophic response in aorta to angiotensin II [84, 85]. These results suggest that despite a reduction in ROS and blood pressure, Nox isoforms other than Nox1 mediate the vascular hypertrophy in response to angiotensin II. Alternatively, other sources of superoxide or other mechanisms not related to ROS may account for this response.

Blood pressure and aortic structure is normal in transgenic mice expressing human Nox1 targeted to smooth muscle [86]. In contrast, infusion of angiotensin II produced a greater increase in blood pressure and vascular hypertrophy in these mice than in nontransgenic mice [86]. These findings suggest that Nox1-derived superoxide specifically in smooth muscle has an important influence on blood pressure and vascular hypertrophy in response to angiotensin II. Although there have been no reports, to our knowledge, of Nox1 genetic polymorphisms associated with human hypertension, these initial experiments in genetically altered Nox1 mice clearly implicate an important role for this Nox isoform in contributing to hypertension in response to angiotensin II.

3.1.2 Nox2

In Nox2 (gp91^{phox})-deficient mice, arterial blood pressure is modestly reduced, suggesting a role for Nox2 in the maintenance of basal blood pressure [37, 87–90]. Chronic administration of angiotensin II in Nox2-deficient mice produced a similar degree of hypertension as in wild-type mice [37, 87–90]. Nox2 deficiency also did not limit hypertension produced by overexpression of a constitutively active renin gene [37], providing additional evidence that Nox2 expression is not necessary for the development of hypertension. In contrast to the lack of effect on blood pressure, Nox2 deficiency produced a reduction in aortic superoxide levels, suggesting that the increase in blood pressure in response to angiotensin II in Nox2-deficient mice occurs independently of the increase in oxidative stress [37, 87–90]. In addition, Nox2 deficiency produces a reduction in the pressor response and vascular oxidative stress in a model of renovascular (2-kidney, 1-clip) hypertension [90].

Endothelial dysfunction produced by renovascular hypertension was less in Nox2 deficient mice compared with controls [90]. A potential limitation related to many of these studies is that examination of superoxide levels and vascular responses were generally restricted to aorta or other large blood vessels that do not contribute to vascular resistance. Thus, it would be of interest to determine the functional effects of these genetic manipulations on vascular structure and function in resistance blood vessels in future studies.

Recently, Nox2 overexpression in endothelium has been achieved in mice, and although an increase in basal superoxide levels in aorta was produced, there was no alteration in blood pressure [33]. In response to angiotensin II, both vascular superoxide and blood pressure were increased in endothelial-specific Nox2 transgenic mice [33]. Interestingly, Nox2 overexpression also was associated with increases in expression of vascular antioxidant enzymes, perhaps reflecting a compensatory response to the increase in oxidative stress. Such findings demonstrate that overexpression of Nox2 targeted to endothelium promotes oxidative stress and hypertension in response to angiotensin II.

3.1.3 Nox4

Nox4 is the only Nox isoform expressed throughout the vessel wall (i.e., endothelium, smooth muscle, and adventitia) [76], but the major site of expression is endothelium [81]. Because Nox4 expression can be induced by various stimuli, including angiotensin II [76], studies of Nox4-deficient and transgenic mice should be of interest. To date, genetic polymorphisms in the Nox4 gene have not, to our knowledge, been identified.

3.2 p22^{phox}

Along with the Nox subunit isoforms, p22^{phox} comprises the membrane-bound component of the vascular NAD(P)H oxidase [73]. p22^{phox} plays an important role in the membrane stabilization of NAD(P)H oxidase, and it serves as a docking site for the cytosolic components [73].

The human p22^{phox} gene has been expressed in smooth muscle in transgenic mice [34, 91]. Expression of p22^{phox} in this model had no effect on baseline blood pressure or vascular structure [34]. However, in response to angiotensin II infusion, p22^{phox} overexpression in smooth muscle resulted in an increase in the development, but not the sustained phase, of hypertension [91]. In addition, expression of p22^{phox} increased smooth muscle ROS and hypertrophy [91]. Although these findings provide important insight into the effect of p22^{phox} in smooth muscle, we are not aware of studies describing the effects of p22^{phox} targeted to endothelium or adventitial fibroblasts, or the generation of mice genetically deficient in p22^{phox}.

SHR exhibit increased expression of p22^{phox} that may be related to functional polymorphisms within the promoter region of the p22^{phox} gene [92–94]. At least five p22^{phox} polymorphisms have been identified in SHR, all of which are associated with increased p22^{phox} expression and NAD(P)H oxidase activity [94]. Similarly, multiple polymorphisms of the p22^{phox} gene have been identified in humans and several have been associated with hypertension [95–98]. These polymorphisms are primarily localized to the promoter region of the p22^{phox} gene, and they are associated with increased translational activity, and increased vascular superoxide and blood pressure in hypertensive individuals [95–98]. Three polymorphisms have been identified in the human p22^{phox} gene, including –930A/G, A640G, and C242T [95–98]. The –930A/G polymorphism is associated with increased expression of p22^{phox} and increased NAD(P)H oxidase in primary hypertension [95]. Of the three known polymorphisms of the p22^{phox} gene, the $-930A/G$ polymorphism has been suggested to be a potential and novel marker for hypertension in humans [95].

3.3   p47^{phox}

The cytosolic NAD(P)H oxidase subunit p47^{phox} is a critical subunit of all NAD(P)H oxidase isoforms [73, 76]. Although p47^{phox} may not be essential for basal NAD(P)H oxidase activity, it plays a critical role in NAD(P)H oxidase activation in response to various stimuli, including angiotensin II [99–103]. Angiotensin II increases the phosphorylation and translocation of p47^{phox} via a cSrc-dependent mechanism, resulting in increased NAD(P)H oxidase activity [100–102].

Baseline blood pressure in p47^{phox}-deficient mice is either normal or increased [104–107]. In one report, systolic blood pressure was markedly higher in homozygous p47^{phox} deficient mice than in controls [106]. The difference in blood pressure in younger p47^{phox} deficient mice could be normalized with either an ACE inhibitor or an angiotensin receptor blocker [106]. The hypertension in these mice seems to be transient because blood pressure normalizes as the mice continue to age, so that blood pressure was not different at 60 weeks of age compared with wild type [106]. Moreover, basal vascular superoxide levels are similar in p47^{phox}-deficient and wild-type mice, suggesting that activation of renin-angiotensin system with p47^{phox} deficiency occurs independent of ROS formation [106]. The pressor response to angiotensin II was blunted in p47^{phox}-deficient mice [104]. Together, these findings suggest an important role for p47^{phox} in baseline blood pressure and in mediating the pressor response to angiotensin II.

Although the exact role of p47^{phox} in activation of NAD(P)H oxidase has yet to be fully elucidated, p47^{phox} expression has multiple effects beyond NAD(P)H oxidase activity [106, 108, 109], thus making it one of the more intriguing subunits of the oxidase. For example, one of the beneficial effects of ascorbic acid is downregulation of p47^{phox} expression and NAD(P)H oxidase in microvascular endothelial cells [109]. Although p47^{phox} is important for activation of some Nox isoforms, we are not aware of linkages between polymorphisms of this subunit of the NAD(P)H oxidase with hypertension.

3.4   p67^{phox}

Although p67^{phox} is not thought to be essential for NAD(P)H oxidase activity, interaction of p67^{phox} with Nox2 has been shown to increase the rate of reduction of flavin adenine dinucleotide, thereby increasing oxidase activity [73]. However, there are no reports, to our knowledge, of p67^{phox} deletion or overexpression in mice. Although polymorphisms have been identified in the p67^{phox} gene [110], there are no reports in which polymorphisms in p67^{phox} are associated with human hypertension.

3.5   Rac1

NAD(P)H oxidase activity is dependent upon interaction with the G protein-Rac [111–113]. To date, three mammalian Racs have been identified: Rac1, Rac2, and Rac3 (Rac1B) [111]. Rac1 is ubiquitously expressed, and is a key component of vascular NAD(P)H oxidase activity, particularly Nox1- and Nox4-containing NAD(P)H oxidases [111]. Rac2 expression is restricted primarily to hematopoitic cells, and it is an essential component of the neutrophil NAD(P)H oxidase [114], whereas Rac3 is expressed primarily in the nervous system [111].

Although Rac1 deficiency in mice is embryonic lethal [115], expression of a constitutively active human Rac1 in smooth muscle has been achieved in mice [38]. Baseline blood pressure is substantially elevated in Rac1 transgenic mice [38]. The hypertension in these mice is reduced by the antioxidant N-acetyl-cysteine, suggesting an important role for ROS [38]. Interestingly, overexpression of Rac1 produces a marked increase in all three isoforms of SOD, perhaps as a compensatory response to increases in ROS [38].

Although no polymorphisms in Rac1 have yet to be associated with hypertension, experiments by using adenovirus for Rac1 have provided support for an important role of Rac1 in hypertension. For example, intracerebroventricular (i.c.v.) administration of a dominant-negative form of Rac1 reduces the pressor response to chronic systemic angiotensin II administration [116]. Conversely, i.c.v. administration of wild-type Rac1 mimics the increase in blood pressure produced by angiotensin II [116]. Together, these data provide strong evidence for a role of Rac1 in the neural regulation of blood pressure.

4   Superoxide Dismutases, Glutathione Peroxidases, and Catalase: Predominant Antioxidant Defense Systems

4.1   Superoxide Dismutases

SOD is the primary antioxidant limiting superoxide accumulation [4]. The three mammalian isoforms of SOD are copper-zinc SOD (CuZnSOD or SOD1), manganese SOD (MnSOD or SOD2), and an extracellular CuZnSOD (EC-SOD or SOD3). GPx and catalase, both of which catalyze the reduction of hydrogen peroxide to water and molecular oxygen, represent the two major enzyme systems that limit accumulation of hydrogen peroxide [4]. Thus, alterations in expression and/or activity of SODs, GPxs, or catalase would be predicted to influence the level of oxidative stress, vascular function, and vascular growth, and hence blood pressure.

4.1.1   CuZnSOD

CuZnSOD is highly expressed in the vessel wall, accounts for the majority of total SOD activity, and it is the isoform of SOD that limits increases in oxidative stress within the cytosolic compartment (including within mitochondria and the nucleus) [4]. There is little, if any, compensation by other isoforms of SOD, or by catalase and GPxs, in CuZnSOD deficient mice [117–119]. Superoxide levels are elevated in these animals under basal conditions in both large arteries and in resistance vessels [120–122]. These increases in superoxide produce reductions in NO-mediated vasodilation [120–123].

Some studies indicate that arterial pressure is similar in CuZnSOD$^{+/+}$ and CuZnSOD$^{+/-}$ mice, but that it is modestly reduced in CuZnSOD$^{-/-}$ mice [120, 122, 124]. There is one report in which blood pressure was normal in CuZnSOD$^{-/-}$ mice [117]. Thus, in general, blood pressure seems to be reduced somewhat with homozygous CuZnSOD deficiency. The findings that blood pressure is relatively normal in CuZnSOD$^{-/-}$ is interesting, because one might predict that blood pressure would be higher in CuZnSOD-deficient mice because superoxide is increased and endothelial dysfunction is present in resistance blood vessels [122]. These findings highlight the complexity of ROS as regulators of blood pressure, and they suggest that although CuZnSOD deficiency increases superoxide and produces endothelial dysfunction, it is not associated with an increase in blood pressure.

In addition to the effects on blood pressure, vascular function and superoxide levels, genetic deficiency in CuZnSOD produces vascular hypertrophy in cerebral arterioles [122]. Interestingly, the degree of hypertrophy in these vessels correlates with the level of CuZnSOD expression so hypertrophy is present in CuZnSOD$^{+/-}$ mice, but increases further in CuZnSOD$^{-/-}$ mice. Studies such as these support the concept that ROS promotes vascular hypertrophy independent of an increase in blood pressure.

Superoxide and hydrogen peroxide have been implicated in mechanisms that produce vascular hypertrophy [122, 126]. Because SODs rapidly convert superoxide to hydrogen peroxide, CuZnSOD deficiency produces a shift in the equilibrium between superoxide and hydrogen peroxide, resulting in an increase in superoxide. These findings suggest that in vivo model hypertrophy occurs in response to increased superoxide (reduced NO, or both). These findings provide evidence that superoxide may be an important determinant of hypertrophy in vivo.

CuZnSOD transgenic mice have been developed using various cDNA constructs and promoters [127–130]. Expression of human CuZnSOD results in an increase in CuZnSOD expression and activity in various tissues, including blood vessels [125, 127, 131, 132]. Under baseline conditions, blood pressure is similar in CuZnSOD transgenic and non-transgenic mice, suggesting that overexpression of CuZnSOD does not influence resting blood pressure [133]. The pressor response to angiotensin II administration in CuZnSOD transgenic mice is reduced by approximately 40% compared with controls [133]. This reduction in blood pressure is accompanied by a reduction in angiotensin II-induced increases in superoxide in aorta [133]. In contrast, vascular hypertrophy in response to angiotensin II was unaffected, suggesting expression of human CuZnSOD was not sufficient to inhibit hypertrophy in this model.

In addition to its effect on blood pressure and vascular ROS levels under basal conditions, expression of human CuZnSOD in transgenic mice protects against increases in superoxide and endothelial dysfunction produced by angiotensin II [123, 133]. Although one might predict that the protective effects of CuZnSOD expression on endothelial function reflect increases in hydrogen peroxide formation, catalase had no effect on vascular responses in CuZnSOD transgenic mice [123].

Although many polymorphisms have been identified in the human CuZnSOD gene, none have been found to be associated with hypertension to our knowledge [134]. Interestingly, individuals with Down's syndrome (Trisomy 21) have lower blood pressure than the general population [135]. Whether this difference is linked to the CuZnSOD gene per se (located on chromosome 21) has not been studied.

4.1.2 MnSOD

MnSOD is the primary isoform of SOD limiting increases in superoxide in mitochondria. Although CuZnSOD is expressed in the intermembrane space, MnSOD is expressed in the mitochondrial matrix and is the sole SOD isoform in this compartment [4, 136]. MnSOD expression is thought to account for up to approximately 10% of total vascular SOD activity [4]. Unlike CuZnSOD and EC-SOD, MnSOD homozygous deficiency is lethal (neonatal) [137, 138], reflecting the critical importance of this SOD isoform in limiting ROS within this subcellular compartment.

Blood pressure is normal in young MnSOD deficient (MnSOD$^{+/-}$) mice [139]. However, MnSOD heterozygous deficiency is associated with the development of mild hypertension with age [139]. Hypertension in old MnSOD$^{+/-}$ mice is associated with increased plasma markers of oxidative stress, suggesting that age-induced hypertension in this model reflects an increase in ROS [139].

A high-salt diet that had no effect on blood pressure in wild-type mice produced an increase in blood pressure in MnSOD$^{+/-}$ mice [139]. The finding that a hypertensive phenotype can be evoked in mice deficient in MnSOD, suggests that normal expression of MnSOD is needed to prevent at least one form of hypertension. These findings also suggest a novel role for mitochondrial-derived ROS in regulating blood pressure. Preliminary studies in MnSOD$^{+/-}$ mice suggest that MnSOD also protects the vasculature from angiotensin II-induced endothelial dysfunction [140].

Several polymorphisms have been identified in the human MnSOD gene [134, 141]. However, the C47T polymorphism is the only one that has been linked to hypertension. This polymorphism is associated with increased MnSOD activity and reductions in ROS and thus may be protective [142].

4.1.3 EC-SOD

As its name implies, EC-SOD is expressed extracellularly and thus should protect NO in transient from endothelium to one of its major sites of action, vascular muscle [4, 143]. EC-SOD contains a heparin-binding domain with high affinity for heparan sulfate proteoglycans on cell surfaces, particularly in basal membranes and the extracellular matrix [143]. Because EC-SOD accounts for a significant portion of total SOD activity, reductions in EC-SOD would be predicted to increase superoxide and have a significant impact on bioavailability of NO.

In most studies, resting blood pressure in EC-SOD$^{-/-}$ mice is normal [43, 144–147], with one study reporting an increase in blood pressure [146]. Basal levels of superoxide in the vasculature have been found to be increased [43, 145, 146] or unchanged [147] in EC-SOD$^{-/-}$ mice, and they seem to vary depending on the blood vessel studied. The effect of EC-SOD deficiency on endothelial- and NO-dependent responses in aorta under baseline conditions has for the most part been found to be minimal [43, 145, 147], suggesting that the role of EC-SOD in protecting NO may not be as great as initially hypothesized.

Several studies have examined the effect of EC-SOD in mouse models of hypertension [43, 145–147]. For example, similar levels of hypertension (despite higher levels of vascular ROS) occur in EC-SOD$^{-/-}$ mice and wild-type controls with reno-vascular (1-kidney, 1-clip) hypertension [43]. In contrast, the level of renovascular hypertension in a 2-kidney, 1-clip model is modestly increased in EC-SOD$^{-/-}$ mice as compared with wild type [145]. Similarly, hypertension produced by angiotensin II administration is greater in EC-SOD$^{-/-}$ mice with a pressor dose, but not with a slow pressor dose, of angiotensin II [146, 147]. Together, these studies suggest that EC-SOD is protective in some models of hypertension. Although these studies have all focused on the effect of homozygous EC-SOD deficiency, complete genetic deficiency in EC-SOD is very rare in humans. Thus, studies in heterozygous EC-SOD-deficient mice would be of value in more fully defining the functional importance of EC-SOD in hypertension.

Although resting blood pressure in EC-SOD transgenic mice is normal [148–151], we are not aware of any studies that have examined the effect of experimental hypertension on vascular function, superoxide levels, or vascular structure in these mice. However, systemic gene transfer of human EC-SOD reduces blood pressure modestly in SHR, but not WKY controls [152]. The reduction in blood pressure with EC-SOD gene transfer also was associated with reductions in vascular oxidative stress. Gene transfer of an EC-SOD gene variant (discussed below) containing a deletion in the heparin-binding domain had no effect on blood pressure in SHR, suggesting that the heparin binding domain is essential for the antihypertensive effect of EC-SOD.

In human hypertension, plasma EC-SOD activity in African-Americans with hypertension is lower than in normotensive controls [153]. Reduced EC-SOD activity in these patients is associated with oxidative stress, including increases in plasma 8-isoprostanes [153]. In humans, the C637G polymorphism of the EC-SOD gene results in a relatively common gene variant in the heparin-binding domain of

EC-SOD that does not reduce EC-SOD activity [134, 154, 155]. This polymorphism is associated with increases in EC-SOD in plasma, but it has not been associated with hypertension because blood pressure is normal in humans with the C637G mutation [155].

4.2 Glutathione Peroxidases

There are five mammalian isoforms of GPx (GPx1–5), and each catalyzes the same reaction—the reduction of hydrogen peroxide and lipid peroxides to water and molecular oxygen and their corresponding lipid alcohols, respectively. GPx1 is ubiquitously expressed in cytosol and mitochondria [156]. The expression of other isoforms of GPx in vascular tissue are limited. For example, GPx2 expression is mainly limited to the gastrointestinal tract, and it is not expressed at all or at much lower levels than GPx1 in the vasculature [156]. GPx3 is primarily expressed in erythrocytes; however, GPx3 message also has been detected in blood vessels [157]. Because GPx1 is considered the predominate GPx in blood vessels, deficiency or overexpression of GPx1 would be predicted to have the most significant effect on cardiovascular function.

Blood pressure under normal conditions or in response to experimental hypertension has not, to our knowledge, been examined in either GPx1 deficient or transgenic mice. Although studies of vascular function in genetically altered GPx1 mice are limited [158–162], pharmacologic depletion of glutathione, a substrate for GPx, in rats produces severe hypertension and can be partially mitigated by antioxidants [163, 164], suggesting an important role for glutathione (and by extrapolation GPx activity) in limiting oxidative stress. Although several polymorphisms in the GPx1 gene have been associated with reductions in GPx1 activity [134], they have not been found to be associated with hypertension in humans.

4.3 Catalase

Like GPx, catalase reduces hydrogen peroxide to water and molecular oxygen. Expression of catalase is for the most part limited to peroxisomes, but the level of expression of catalase is highly tissue specific [156]. For example, expression of catalase is high in red blood cells and in the liver and relatively low in brain and connective tissues [156]. Several groups have produced mice that globally express human catalase and specifically in smooth muscle [126, 165, 166]. Although baseline blood pressure was similar in mice expressing human catalase, the pressor response to angiotensin II were markedly reduced in these animals compared with non-transgenic controls [165].

Although mice deficient in catalase have been produced [167, 168], the effect of catalase deficiency on resting blood pressure or in response to experimental

hypertension has not yet been explored. Several polymorphisms have been identified in the human catalase gene, some of which are associated with severe reductions
in catalase expression and activity [134]; however, association of these polymorphisms with hypertension do not seem to be very strong [169, 170].

5 Summary

As presented, there is strong evidence to support a role for oxidative stress in both
animal models and human hypertension. NAD(P)H oxidase is a major source of
oxidative stress in the vasculature. Expression and activity of NAD(P)H oxidase
are increased in hypertension, particularly that produced by angiotensin II. Genetic
manipulation of NAD(P)H oxidase subunits is associated with alterations in ROS
levels, changes in vascular structure and function, and it has significant effects on
resting blood pressure. Moreover, the degree of hypertension produced in experimental

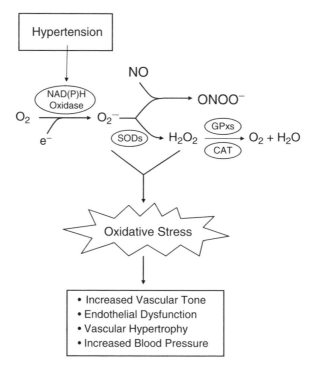

Fig. 13.1 Hypertension is associated with oxidative stress (i.e., increased vascular superoxide
and hydrogen peroxide). Oxidative stress contributes to vascular alterations, including endothelial
dysfunction and hypertrophy. In addition to being a consequence of hypertension, oxidative stress
may further promote/contribute to hypertension. SODs, superoxide dismutases; GPxs, glutathione
peroxidases; CAT, catalase, NO, nitric oxide, ONOO-, peroxynitrite; $O_2^{\cdot-}$, superoxide; O_2, oxygen;
H_2O_2, hydrogen peroxide; H_2O, water

models, such as angiotensin II-induced hypertension, can be influenced by changes in NAD(P)H oxidase subunit expression.

Although NAD(P)H oxidase represents a major source of oxidative stress, expression and activity of enzymes such as SODs and GPxs are important in limiting oxidative stress. Genetic overexpression and deletion of SOD or GPx expression and/or activity are associated with alterations in vascular levels of ROS. Therefore, changes in antioxidant levels can influence vascular structure and function and also blood pressure. Together, expression of prooxidant and antioxidant enzymes have profound effects on oxidative stress within the vasculature and alterations in the balance of these opposing systems can directly contribute to or limit the development of hypertension (Fig. 13.1).

Acknowledgements This work was supported by National Institutes of Health Grants NS-24621, HL-38901, and HL-62984 as well as support from the American Heart Association Beginning Grant-in-Aid (0565486Z).

References

1. Commoner B, Townsend J, Pake GE. Free radicals in biological materials. Nature 1954;174:689–691.
2. Valko M, Leibfritz D, Moncol J, Cronin MT, Mazur M, Telser T. Free radicals and antioxidants in normal physiological functions and human disease. Int J Biochem Cell Biol 2007;39:44–84.
3. Darley-Usmar V, Halliwell B. Blood radicals: reactive nitrogen species, reactive oxygen species, transition metal ions, and the vascular system. Pharm Res 1996;13:649–662.
4. Faraci FM, Didion SP. Vascular protection: superoxide dismutase isoforms in the vessel wall. Arterioscler Thromb Vasc Biol 2004;24:1367–1373.
5. Wei EP, Kontos HA, Christman CW, DeWitt DS, Povlishock JT. Superoxide generation and reversal of acetylcholine-induced cerebral arteriolar dilation after acute hypertension. Circ Res 1985;57:781–787.
6. Luscher TF, Vanhoutte PM. Endothelium-dependent contractions to acetylcholine in the aorta of the spontaneously hypertensive rat. Hypertension 1986;8:344–348.
7. Mayhan WG, Faraci FM, Heistad DD. Impairment of endothelium-dependent responses of cerebral arterioles in chronic hypertension. Am J Physiol 1987;253:H1435–H1440.
8. Auch-Schwelk W, Katusic ZS, Vanhoutte PM. Contractions to oxygen-derived free radicals are augmented in aorta of the spontaneously hypertensive rat. Hypertension 1989;13:859–864.
9. Rubanyi GM, Vanhoutte PM. Oxygen-derived free radicals, endothelium, and responsiveness of vascular smooth muscle. Am J Physiol 1986;250:H815–H821.
10. Mayhan WG. Impairment of endothelium-dependent dilatation of basilar artery during chronic hypertension. Am J Physiol 1990;259:H1455–H1462.
11. Yang D, Feletou M, Boulanger CM, Wu HF, Levens N, Zhang JN, Vanhoutte PM. Oxygen-derived free radicals mediate endothelium-dependent contractions to acetylcholine in aortas from spontaneously hypertensive rats. Br J Pharmacol 2002;136:104–110.
12. Ceriello A, Giugliano D, Quatraro A, Lefebvre PJ. Anti-oxidants show an anti-hypertensive effect in diabetic and hypertensive subjects. Clin Sci 1991;81:739–742.
13. Nakazono K, Watanabe N, Matsuno J, Saski J, Sato T, Inoue M. Does superoxide underlie the pathogenesis of hypertension? Proc Natl Acad Sci U S A 1991;88:10045–10048.
14. Pedro-Botet J, Covas MI, Martin S, Rubies-Prat J. Decreased endogenous antioxidant enzymatic status in essential hypertension. J Hum Hypertens 2000;14:343–345.

15. Redon J, Oliva MR, Tormos C et al. Antioxidant activities and oxidative stress byproducts in human hypertension. Hypertension 2003;41:1096–1101.
16. Chaves FJ, Mansego ML, Blesa S et al. Inadequate cytoplasmic antioxidant enzymes response contributes to the oxidative stress in human hypertension. Am J Hypertens 2007;20:62–69.
17. Saez GT, Tormos C, Giner V, Chaves J, Lozano JV, Iradi A, Redon J. Factors related to the impact of antihypertensive treatment in antioxidant activities and oxidative stress by-products in human hypertension. Am J Hypertens 2004;17:809–816
18. Ulker S, McMaster D, McKeown PP, Bayraktutan U. Impaired activities of antioxidant enzymes elicit endothelial dysfunction in spontaneous hypertensive rats despite enhanced vascular nitric oxide generation. Cardiovasc Res 2003;59:488–500.
19. Zhang Y, Jang R, Mori TA, Croft KD, Schyvens CG, McKenzie KU, Whitworth JA. The anti-oxidant Tempol reverses and partially prevents adrenocorticotrophic hormone-induced hypertension in the rat. J Hypertens 2003;21:1513–1518.
20. Laursen JB, Rajagopalan S, Galis Z, Tarpey M, Freeman BA, Harrison DG. Role of super-oxide in angiotensin II-induced but not catecholamine-induced hypertension. Circulation 1997;95:588–593.
21. Rey FE, Cifuentes ME, Kiarash A, Quinn MT, Pagano PJ. Novel competitive inhibitor of NAD(P)H oxidase assembly attenuates vascular $O_2^{\cdot-}$ and systolic blood pressure in mice. Circ Res 2001;89:408–414.
22. Ortiz MC, Manriquez MC, Romero JC, Juncos LA. Antioxidants block angiotensin II-induced increases in blood pressure and endothelin. Hypertension 2001;38:655–659.
23. Nishiyama A, Fukui T, Fujisawa Y, Rahman M, Tian RX, Kimura S, Abe Y. Systemic and regional hemodynamic responses to Tempol in angiotensin II-infused hypertensive rats. Hypertension 2001;37:77–83.
24. Wang HD, Xu S, Johns DG, Du Y, Quinn MT, Cayatte AJ, Cohen RA. Role of NADPH oxi-dase in the vascular hypertrophic and oxidative stress response to angiotensin II in mice. Circ Res 2001;88:947–953.
25. Dobrian AD, Schriver SD, Prewitt RL. Role of angiotensin II and free radicals in blood pres-sure regulation in a rat model of renal hypertension. Hypertension 2001;38:361–366.
26. Meng S, Roberts LJ II, Cason GW, Curry TS, Manning RD Jr. Superoxide dismutase and oxidative stress in Dahl salt-sensitive and -resistant rats. Am J Physiol 2002;283: R732–R738.
27. Swei A, Lacy F, DeLano FA, Schmid-Schonbein GW. Oxidative stress in the Dahl hyperten-sive rat. Hypertension 1997;30:1628–1633.
28. Schnackenberg CG, Welch WJ, Wilcox CS. Normalization of blood pressure and renal vas-cular resistance in SHR with a membrane-permeable superoxide dismutase mimetic: role of nitric oxide. Hypertension 1998;32:59–64.
29. Schnackenberg CG, Wilcox CS. Two-week administration of tempol attenuates both hypertension and renal excretion of 8-iso prostaglandin F_2-α. Hypertension 1999; 33:424–428.
30. Yanes L, Romero D, Iliescu R et al. Systemic arterial pressure response to two weeks of Tempol therapy in SHR: involvement of NO, the RAS, and oxidative stress. Am J Physiol 2005;288:R903–R908.
31. Park JB, Touyz RM, Chen X, Schiffrin EL. Chronic treatment with a superoxide dismutase mimetic prevents vascular remodeling and progression of a hypertension in salt loaded stroke-prone spontaneously hypertensive rats. Am J Hypertens 2002;15:78–84.
32. Uddin M, Yang H, Shi M, Polley-Mandal M, Guo Z. Elevation of oxidative stress in the aorta of genetically hypertensive mice. Mech Ageing Dev 2003;124:811–817.
33. Bendall JK, Rinze R, Adlam D, Tatham AL, de Bono J, Channon KM. Endothelial Nox2 overexpression potentiates vascular oxidative stress and hemodynamic response to angiotensin II. Studies in endothelial-targeted Nox2 transgenic mice. Circ Res 2007;100:1016–1025.
34. Laude K, Cai H, Fink B et al. Hemodynamic and biochemical adaptations to vascular smooth muscle overexpression of p22[phox] in mice. Am J Physiol 2005;288:H7–H12.

35. Didion SP, Ryan MJ, Baumbach GL, Sigmund CD, Faraci FM. Superoxide contributes to vascular dysfunction in mice that express human renin and angiotensinogen. Am J Physiol 2002;283:H1569–H1576.
36. Baumbach GL, Sigmund CD, Faraci FM. Cerebral arteriolar structure in mice overexpressing human renin and angiotensinogen. Hypertension 2003;41:50–55.
37. Touyz RM, Mercure C, He Y et al. Angiotensin II-dependent chronic hypertension and cardiac hypertrophy are unaffected by gp91phox-containing NADPH oxidase. Hypertension 2005;45:530–537.
38. Hassanain HH, Gregg D, Marcelo ML et al. Hypertension caused by transgenic overexpression of Rac1. Antioxid Redox Signal 2007;9:91–100.
39. Jin L, Beswick RA, Yamamoto T et al. Increased reactive oxygen species contributes to kidney injury in mineralocorticoid hypertensive rats. J Physiol Pharmacol 2006;57:343–357.
40. Beswick RA, Dorrance AM, Leite R, Webb RC. NADH/NADPH oxidase and enhanced superoxide production in the mineralocorticoid hypertensive rat. Hypertension 2001;38:1107–1111.
41. Somers MJ, Mavromatis K, Galis ZS, Harrison DG. Vascular superoxide production and vasomotor function in hypertension induced by deoxycorticosterone acetate-salt. Circulation 2000;101:1722–1728.
42. Heitzer T, Wenzel U, Hink U et al. Increased NAD(P)H oxidase-mediated superoxide production in renovascular hypertension: evidence for an involvement of protein kinase C. Kidney Int 1999;55:252–260.
43. Jung O, Marklund SL, Xia N, Busse R, Brandes RP. Inactivation of extracellular superoxide dismutase contributes to the development of high-volume hypertension. Arterioscler Thromb Vasc Biol 2007;27:470–477.
44. Lacy F, O'Connor DT, Schmid-Schonbein GW. Plasma hydrogen peroxide production in hypertensives and normotensive subjects at genetic risk of hypertension. J Hypertens 1998;16:291–303.
45. Lacy F, Kailasam MT, O'Connor DT, Schmid-Schonbein GW, Parmer RJ. Plasma hydrogen peroxide production in human essential hypertension: role of heredity, gender, and ethnicity. Hypertension 2000;36:878–884.
46. Kumar KV, Das UN. Are free radicals involved in the pathobiology of human essential hypertension? Free Radic Res Commun 1993;19:59–66.
47. Fortuno A, Olivan S, Beloqui O, San Jose G, Moreno MU, Diez J, Zalba G. Association of increased phagocytic NADPH oxidase-dependent superoxide production with diminished nitric oxide generation in essential hypertension. J Hypertens 2004;22:2169–2175.
48. Garcia CE, Kilcoyne CM, Cardillo C, Cannon RO III, Quyyumi AA, Panza JA. Effect of copper-zinc superoxide dismutase on endothelium-dependent vasodilation in patients with essential hypertension. Hypertension 1995;26:863–868.
49. Schiffrin EL, Deng LY, Larochelle P. Progressive improvement in the structure of resistance arteries of hypertensive patients after treatment with an angiotensin I-converting enzyme inhibitor comparison with effects of a β-blocker. Am J Hypertens 1995;8:229–236.
50. Schiffrin EL, Deng LY. Structure and function of resistance arteries of hypertensive patients treated with a β-blocker or a calcium channel antagonist. J Hypertens 1996;14:1247–1255.
51. Rizzoni D, Muiesan ML, Porteri E et al. Effects of long-term antihypertensive treatment with lisinopril on resistance arteries in hypertensive patients with left ventricular hypertrophy. J Hypertens 1997;15:197–204.
52. Minuz P, Patrignani P, Gaino S et al. Increased oxidative stress and platelet activation in patients with hypertension and renovascular disease. Circulation 2002;106:2800–2805.
53. Higashi Y, Sasaki S, Nakagawa K, Matsuura H, Oshima T, Chayama K. Endothelial function and oxidative stress in renovascular hypertension. N Engl J Med 2002;346:1954–1962.
54. Rizzoni D, Porteri E, Castellano M et al. Vascular hypertrophy and remodeling in secondary hypertension. Hypertension 1996;28:785–790.
55. Rizzoni D, Porteri E, Guefi D et al. Cellular hypertrophy in subcutaneous small arteries of patients with renovascular hypertension. Hypertension 2000;35:931–935.

56. Fennell JP, Brosnan MJ, Frater AJ et al. Adenovirus-mediated overexpression of extracellular superoxide dismutase improves endothelial dysfunction in a rat model of hypertension. Gene Ther 2002;9:110–117.

57. Alexander MY, Brosnan MJ, Hamilton CA et al. Gene transfer of endothelial nitric oxide synthase but not Cu/Zn superoxide dismutase restores nitric oxide availability in the SHRSP. Cardiovasc Res 2000;47:609–617.

58. Griendling KK, Minieri CA, Ollerenshaw JD, Alexander RW. Angiotensin II stimulates NADH and NADPH oxidase activity in cultured vascular smooth muscle cells. Circ Res 1994;74:1141–1148.

59. Cifuentes ME, Rey FE, Carretero OA, Pagano PJ. Upregulation of p67(phox) and gp91(phox) in aortas from angiotensin II-infused mice. Am J Physiol 2000;279:H2234–H2240.

60. Pagano PJ, Clark JK, Cifuentes-Pagano ME, Clark SM, Callis GM, Quinn MT. Localization of a constitutively active, phagocyte-like NADPH oxidase in rabbit aortic adventitia: enhancement by angiotensin II. Proc Nat Acad Sci U S A 1997;94:14483–14488.

61. Pagano PJ, Chanock SJ, Siwik DA, Colucci WS, Clark JK. Angiotensin II induces p67phox mRNA expression and NADPH oxidase superoxide generation in rabbit aortic adventitial fibroblasts. Hypertension 1998;32:331–337.

62. Bleeke T, Zhang H, Madamanchi N, Patterson C, Faber JE. Catecholamine-induced vascular wall growth is dependent on generation of reactive oxygen species. Circ Res 2004;94:37–45.

63. Ghiadoni L, Magagna A, Versari D et al. Different effect of antihypertensive drugs on conduit artery endothelial function. Hypertension 2003;41:1281–1286.

64. Ghiadoni L, Huang Y, Magagna A, Burali S, Taddei S, Salvetti A. Effect of acute blood pressure reduction on endothelial function in the brachial artery of patients with essential hypertension. J Hypertens 2001;19:547–551.

65. Cohuet G, Struijker-Boudier H. Mechanisms of target organ damage caused by hypertension: therapeutic potential. Pharmacol Ther 2006;111:81–98.

66. Thuillez C, Richard V. Targeting endothelial dysfunction in hypertensive subjects. J Hum Hypertens 2005;19:S21–S25.

67. Zimmerman MC, Lazartigues E, Lang JA et al. Superoxide mediates the actions of angiotensin II in the central nervous system. Circ Res 2002;91:1038–1045.

68. Makino A, Skelton MM, Zou AP, Cowley AW Jr. Increased renal medullary H_2O_2 leads to hypertension. Hypertension 2003;42:25–30.

69. Makino A, Skelton MM, Zou AP, Roman RJ, Cowley AW Jr. Increased renal medullary oxidative stress produces hypertension. Hypertension 2002;39:667–672.

70. Wilcox CS. Oxidative stress and nitric oxide deficiency in the kidney: a critical link to hypertension. Am J Physiol 2005; 289:R913–R935.

71. Zimmerman MC, Davisson RL. Redox signaling in central neural regulation of cardiovascular function. Prog Biophys Mol Biol 2004;84:125–149.

72. Geiszt M. NADPH oxidases: new kids on the block. Cardiovasc Res 2006;71:289–299.

73. Bedard K, Krause KH. The Nox family of ROS-generating NADPH oxidases: physiology and pathophysiology. Physiol Rev 2007;87:245–313.

74. Zalba G, San Jose G, Moreno MU et al. Oxidative stress in arterial hypertension: role of NAD(P)H oxidase. Hypertension 2001;38:1395–1399.

75. Fortuno A, San Jose G, Moreno MU, Diez J, Zalba G. Oxidative stress and vascular remodelling. Exp Physiol 2005;90:457–462.

76. Lassegue B, Griendling KK. Reactive oxygen species in hypertension: an update. Am J Hypertens 2004;17:852–860.

77. Guzik TJ, Sadowski J, Kapelak B et al. Systemic regulation of vascular NAD(P)H oxidase activity and Nox isoform expression in human arteries and veins. Arterioscler Thromb Vasc Biol 2004;24:1614–1620.

78. Hamilton CA, Brosnan MJ, Al-Benna S, Berg G, Dominiczak AF. NAD(P)H oxidase inhibition improves endothelial function in rat and human blood vessels. Hypertension 2002;40:755–762.

79. Brosnan MJ, Hamilton CA, Graham D, Lygate CA, Jardine E, Dominiczak AF. Irbesartan lowers superoxide levels and increases nitric oxide bioavailability in blood vessels from spontaneously hypertensive stroke-prone rats. J Hypertens 2002;20:281–286.

80. Rodrigo E, Maeso R, Munoz-Garcia R et al. Endothelial dysfunction in spontaneously hypertensive rats: consequences of chronic treatment with losartan or captopril. J Hypertens 1997;15:613–618.

81. Ago T, Kitazono T, Ooboshi H et al. Nox4 as the major catalytic component of an endothelial NAD(P)H oxidase. Circulation 2004;109:227–233.

82. Ago T, Kitazono T, Kuroda J et al. NAD(P)H oxidases in rat basilar artery endothelial cells. Stroke 2005;36:1040–1046.

83. Miller AA, Drummond GR, Schmidt HH, Sobey CG. NADPH oxidase activity and function are profoundly greater in cerebral versus systemic arteries. Circ Res 2005;97:1055–1062.

84. Matsuno K, Yamada H, Iwata K et al. Nox1 is involved in angiotensin II-mediated hypertension: a study in Nox1-deficient mice. Circulation 2005;112:2677–2685.

85. Gavazzi G, Banfi B, Deffert C, Fiette L, Schappi M, Herrmann F, Krause KH. Decreased blood pressure in NOX1-deficient mice. FEBS Lett 2006;580:497–504.

86. Dikalova A, Clempus R, Lassegue B et al. Nox1 overexpression potentiates angiotensin II-induced hypertension and vascular smooth muscle hypertrophy in transgenic mice. Circulation 2005;112:2668–2676.

87. Wang HD, Xu S, Johns DG, Du Y, Quinn MT, Cayatte AJ, Cohen RA. Role of NADPH oxidase in the vascular hypertrophic response and oxidative stress response to angiotensin II in mice. Circ Res 2001;88:947–953.

88. Byrne JA, Grieve DJ, Bendall JK et al. Contrasting roles of NADPH oxidase isoforms in pressure-overload versus angiotensin II-induced cardiac hypertrophy. Circ Res 2003;93: 802–804.

89. Bendall JK, Cave AC, Heymes C, Gall N, Shah AM. Pivitol role of a gp91(phox) -containing NADPH oxidase in angiotensin II-induced cardiac hypertrophy in mice. Circulation 2002;105;293–296.

90. Jung O, Schreiber JG, Geiger H, Pedrazzini T, Busse R, Brandes RP. gp91^{phox} -containing NADPH oxidase mediates endothelial dysfunction in renovascular hypertension. Circulation 2004;109:1795–1801.

91. Weber DS, Rocic P, Mellis AM et al. Angiotensin II-induced hypertrophy is potentiated in mice overexpressing p22^{phox} in vascular smooth muscle. Am J Physiol 2005;288:H37–H42.

92. Fukui T, Ishizaka N, Rajagopalan S et al. p22^{phox} mRNA expression and NADPH oxidase activity are increased in aortas from hypertensive rats. Circ Res 1997;80:45–51.

93. Zalba G, San Jose G, Moreno MU, Fortuno A, Diez J. NADPH oxidase-mediated oxidative stress: genetic studies of the p22(phox) gene in hypertension. Antioxid Redox Signal 2005;7:1327–1336.

94. Zalba G, San Jose G, Beaumont FJ, Fortuno MA, Fortuno A, Diez J. Polymorphisms and promoter overactivity of the p22(phox) gene in vascular smooth muscle cells from spontaneously hypertensive rats. Circ Res 2001;88:217–222.

95. San Jose G, Moreno MU, Olivan S, Beloqui O, Fortuno A, Diez J, Zalba G. Functional effect of the p22^{phox} -930A/G polymorphism on p22^{phox} expression and NADPH oxidase activity in hypertension. Hypertension 2004;44:163–169.

96. Moreno MU, San Jose G, Orbe J, Paramo JA, Beloqui O, Diez J, Zalba G. Preliminary characterization of the promoter of the human p22(phox) gene: identification of a new polymorphism associated with hypertension. FEBS Lett 2003;542:27–31.

97. Castejon AM, Bracero J, Hoffmann IS, Alfieri AB, Cubeddu LX. NAD(P)H oxidase p22phox gene C242T polymorphism, nitric oxide production, salt sensitivity and cardiovascular risk factors in Hispanics. J Hum Hypertens 2006;20:772–779.

98. Kokubo Y, Iwai N, Tago N et al. Association analysis between hypertension and CYBA, CLCNKB, and KCNMB1 functional polymorphisms in the Japanese population–the Suita study. Circ J 2005;69:138–142.

99. Lavigne MC, Malech HL, Holland SM, Leto TL. Genetic demonstration of p47^{phox}-dependent superoxide anion production in murine vascular smooth muscle cells. Circulation 2001;104:79–84.

100. Touyz RM, Yao G, Schiffrin EL. c-Src induces phosphorylation and translocation of p47^{phox}: role in superoxide generation by angiotensin II in human vascular smooth muscle cells. Arterioscler Thromb Vasc Biol 2003;23:981–987.

101. Li JM, Shah AM. Mechanism of endothelial cell NADPH oxidase activation by angiotensin II. Role of the p47^{phox} subunit. J Biol Chem 2003;278:12094–12100.

102. Li JM, Mullen AM, Yun S, Wientjes F, Brouns GY, Thrasher AJ, Shah AM. Essential role of the NADPH oxidase subunit p47(phox) in endothelial cell superoxide production in response to phorbol ester and tumor necrosis factor-alpha. Circ Res 2002;90:143–150.

103. Li JM, Wheatcroft S, Fan LM, Kearney MT, Shah AM. Opposing roles of p47^{phox} in basal versus angiotensin II-stimulated alterations in vascular O$_2$$^{\cdot-}$ production, vascular tone, and mitogen-activated protein kinase activation. Circulation 2004;109:1307–1313.

104. Landmesser U, Cai H, Dikalov S et al. Role of p47(phox) in vascular oxidative stress and hypertension caused by angiotensin II. Hypertension 2002;40:511–515.

105. Cai H, Li Z, Dikalov S et al. NAD(P)H oxidase-derived hydrogen peroxide mediates endothelial nitric oxide production in response to angiotensin II. J Biol Chem 2002;277:48311–48317.

106. Grote K, Ortmann M, Salguero G et al. Critical role for p47^{phox} in renin-angiotensin system activation and blood pressure regulation. Cardiovasc Res 2006;71:596–605.

107. Hsich E, Segal BH, Pagano PJ et al. Vascular effects following homozygous disruption of p47(phox): an essential component of NADPH oxidase. Circulation 2000;101:1234–1236.

108. Brandes RP, Miller FJ, Beer S et al. The vascular NADPH oxidase subunit p47^{phox} is involved in redox-mediated gene expression. Free Radic Biol Med 2002;32:1116–1122.

109. Wu F, Schuster DP, Tyml K, Wilson JX. Ascorbate inhibits NADPH oxidase subunit p47^{phox} expression in microvascular endothelial cells. Free Radic Biol Med 2007;42:124–131.

110. Kenney RT, Malech HL, Epstein ND, Roberts RL, Leto TL. Characterization of the p67phox gene: genomic organization and restriction fragment length polymorphism analysis for prenatal diagnosis in chronic granulomatous disease. Blood 1993;82:3739–3744.

111. Hordijk PL. Regulation of NADPH oxidases: the role of Rac proteins. Circ Res 2006; 98:453–462.

112. Heyworth PG, Bohl BP, Bokoch GM, Curnutte JT. Rac translocates independently of the neutrophil NADPH oxidase components p47^{phox} and p67^{phox}. Evidence for its interaction with flavocytochrome b558. J Biol Chem 1994;269:30749–30752.

113. Sundaresan M, Yu ZX, Ferrans VJ et al. Regulation of reactive-oxygen-species generation in fibroblasts by Rac1. Biochem J 1996;318:379–382.

114. Roberts AW, Kim C, Zhen L et al. Deficiency of the hematopoietic cell-specfic Rho family GTPase Rac2 is characterized by abnormalities in neutrophil function and host defense. Immunity 1999;10:183–196.

115. Sugihara K, Nakatsuji N, Nakamura K et al. Rac1 is required for the formation of three germ layers during gastrulation. Oncogene 1998;17:3427–3433.

116. Zimmerman MC, Dunlay RP, Lazartigues E et al. Requirement for Rac1-dependent NADPH oxidase in the cardiovascular and dipsogenic actions of angiotensin II in the brain. Circ Res 2004;95:532–539.

117. Morikawa K, Shimokawa H, Matoba T et al. Pivotal role of Cu,Zn-superoxide dismutase in endothelium-dependent hyperpolarization. J Clin Invest 2003;112:1871–1879.

118. Ho YS, Gargano M, Cao J, Bronson RT, Heimler I, Hutz RJ. Reduced fertility in female mice lacking copper-zinc superoxide dismutase. J Biol Chem 1998;273:7765–7769.

119. Carlsson LM, Jonsson J, Edlund T, Marklund SL. Mice lacking extracellular superoxide dismutase are more sensitive to hyperoxia. Proc Natl Acad Sci U S A 1995;92:6264–6268.

120. Didion SP, Ryan MJ, Didion LA, Fegan PE, Sigmund CD, Faraci FM. Increase superoxide and vascular dysfunction in CuZnSOD-deficient mice. Circ Res 2002;91:938–944.

121. Cooke CL, Davidge ST. Endothelial-dependent vasodilation is reduced in mesenteric arteries from superoxide dismutase knockout mice. Cardiovasc Res 2003;60:635–642.

122. Baumbach GL, Didion SP, Faraci FM. Hypertrophy of cerebral arterioles in mice deficient in expression of the gene for CuZn superoxide dismutase. Stroke 2006;37:1850–1855.
123. Didion SP, Kinzenbaw DA, Faraci FM. Critical role for CuZn-superoxide dismutase in preventing angiotensin II-induced endothelial dysfunction. Hypertension 2005;46:1147–1153.
124. Didion SP, Kinzenbaw DA, Schrader LI, Faraci FM. Heterozygous CuZn superoxide dismutase deficiency produces a vascular phenotype with aging. Hypertension 2006;48:1072–1079.
125. Didion SP, Faraci FM. Ceramide-induced impairment of endothelial function is prevented by CuZn superoxide dismutase overexpression. Arterioscler Thromb Vasc Biol 2005;25: 90–95.
126. Zhang Y, Griendling KK, Dikalova A, Owens GK, Taylor WR. Vascular hypertrophy in angiotensin II-induced hypertension is mediated by vascular smooth muscle cell-derived H_2O_2. Hypertension 2005;46:732–737.
127. Epstein CJ, Avraham KB, Lovett M et al. Transgenic mice with increased Cu/Zn-superoxide dismutase activity: animal model of dosage effects in Down syndrome. Proc Natl Acad Sci U S A 1987;84:8044–8048.
128. Ceballos I, Nicole A, Briand P et al. Expression of human Cu-Zn superoxide dismutase gene in transgenic mice: model for gene dosage effect in Down Syndrome. Free Radic Res Commun 1991;12–13 Pt 2:581–589.
129. Ceballos-Picot I, Nicole A, Briand P et al. Neuronal-specific expression of human copper-zinc superoxide dismutase gene in transgenic mice: animal model of gene dosage effects in Down's syndrome. Brain Res 1991;552:198–214.
130. Wang P, Chen H, Qin H et al. Overexpression of human copper, zinc-superoxide dismutase (SOD1) prevents postischemic injury. Proc Natl Acad Sci U S A 1998;95:4556–4560.
131. Przedborski S, Jackson-Lewis V, Kostic V, Carlson E, Epstein CJ, Cadet JL. Superoxide dismutase, catalase, and glutathione peroxidase activities in copper/zinc-superoxide dismutase transgenic mice. J Neurochem 1992;58:1760–1767.
132. Didion SP, Kinzenbaw DA, Fegan PE, Didion LA, Faraci FM. Overexpression of CuZn-SOD prevents lipopolysaccharide-induced endothelial expression. Stroke 2004;35:1963–1967.
133. Wang HD, Johns DG, Xu S, Cohen RA. Role of superoxide anion in regulating pressor and vascular hypertrophic response to angiotensin II. Am J Physiol 2002;282:H1697–H1702.
134. Forsberg L, de Faire U, Morgenstern R. Oxidative stress, human genetic variation, and disease. Arch Biochem Biophys 2001;389;84–93.
135. Morrison RA, McGrath A, Davidson G, Brown JJ, Murray GD, Lever AF. Low blood pressure in Down's syndrome; a link with Alzheimer's disease? Hypertension 1996;28:569–575.
136. Macmillan-Crow LA, Cruthirds DL. Invited review: manganese superoxide dismutase in disease. Free Radic Res 2001;34:325–336.
137. Lebovitz RM, Zhang H, Vogel H et al. Neurodegeneration, myocardial injury, and perinatal death in mitochondrial superoxide dismutase-deficient mice. Proc Natl Acad Sci U S A 1996;93:9782–9787.
138. Li Y, Huang TT, Carlson EJ et al. Dilated cardiomyopathy and neonatal lethality in mutant mice lacking manganese superoxide dismutase. Nat Genet 1995;11:376–381.
139. Rodriguez-Iturbe B, Sepassi L, Quiroz Y, Ni Z, Wallace DC, Vaziri ND. Association of mitochondrial SOD deficiency with salt-sensitive hypertension and accelerated renal senescence. J Appl Physiol 2007;102:255–260.
140. Chrissobolis S, Didion SP, Faraci FM. Protective role of manganese superoxide dismutase against angiotensin II-induced, Nox2-dependent cerebral endothelial dysfunction. (Abstract). FASEB J 2007;21:A1262–A1263.
141. Shao J, Chen L, Marrs B et al. SOD2 polymorphisms: unmasking the effect of polymorphisms on splicing. BMC Med Genet 2007;8:7.
142. Hsueh YM, Lin P, Chen HW et al. Genetic polymorphisms of oxidative and antioxidant enzymes and arsenic-related hypertension. J Toxicol Environ Health A 2005;68:1471–1484.
143. Stralin P, Karlsson K, Johansson BO, Marklund SL. The interstitium of the human arterial wall contains very large amounts of extracellular superoxide dismutase. Arterioscler Thromb Vasc Biol 1995;15:2032–2036.

144. Jonsson LM, Rees DD, Edlund T, Marklund SL. Nitric oxide and blood pressure in mice lacking extracellular-superoxide dismutase. Free Radic Res 2002;36:755–758.

145. Jung O, Marklund SL, Geiger H, Pedrazzini T, Busse R, Brandes RP. Extracellular superoxide dismutase is a major determinant of nitric oxide bioavailability: in vivo and ex vivo evidence from ecSOD-deficient mice. Circ Res 2003;93:622–629.

146. Welch WJ, Chabrashvili T, Solis G et al. Role of extracellular superoxide dismutase in the mouse angiotensin slow pressor response. Hypertension 2006;48:934–941.

147. Gongora MC, Qin Z, Laude K et al. Role of extracellular superoxide dismutase in hypertension. Hypertension 2006;48:473–481.

148. Oury TD, Ho YS, Piantadosi CA, Crapo JD. Extracellular superoxide dismutase, nitric oxide, and central nervous system O_2 toxicity. Proc Natl Acad Sci U S A 1992;89:9715–9719.

149. Sheng H, Kudo M, Mackensen GB, Pearlstein RD, Crapo JD, Warner DS. Mice overexpressing extracellular superoxide dismutase have increased resistance to global cerebral ischemia. Exp Neurol 2000;163:392–398.

150. Demchenko IT, Oury TD, Crapo JD, Piantadosi CA. Regulation of the brain's vascular responses to oxygen. Circ Res 2002;91:1031–1037.

151. McGirt MJ, Parra A, Sheng H et al. Attenuation of cerebral vasospasm after subarachnoid hemorrhage in mice overexpressing extracellular superoxide dismutase. Stroke 2002; 33:2317–2323.

152. Chu Y, Iida S, Lund DD et al. Gene transfer of extracellular superoxide dismutase reduces arterial pressure in spontaneously hypertensive rats: role of heparin-binding domain. Circ Res 2003;92:461–468.

153. Zhou LC, Xiang W, Potts J et al. Reduction in extracellular superoxide dismutase activity in African-American patients with hypertension. Free Radic Biol Med 2006;41:1384–1391.

154. Sandstrom J, Nilsson P, Karlsson K, Marklund SL. 10-fold increase in human plasma extracellular superoxide dismutase content caused by a mutation in heparin-binding domain. J Biol Chem 1994;269:19163–19166.

155. Marklund SL, Nilsson P, Israelsson K, Schampi I, Peltonen M, Asplund K. Two variants of extracellular-superoxide dismutase: relationship to cardiovascular risk factors in an unselected middle-aged population. J Intern Med 1997;242:5–14.

156. Leopold JA, Loscalzo J. Oxidative enzymopathies and vascular disease. Arterioscler Thromb Vasc Biol 2005;25:1332–1340.

157. 't Hoen PA, Van der Lans CA, Van Eck M, Bijsterbosch MK, Van Berkel TJ, Twisk J. Aorta of ApoE-deficient mice response to atherogenic stimuli by a prelesional increase and subsequent decrease in the expression of antioxidant enzymes. Circ Res 2003;93:262–269.

158. Ho YS, Magnenat JL, Bronson RT et al. Mice deficient in cellular glutathione peroxidase develop normally and show no increased sensitivity to hyperoxia. J Biol Chem 1997;272:16644–16651.

159. Forgione MA, Weiss N, Heydrick S et al. Cellular glutathione peroxidase deficiency and endothelial dysfunction. Am J Physiol 2002;282:H1255–H1261.

160. Forgione MA, Cap A, Liao R et al. Heterozygous cellular glutathione peroxidase deficiency in the mouse: abnormalities in vascular and cardiac function and structure. Circulation 2002;106:1154–1158.

161. Dayal S, Brown KL, Weydert CJ et al. Deficiency of glutathione peroxidase-1 sensitizes hyperhomocysteinemic mice to endothelial dysfunction. Arterioscler Thromb Vasc Biol 2002;22:1996–2002.

162. Yoshida T, Watanabe M, Engelman DT et al. Transgenic mice overexpressing glutathione peroxidase are resistant to myocardial ischemia reperfusion injury. J Mol Cell Cardiol 1996;28:1759–1767.

163. Vaziri ND, Wang XQ, Oveisi F, Rad B. Induction of oxidative stress by glutathione depletion causes severe hypertension in normal rats. Hypertension 2000;36;142–146.

164. Ford RJ, Graham DA, Denniss SG, Quadrilatero J, Rush JW. Glutathione depletion in vivo enhances contraction and attenuates endothelium-dependent relaxation of isolated rat aorta. Free Radic Biol Med 2006;40:670–678.

165. Chen X, Liang H, Van Remmen H, Vijg J, Richardson A. Catalase transgenic mice: characterization and sensitivity to oxidative stress. Arch Biochem Biophys 2004;422:197–210.
166. Yang H, Shi MJ, Van Remmen H, Chen XL, Vijg J, Richardson A, Guo ZM. Reduction of pressor response to vasoconstrictor agents by overexpression of catalase in mice. Am J Hypertens 2003;16:1–5.
167. Kobayashi M, Sugiyama H, Wang DH et al. Catalase deficiency renders remnant kidneys more susceptible to oxidant tissue injury and renal fibrosis in mice. Kidney Int 2005;68:1018–1031.
168. Jiang Z, Akey JM, Shi J et al. A polymorphism in the promoter region of catalase is associated with blood pressure levels. Hum Genet 2001;109:95–98.
169. Zhou XF, Cui J, DeStefano AL et al. Polymorphisms in the promoter region of catalase gene and essential hypertension. Dis Markers 2005;21:3–7.
170. Ho YS, Xiong Y, Ma W, Spector A, Ho DS. Mice lacking catalase develop normally but show differential sensitivity to oxidant tissue injury. J Biol Chem 2004;279:32804–32812.

14
Aging and Cardiac Ischemia—Mitochondria and Free Radical Considerations

Paul S. Brookes and David L. Hoffman

Summary Acute myocardial infarction (MI, heart attack) kills 200,000 people annually in the United States, and the number 1 risk factor for fatal MI is age. The biological phenomenon underlying MI is ischemia-reperfusion (IR) injury, i.e., damage to tissue that occurs during and immediately after an MI. In this chapter, following a brief primer on mitochondrial ROS generation, and the underlying pathologic mechanisms of mitochondrial dysfunction in IR injury, we then break down the effects of aging on MI risk into three parts. First, changes in risk factors for MI with aging, most of which are external to the heart (e.g., atherosclerosis). Second, changes in cardiac mitochondrial function with age. Third, changes in the response of cardiac mitochondria to MI. Last, the role of mitochondria in protecting the heart from ischemia is introduced, and the possibility that increased MI risk with aging originates from a degeneration of protective signaling pathways is raised. We conclude with an outlook on the therapeutic opportunities, both current and developing, for the treatment of increased MI risk, in aged human populations.

Keywords Ischemia-reperfusion injury, mitochondria, reactive oxygen species, aging, electron transport chain, cardioprotection.

1 Cardiac Mitochondria and Metabolism

Because the overall theme of the current volume is the free radical theory of aging, with a particular focus on mitochondria-derived free radicals, an important introductory role to this chapter on cardiac ischemic injury is served by describing aspects of mitochondrial metabolism that are unique to the heart.

The heart is a highly oxygen-dependent tissue, evidenced by the ultrastructure of the tissue itself, which is greatly vascularized with a large endothelial surface area. Although the heart typically accounts for only 0.5% of body weight in humans, it accounts for 10% of whole body O_2 consumption. In addition the O_2 consumption of the heart can be expanded over a large dynamic range under intense exercise [1]. The heart also exhibits a very large ateriovenous O_2 gradient,

From: *Aging Medicine: Oxidative Stress in Aging: From Model Systems to Human Diseases* 253
Edited by: S. Miwa, K.B. Beckman, and F.L. Muller © Humana Press, Totowa, NJ

indicating almost complete O_2 extraction, with O_2 diffusion to mitochondria facilitated by myoglobin.

The heart obtains the majority of its large ATP requirement from mitochondrial oxidative phosphorylation. This dependence on mitochondria is highlighted by observations that approximately 35% of the volume of a typical cardiomyocyte is occupied by mitochondria. The major ATP-consuming reactions in cardiomyocytes are myosin cross-bridge cycling and ion homeostasis (mostly Ca^{2+}). Consistent with these two major functions, cardiac mitochondria are thought to be stratified into two distinct populations: subsarcolemmal (SSM) and intermyofibrillar (IFM), which each exhibit unique bioenergetic properties [2]. The relative importance of SSM versus IFM in the pathology of the aging heart is an area of much scientific interest.

The primary energy source of healthy heart is fatty acid β-oxidation, although, as will be discussed later (Sections 3 and 5), perturbations in mitochondrial substrate choice are prevalent in several pathologic situations. A recent comparative proteomics study examined mitochondria from a variety of tissues, and the study found cardiac mitochondria to be highly enriched in the enzymes of β-oxidation, the respiratory chain, and the "lower-half" of the tricarboxylic acid cycle [3, 4]. In addition, it has been shown that heart mitochondria express a functional monocarboxylate carrier, and this has led to speculation that heart mitochondria may be capable of using lactate as an energy source [5, 6].

2   Cardiac Mitochondrial Generation of Reactive Oxygen Species (ROS)

The generation of ROS by heart mitochondria has received much attention, and index values for H_2O_2 generation rates are given in Table 14.1 for mitochondria respiring on a variety of substrates. For comparison, values from liver, skeletal muscle and brain mitochondria are also shown.

When considering mitochondrial ROS generation, most research has focused on complexes I and III of the respiratory chain, although it should be noted that there are at least nine sources of ROS within mitochondria (for review, see [7]). In heart mitochondria, the electron transfer flavoprotein of β-oxidation is likely to be a significant source of ROS, given the reliance of the heart on fat oxidation for energy. The role of this protein in cardiac ROS-mediated pathology has received little attention to date.

Another factor that may influence cardiac mitochondrial ROS generation is the availability of antioxidants within the myocyte. Interestingly, it has been reported that cardiac mitochondria may contain catalase [8]. The reason for this is not clear (catalase is not found in mitochondria of other tissues), although the high content of iron in myocytes (albeit bound to heme in myoglobin) would suggest that excessive H_2O_2 would be more dangerous than in other tissues, due to the possibility of $OH^·$ generation via Fenton/Haber–Weiss chemistry. The prevalence of other

Table 14.1 Baseline ROS generation rates from mitochondria under various combinations of substrates and inhibitors, measured by several different methods. See text for details ROS generation rates are given in units of pmols/min/mg mitochondrial protein

Linked substrate/tissue	Heart	Liver	Brain	Skeletal	Reference	Method
Complex I - no rotenone	100	100		100	[70]	Homovanillic acid
			577	288	[71]	Epinephrine to epichrome
			20		[72]	Epinephrine to epichrome
			49.3		[73]	Amplex red
			70		[74]	Amplex red
		170			[75]	DCF
			80		[76]	Amplex red
		200			[77]	DCF
	25				[34]	Amplex red
		250			D.L.H. (unpublished]	Amplex red
	1,800				[78]	Review
				70	[79]	Amplex red
				200	[80]	Amplex red
Complex I + rotenone			577	288	[71]	Epinephrine to epichrome
			530		[72]	Epinephrine to epichrome
			275		[73]	Amplex red
			300		[34]	Amplex red
	4,200				[78]	Review
	1,700				[81]	Adrenochrome
				610	[79]	Amplex red
	250	200		300	[70]	Homovanillic acid
				350	[80]	Amplex red
Complex I + antimycin a			250		[73]	Amplex red
Complex II - no rotenone			1,237	1,718	[71]	Epinephrine to epichrome
			660		[72]	Epinephrine to epichrome
			280		[73]	Amplex red
				2,600	[82]	PHPA
	400				[75]	DCF
		200			[77]	DCF
		420			[83]	Cyt c peroxidase
			1700		[84]	Amplex red
complex II + rotenone		300			[85]	Scopoletin
	150				[34]	Amplex red
		100			[13]	Amplex red
		770			[86]	Cyt c reduction ± SOD
Complex II + antimycin A			50		[73]	Amplex red
			1,200		[73]	Amplex red
	1,750				[34]	Amplex red
				300	[80]	Amplex red
	500	100		300	[70]	Homovanillic acid

antioxidants in the heart is relatively under studied, although heart mitochondria (like all mitochondria) contain high (millimolar) levels of glutathione. Thioredoxin also is expressed widely in the heart, and superoxide dismutase (SOD) is also critical for cardiac function, because MnSOD null die prematurely of cardiac complications [9, 10].

The generation of ROS by the two different subpopulations of heart mitochondria is relatively unclear, and one important factor in this regard is O_2 gradients across the cell [11]. Several laboratories have reported on the possibility that mitochondrial ROS generation may increase as O_2 levels decrease [12], but O_2 is the substrate for all ROS generation, and we recently used a steady-state $[O_2]$ respiration system to demonstrate that isolated mitochondrial ROS generation follows a hyperbolic function of $[O_2]$ (i.e., less ROS at low $[O_2]$, saturating at high $[O_2]$) [13]. Thus, it can be hypothesized that mitochondria closer to the cell membrane and the vasculature (i.e., SSM) may generate higher levels of ROS than those further inside the cell and subject to lower O_2 availability. However, in contrast to this hypothesis, in vitro studies on isolated SSM and IFM have shown that the latter contain higher amounts of cytochromes and thus make higher amounts of ROS [14, 15]. These two parameters (cytochrome content vs O_2 availability in vivo) may effectively cancel each other out.

3   Pathologic Mechanisms in Ischemia-Reperfusion (IR) Injury

The overall mechanisms of tissue damage during IR injury can be divided into three temporal stages: (1) Damage occurring during the ischemia itself. (2) Damage occurring immediately upon reperfusion. (3) Damage occurring at a later time (hours to days) after reperfusion, usually associated with an inflammatory response. The role of mitochondria in mediating 1 and 2 has been demonstrated [16], although evidence of a role for mitochondria in 3 is still somewhat unclear.

During IR injury, the major biochemical participants can be summarized as Ca^{2+}, ROS, and pH. The complex interplay between these key factors has been reviewed extensively previously [16], and so herein a brief summary and reference to Fig. 14.1 serve to introduce the topic. The development of ischemia (lack of O_2 and substrates due to occlusion of an upstream blood vessel) inhibits mitochondrial oxidative phosphorylation, leading to lower ATP availability and a buildup of lactate from anaerobic metabolism. Low ATP impairs Ca^{2+} export via the, sarco(endo)plasmic reticulum Ca^{2+} ATPase pump, leading to elevated cytosolic $[Ca^{2+}]$, whereas acidic pH from lactic acidosis leads to a secondary overload of cytosolic $[Na^+]$ via the Na^+/H^+ exchanger (NHE). In responding to this Na^+ overload, reverse operation of the Na^+/Ca^{2+} exchanger then further elevates cytosolic $[Ca^{2+}]$. During ischemia, at least some of this elevated cytosolic Ca^{2+} accumulates inside mitochondria [17]. However, the prevailing acidic pH inhibits opening of the permeability transition (PT) pore that would normally occur in response to such a Ca^{2+} accumulation [18]. Upon reperfusion, several factors converge to trigger opening of the PT pore [19]. Re-energization of mitochondria enhances their ability to accumulate Ca^{2+}, resulting in $[Ca^{2+}]_m$ overload. A surge of ROS generation also occurs after reintroduction of O_2 to the tissue. Furthermore, the normalization of cellular pH removes the acidic-pH brake on the PT pore. Thus, there is widespread agreement that the PT pore stays

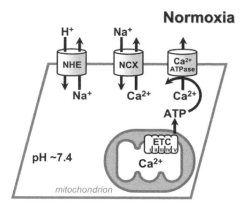

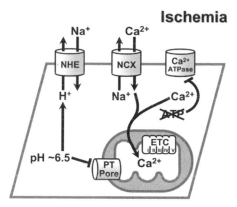

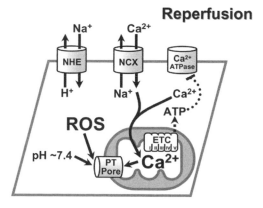

Fig. 14.1 Pathologic events in cardiac IR injury. See text for details

closed during ischemia and opens at reperfusion. Opening of the PT pore results in cytochrome c release and triggering of the intrinsic apoptotic cascade. In addition, the release of cytochrome c interrupts the normal function of the respiratory chain, and owing to the heavy reliance of myocardium on mitochondrially derived ATP, this results in a profound bioenergetic defect that negatively impacts cardiac contractile function [16].

Several cardioprotective agents are known to elicit their effects by inhibiting the events outlined in Fig. 14.1. These agents includes PT pore inhibitors such as cyclosporin A and sanglifehrin A [20], NHE inhibitors such as cariporide [21, 22], mild uncoupling of mitochondria either by chemical uncouplers or changes in expression of uncoupling proteins [23–25], antioxidants such as 2-mercaptopropionyl glycine [26], and the recent discovery that ischemic postconditioning (staccato reperfusion) serves to maintain acidic pH into the reperfusion window [27].

In addition to the events described above, several proteins and other biomolecules within the mitochondrion are subject to oxidative damage after IR injury [28]. These include complexes I, II, III, and IV of the respiratory chain, the ATP synthase (complex V), the adenine-nucleotide translocase, the phospholipid cardiolipin, and tricarboxylic acid cycle enzymes (aconitase, α–ketoglutarate dehydrogenase) [23–25, 29–33]. Together with the energetic defect caused by cytochrome c release (see above), these events together contribute to a prolonged defect in energy metabolism at the mitochondrial level that can last well beyond the immediate reperfusion period. Interestingly, we also showed recently that oxidative posttranslational modification of complex I in IR injury can increase the ability of the complex to generate ROS [34].

4   Changes in Risk Factors and Underlying Causes of IR Injury during Aging

In considering the impact of aging on cardiac IR injury, a distinction should be drawn between IR injury itself (i.e., the events outlined in Fig. 14.1), and the risk factors that lead to an increased incidence of IR injury with aging, i.e., the risk factors for a stoppage of blood flow to the heart.

The primary risk factors for cardiovascular disease are the "usual suspects": family history, diet, sedentary lifestyle, smoking, cholesterol, diabetes, and oxidative stress. All of these factors contribute to the development of atherosclerosis [35]. Although atherosclerosis was once considered a disease of the elderly, its prevalence among the young is increasing, to the extent that atherosclerotic lesions have now been found in children as young as 12 [36].

Additional risk factors that change with age include decreased activities of nitric-oxide synthases (especially endothelial nitric-oxide synthase) [37], of which activation has been shown to be cardioprotective [38, 39]. Also, arterial stiffness and dilation; prolonged repolarization of the cardiomyocyte action potential, and decreased sensitivity in β-adrenergic signaling; which as been shown to contribute to decreased cardiac output under higher workloads [40].

In addition to the simple observation that the likelihood of suffering an acute cardiac ischemic event (usually due to plaque rupture) increases with age, it also should be noted that the amount of damage suffered as a result of such an event also increases with age, i.e., there are changes at the cardiomyocyte level that make these cells more susceptible to injury upon encountering IR. As mentioned, aged cardiomyocytes have been shown to exhibit a prolonged action potential, which contributes to a decreased sarcoplasmic reticulum Ca^{2+} pumping rate (decreased Ca^{2+} handling efficiency). This alteration in Ca^{2+} handling can increase cytosolic concentrations of Ca^{2+} (increasing the likelihood of a Ca^{2+} overload during IR injury) and place added strain on the excitation-contraction-coupling machinery by varying the amplitude and pattern of Ca^{2+} stimulation. Adding to this, alterations in various myofilament proteins, specifically the Ca^{2+}-activated ATPase, also have been shown to influence Ca^{2+} handling in the aging myocyte [40]. Furthermore, it is known that the activity of NHE increases with aging [41], and this is thought to be responsible for the apparently greater sensitivity of aged hearts to NHE inhibitors such as cariporide [42]. All of these effects of aging, combined with aging-associated mitochondrial damage (discussed below), influence Ca^{2+} homeostasis within the cardiomyocyte, rendering it more susceptible to injury during an ischemic event.

5   Changes in Cardiac Mitochondria during Aging

As mentioned in Section 2, cardiac mitochondria generate ROS, and over time these ROS and other factors associated with aging can lead to damage of mitochondrial proteins, resulting in mitochondrial dysfunction. Notably, of the two subpopulations of mitochondria found in the heart (see Section 1), IFM seem to be far more sensitive to aging than SSM [15, 43–47].

It has been shown that in IFM, aging decreases the activity of complexes III and IV of the electron transport chain (ETC) [43, 46–48]. Although the exact cause of this decrease, or the specific mechanisms of damage unique to IFM versus SSM are not known, it has been hypothesized that IFM have increased numbers of respiratory cytochromes, and increased ETC activity compared with SSM. Indeed, IFM may be more sensitive to age-induced damage because they have a greater target density; alternatively they may produce larger quantities of ROS, or they may lack the means to detoxify ROS (e.g., SOD levels) compared with than their relatively unaffected SMM counterparts [49].

5.1   Aging and Complex III

Ubiquinol cytochrome c oxidoreductase (complex III) is one of the major sources of ROS (in addition to complex I) within the ETC. As electrons enter the ETC from complex I and II, they are passed to complex III via the membrane-bound electron carrier Co-enzyme Q (ubiquinone/ubiquinol). At the Q_o site of complex III, electrons are passed from fully reduced ubiquinol (QH_2), generating a ubisemiquinone radical ($QH^·$), which can interact with molecular oxygen to form superoxide ($O_2^{·-}$), although this is a minor reaction, and normally electrons are passed from $QH^·$ on to the b-cytochromes of complex III, resulting in the regeneration of ubiquinone (Q).

Because of its ability to generate ROS, it is not surprising that the effects of aging have been linked to increased ROS production, due to damage of this complex. The effects of aging have been shown to damage the Q_o site in complex III, proximal to the Rieske iron sulfur protein (RISP) [46]. The RISP is a 2Fe-2S protein that functions as an electron shuttle from the Q_o site to cytochrome c_1 [50]. Generation of $O_2^{·-}$ can have drastic effects on the mitochondria, through either mutations in mitochondrial DNA, the formation of lipid radicals, or oxidation of mitochondrial proteins [7, 16]. All of these can further augment the generation of $O_2^{·-}$ and increase mitochondrial dysfunction.

5.2   Aging and Complex IV

Damage to cytochrome c oxidase (complex IV) during aging was initially thought to involve alterations in the level of cytochromes within the complex [2, 51], although further studies measuring the activity of purified enzyme in aged and control samples suggested that this was not the case [49]. Interestingly, studies measuring complex IV activity in permeabilized mitochondria noted that the addition of phospholipid liposomes containing cardiolipin reversed the age-related loss of activity [52, 53]. Cardiolipin is a mitochondrial phospholipid thought to be uniquely localized to the outer surface of the inner membrane. Its location and strong negative charge are the basis for its proposed role in membrane tethering of the mobile electron carrier cytochrome c, which is positively charged due to its high content of lysine residues. Thus, alterations in the structure or quantity of cardiolipin can affect the activity of complex IV by decreasing the influx of electrons from cytochrome c. In addition, it is now recognized that breakage of the strong interaction between cytochrome c and the inner membrane may be an important regulatory step in the process of cytochrome c release, during cell death pathways [54, 55].

The proposal of a link between cardiolipin and cardiac mitochondrial aging has promoted much interest in this field. However, it has been revealed that there is no change in the composition of the acyl chains of cardiolipin (CL) during aging [56, 57]. Thus, it is suggested that the loss of "functional" CL at the inner membrane in aging is a true loss in the amount of CL, and not due to changes in the properties of individual CL molecules in situ [14].

Because complex IV is the terminal site of the electron transport chain, it could be hypothesized that inhibition at this site may cause a "buildup" of electrons in the proximal respiratory chain and enhance ROS generation (e.g., at the Q_o site of complex III). However, such a hypothesis is not supported by empirical observations regarding the effects of complex IV inhibitors (e.g., cyanide, azide, nitric oxide) on ROS generation (reviewed in [7]).

6   Response of Cardiac Mitochondria to IR Injury: Changes during Aging

The susceptibility to damage during IR injury, increases with aging [41, 52]. This may be due to preexisting mitochondrial damage caused by aging, as described above. However, there are some examples which suggest this is not the case.

First, if underlying mitochondrial damage were a cause of increased damage sensitivity in IR injury, it might be predicted that damage in IR would be primarily in IFM, because IFM bear the overwhelming burden of aging-related damage [14]. This is not the case, because both IFM and SSM seem to be targets of IR injury-induced damage in young and old hearts alike [52].

Second, inactivation of the complex III occurs during aging [52], and it has been shown that genetic ablation of the RISP component of complex III causes a decrease in mitochondrial ROS generation in response to hypoxia [12, 58, 59]. Thus, it could be predicted that loss of complex III in aging would lead to lower ROS in IR injury. This is opposite to the actual experimental observation of greater ROS generation in mitochondria from aged hearts [46]. In addition, although some generation of ROS may indeed occur during ischemia/hypoxia [60], the majority of ROS generation during IR typically occurs upon reperfusion [61], and studies performed in this laboratory have shown that under tightly controlled O_2 tensions, mitochondrial ROS production decreases as O_2 levels drop [14].

Third, it is known that the sensitivity of SSM to Ca^{2+} overload and PT pore opening is far greater than that of IFM [62]. Thus, it would be expected that SSM would be more susceptible to damage in IR injury, and yet IFM are in fact more sensitive in aging. Thus, the patterns of damage to IFM versus SSM seem opposite in response to either aging or IR injury, with IFM dominating the former and SMM the latter. The role of preexisting age-related mitochondrial defects, in determining the response of mitochondria to IR injury, is both complicated and sometimes paradoxical.

In addition to changes in mitochondrial function in aging, that may affect IR injury response, several of the key pathologic mechanisms of IR injury also may be perturbed in aging. An example of this is increased levels of accumulated cytosolic Ca^{2+} in aged hearts compared with younger controls [41], thus increasing the probability for opening of the PT pore. In addition, the amount of SOD, an enzyme responsible for converting O_2^{-} to the less reactive H_2O_2, has been shown to decrease during aging [48]. Such as loss of SOD occurring at the cardiac level, could explain the observed increased risk of ROS-induced ROS production in aged hearts [63].

7 Efficacy of Cardioprotection in Aging

How much of the decline in recovery from IR injury observed during aging, is a gain of damage versus a loss of protective signaling pathways? To answer this question, it is necessary to give a brief overview of cardioprotection.

One of the most widely studied cardioprotective mechanisms is that of ischemic preconditioning (IPC), in which short nonlethal periods of IR (typically three to five cycles of 3–5 min each) can significantly protect the heart from subsequent prolonged IR injury [64]. Figure 14.2 shows an overview of various cardioprotective signaling pathways, highlighting that nearly all of these pathways converge on the mitochondrion, and the inhibition of PT pore opening, as an end effector. Notably, several widely popular clinical protective interventions also converge on these same signaling pathways.

The efficacy of IPC decreases with aging and the underlying mechanisms for this decrease are currently unclear [41, 63, 65]. In addition, several pharmacologic

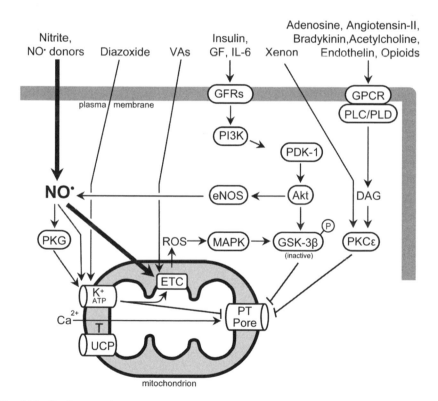

Fig. 14.2 Cardioprotective signaling pathways converging on the mitochondrion. Three main signaling pathways have been defined in cardioprotective signaling, namely the insulin → Akt pathway, the G protein-coupled receptor → protein kinase (PK)C pathway, and the nitric oxide → PKG → mK$^+_{ATP}$ channel pathway. Several components of these pathways are known to be inactivated during aging (see text for details)

agents that mimic IPC (e.g., volatile anesthetics) are also less effective in aged patient populations [66, 67]. Perturbations to several cardioprotective signaling pathways have been observed during aging. One such example is a defect in the cardioprotective signal transduction pathway that leads to activation of the mitochondrial K_{ATP} channel [68]. Another example is the insulin → Akt signaling pathway, which is known to decrease significantly with age, probably related to the increased incidence of insulin resistance/type 2 diabetes/metabolic syndrome with aging [69]. Another example is the nitric-oxide signaling pathway, which has been shown to decline with aging [37].

Interestingly, this final example also serves to highlight the potential therapeutic avenues that are available to ameliorate the decline in functional recovery of the heart from IR injury, which occurs during aging. Nitric oxide is a broad cardioprotective agent, acting at a number of molecular targets within cardiomyocytes (including several in mitochondria) to afford protection from IR injury [38, 39]. Thus, increasing NO levels may be an important therapeutic in aging.

8   Concluding Remarks

In considering the mitochondrial free radical theory of aging, it is sometimes difficult to delineate between aging itself, and the various diseases of aging. It is hoped that the proceeding chapter highlighted that the myocardium, in the pathologic setting of IR injury, is both a disease condition, and a microcosm of the aging process. Further delineation of the changes in mitochondrial function in IR injury alone, in aging, and in both, will lead to the identification of novel targets for therapeutic approaches. Given the essential nature of cardiac function for viability of most organisms, it is hoped that such therapies may be broadly considered as life span extenders.

Acknowledgements   Work in the laboratory of P.S.B. is funded by National Institutes of Health grant HL-071158.

References

1. Neely JR, Liebermeister H, Battersby EJ, Morgan HE. Effect of pressure development on oxygen consumption by isolated rat heart. Am J Physiol 1967;212(4):804–14.
2. Palmer JW, Tandler B, Hoppel CL. Biochemical properties of subsarcolemmal and interfibrillar mitochondria isolated from rat cardiac muscle. J Biol Chem 1977;252(23):8731–9.
3. Johnson DT, Harris RA, Blair PV, Balaban RS. Functional consequences of mitochondrial proteome heterogeneity. Am J Physiol 2007;292(2):C698–707.
4. Johnson DT, Harris RA, French S et al. Tissue heterogeneity of the mammalian mitochondrial proteome. Am J Physiol 2007;292(2):C689–97.
5. Brooks GA, Brown MA, Butz CE, Sicurello JP, Dubouchaud H. Cardiac and skeletal muscle mitochondria have a monocarboxylate transporter MCT1. J Appl Physiol 1999;87(5):1713–8.
6. Chatham JC. Lactate–the forgotten fuel! J Physiol 2002;542(2):333.

7. Brookes P. Mitochondrial production of oxidants and their role in the regulation of cellular processes. In: Gibson G, Dienel G, eds. Handbook of neurochemistry and molecular neurobiology. New York: Springer; 2007.

8. Radi R, Turrens JF, Chang LY, Bush KM, Crapo JD, Freeman BA. Detection of catalase in rat heart mitochondria. J Biol Chem 1991;266(32):22028–34.

9. Lebovitz RM, Zhang H, Vogel H et al. Neurodegeneration, myocardial injury, and perinatal death in mitochondrial superoxide dismutase-deficient mice. Proc Natl Acad Sci U S A 1996;93(18):9782–7.

10. Li Y, Huang TT, Carlson EJ et al. Dilated cardiomyopathy and neonatal lethality in mutant mice lacking manganese superoxide dismutase. Nat Genet 1995;11(4):376–81.

11. Takahashi E, Endoh H, Doi K. Visualization of myoglobin-facilitated mitochondrial O(2) delivery in a single isolated cardiomyocyte. Biophys J 2000;78(6):3252–9.

12. Chandel NS, McClintock DS, Feliciano CE et al. Reactive oxygen species generated at mitochondrial complex III stabilize hypoxia-inducible factor-1alpha during hypoxia: a mechanism of O_2 sensing. J Biol Chem 2000;275(33):25130–8.

13. Hoffman DL, Salter JD, Brookes PS. Response of mitochondrial reactive oxygen species generation to steady-state oxygen tension: implications for hypoxic cell signaling. Am J Physiol 2007;292(1):H101–8.

14. Lesnefsky EJ, Hoppel CL. Oxidative phosphorylation and aging. Ageing Res Rev 2006;5(4):402–33.

15. Lesnefsky EJ, Hoppel CL. Ischemia-reperfusion injury in the aged heart: role of mitochondria. Arch Biochem Biophys 2003;420(2):287–97.

16. Brookes PS, Yoon Y, Robotham JL, Anders MW, Sheu SS. Calcium, ATP, and ROS: a mitochondrial love-hate triangle. Am J Physiol 2004;287(4):C817–33.

17. Riess ML, Camara AK, Novalija E, Chen Q, Rhodes SS, Stowe DF. Anesthetic preconditioning attenuates mitochondrial Ca2+ overload during ischemia in guinea pig intact hearts: reversal by 5-hydroxydecanoic acid. Anesth Analg 2002;95(6):1540–6, table of contents.

18. Kim JS, Jin Y, Lemasters JJ. Reactive oxygen species, but not Ca2+ overloading, trigger pH- and mitochondrial permeability transition-dependent death of adult rat myocytes after ischemia-reperfusion. Am J Physiol 2006;290(5):H2024–34.

19. Griffiths EJ, Halestrap AP. Mitochondrial non-specific pores remain closed during cardiac ischaemia, but open upon reperfusion. Biochem J 1995;307(1):93–8.

20. Clarke SJ, McStay GP, Halestrap AP. Sanglifehrin A acts as a potent inhibitor of the mitochondrial permeability transition and reperfusion injury of the heart by binding to cyclophilin-D at a different site from cyclosporin A. J Biol Chem 2002;277(38):34793–9.

21. Chakrabarti S, Hoque AN, Karmazyn M. A rapid ischemia-induced apoptosis in isolated rat hearts and its attenuation by the sodium-hydrogen exchange inhibitor HOE 642 (cariporide). J Mol Cell Cardiol 1997;29(11):3169–74.

22. Portman MA, Panos AL, Xiao Y, Anderson DL, Ning X. HOE-642 (cariporide) alters pH(i) and diastolic function after ischemia during reperfusion in pig hearts in situ. Am J Physiol 2001;280(2):H830–4.

23. Ganote CE, Armstrong SC. Effects of CCCP-induced mitochondrial uncoupling and cyclosporin A on cell volume, cell injury and preconditioning protection of isolated rabbit cardiomyocytes. J Mol Cell Cardiol 2003;35(7):749–59.

24. Hoerter J, Gonzalez-Barroso MD, Couplan E et al. Mitochondrial uncoupling protein 1 expressed in the heart of transgenic mice protects against ischemic-reperfusion damage. Circulation 2004;110(5):528–33.

25. Minners J, van den Bos EJ, Yellon DM, Schwalb H, Opie LH, Sack MN. Dinitrophenol, cyclosporin A, and trimetazidine modulate preconditioning in the isolated rat heart: support for a mitochondrial role in cardioprotection. Cardiovasc Res 2000;47(1):68–73.

26. Nadtochiy SM, Burwell LS, Brookes PS. Cardioprotection and mitochondrial S-nitrosation: effects of S-nitroso-2-mercaptopropionyl glycine (SNO-MPG) in cardiac ischemia-reperfusion injury. J Mol Cell Cardiol 2007;42:812–25.

27. Cohen MV, Yang XM, Downey JM. The pH hypothesis of postconditioning: staccato reperfusion reintroduces oxygen and perpetuates myocardial acidosis. Circulation 2007;115(14):1895–903.
28. Lucas DT, Szweda LI. Cardiac reperfusion injury: aging, lipid peroxidation, and mitochondrial dysfunction. Proc Natl Acad Sci U S A 1998;95(2):510–4.
29. Nadtochiy SM, Tompkins AJ, Brookes PS. Different mechanisms of mitochondrial proton leak in ischaemia/reperfusion injury and preconditioning: implications for pathology and cardioprotection. Biochem J 2006;395(3):611–8.
30. Paradies G, Ruggiero FM, Petrosillo G, Quagliariello E. Peroxidative damage to cardiac mitochondria: cytochrome oxidase and cardiolipin alterations. FEBS Lett 1998;424(3):155–8.
31. Rouslin W, Millard RW. Mitochondrial inner membrane enzyme defects in porcine myocardial ischemia. Am J Physiol 1981;240(2):H308–13.
32. Rouslin W. Mitochondrial complexes I, II, III, IV, and V in myocardial ischemia and autolysis. Am J Physiol 1983;244(6):H743–8.
33. Tretter L, Adam-Vizi V. Inhibition of Krebs cycle enzymes by hydrogen peroxide: A key role of α-ketoglutarate dehydrogenase in limiting NADH production under oxidative stress. J Neurosci 2000;20(24):8972–9.
34. Tompkins AJ, Burwell LS, Digerness SB, Zaragoza C, Holman WL, Brookes PS. Mitochondrial dysfunction in cardiac ischemia-reperfusion injury: ROS from complex I, without inhibition. Biochim Biophys Acta 2006;1762(2):223–31.
35. Hirsch AT, Haskal ZJ, Hertzer NR et al. ACC/AHA 2005 Practice Guidelines for the management of patients with peripheral arterial disease (lower extremity, renal, mesenteric, and abdominal aortic): a collaborative report from the American Association for Vascular Surgery/Society for Vascular Surgery, Society for Cardiovascular Angiography and Interventions, Society for Vascular Medicine and Biology, Society of Interventional Radiology, and the ACC/AHA Task Force on Practice Guidelines (Writing Committee to Develop Guidelines for the Management of Patients With Peripheral Arterial Disease): endorsed by the American Association of Cardiovascular and Pulmonary Rehabilitation; National Heart, Lung, and Blood Institute; Society for Vascular Nursing; TransAtlantic Inter-Society Consensus; and Vascular Disease Foundation. Circulation 2006;113(11):e463–654.
36. Groner JA, Joshi M, Bauer JA. Pediatric precursors of adult cardiovascular disease: noninvasive assessment of early vascular changes in children and adolescents. Pediatrics 2006;118(4):1683–91.
37. d'Alessio P. Aging and the endothelium. Exp Gerontol 2004;39(2):165–71.
38. Cohen MV, Yang XM, Downey JM. Nitric oxide is a preconditioning mimetic and cardioprotectant and is the basis of many available infarct-sparing strategies. Cardiovasc Res 2006;70(2):231–9.
39. Jones SP, Bolli R. The ubiquitous role of nitric oxide in cardioprotection. J Mol Cell Cardiol 2006;40(1):16–23.
40. Lakatta EG. Cardiovascular regulatory mechanisms in advanced age. Physiol Rev 1993;73(2):413–67.
41. Tani M, Suganuma Y, Hasegawa H et al. Decrease in ischemic tolerance with aging in isolated perfused Fischer 344 rat hearts: relation to increases in intracellular Na+ after ischemia. J Mol Cell Cardiol 1997;29(11):3081–9.
42. Besse S, Tanguy S, Boucher F et al. Cardioprotection with cariporide, a sodium-proton exchanger inhibitor, after prolonged ischemia and reperfusion in senescent rats. Exp Gerontol 2004;39(9):1307–14.
43. Chen Q, Moghaddas S, Hoppel CL, Lesnefsky EJ. Reversible blockade of electron transport during ischemia protects mitochondria and decreases myocardial injury following reperfusion. J Pharmacol Exp Ther 2006;319(3):1405–12.
44. Fannin SW, Lesnefsky EJ, Slabe TJ, Hassan MO, Hoppel CL. Aging selectively decreases oxidative capacity in rat heart interfibrillar mitochondria. Arch Biochem Biophys 1999;372(2):399–407.

45. Lesnefsky EJ, Gudz TI, Moghaddas S et al. Aging decreases electron transport complex III activity in heart interfibrillar mitochondria by alteration of the cytochrome c binding site. J Mol Cell Cardiol 2001;33(1):37–47.
46. Moghaddas S, Hoppel CL, Lesnefsky EJ. Aging defect at the QO site of complex III augments oxyradical production in rat heart interfibrillar mitochondria. Arch Biochem Biophys 2003;414(1):59–66.
47. Moghaddas S, Stoll MS, Minkler PE, Salomon RG, Hoppel CL, Lesnefsky EJ. Preservation of cardiolipin content during aging in rat heart interfibrillar mitochondria. J Gerontol A Biol Sci Med Sci 2002;57(1):B22–8.
48. Darr D, Fridovich I. Adaptation to oxidative stress in young, but not in mature or old, Caenorhabditis elegans. Free Radic Biol Med 1995;18(2):195–201.
49. Fridovich I. Mitochondria: are they the seat of senescence? Aging Cell 2004;3(1):13–6.
50. Rieske JS, Hansen RE, Zaugg WS. Studies on the electron transfer system. 58. Properties of a new oxidation-reduction component of the respiratory chain as studied by electron paramagnetic resonance spectroscopy. J Biol Chem 1964;239:3017–22.
51. Hoppel CL, Tandler B, Parland W, Turkaly JS, Albers LD. Hamster cardiomyopathy. A defect in oxidative phosphorylation in the cardiac interfibrillar mitochondria. J Biol Chem 1982;257(3):1540–8.
52. Paradies G, Ruggiero FM, Petrosillo G, Quagliariello E. Age-dependent decrease in the cytochrome c oxidase activity and changes in phospholipids in rat-heart mitochondria. Arch Gerontol Geriatr 1993;16(3):263–72.
53. Paradies G, Ruggiero FM, Petrosillo G, Quagliariello E. Age-dependent decline in the cytochrome c oxidase activity in rat heart mitochondria: role of cardiolipin. FEBS Lett 1997;406(1–2):136–8.
54. Ott M, Robertson JD, Gogvadze V, Zhivotovsky B, Orrenius S. Cytochrome c release from mitochondria proceeds by a two-step process. Proc Natl Acad Sci U S A 2002;99(3):1259–63.
55. Tyurin VA, Tyurina YY, Osipov AN et al. Interactions of cardiolipin and lyso-cardiolipins with cytochrome c and tBid: conflict or assistance in apoptosis. Cell Death Differ 2007;14(4):872–5.
56. Hansford RG, Tsuchiya N, Pepe S. Mitochondria in heart ischaemia and aging. Biochem Soc Symp 1999;66:141–7.
57. Pepe S. Mitochondrial function in ischaemia and reperfusion of the ageing heart. Clin Exp Pharmacol Physiol 2000;27(9):745–50.
58. Brunelle JK, Bell EL, Quesada NM et al. Oxygen sensing requires mitochondrial ROS but not oxidative phosphorylation. Cell Metab 2005;1(6):409–14.
59. Guzy RD, Hoyos B, Robin E et al. Mitochondrial complex III is required for hypoxia-induced ROS production and cellular oxygen sensing. Cell Metab 2005;1(6):401–8.
60. Duranteau J, Chandel NS, Kulisz A, Shao Z, Schumacker PT. Intracellular signaling by reactive oxygen species during hypoxia in cardiomyocytes. J Biol Chem 1998;273(19):11619–24.
61. Honda HM, Korge P, Weiss JN. Mitochondria and ischemia/reperfusion injury. Ann N Y Acad Sci 2005;1047:248–58.
62. Palmer JW, Tandler B, Hoppel CL. Heterogeneous response of subsarcolemmal heart mitochondria to calcium. Am J Physiol 1986;250(5 Pt 2):H741–8.
63. Juhaszova M, Rabuel C, Zorov DB, Lakatta EG, Sollott SJ. Protection in the aged heart: preventing the heart-break of old age? Cardiovasc Res 2005;66(2):233–44.
64. Murry CE, Jennings RB, Reimer KA. Preconditioning with ischemia: a delay of lethal cell injury in ischemic myocardium. Circulation 1986;74(5):1124–36.
65. Devlin W, Cragg D, Jacks M, Friedman H, O'Neill W, Grines C. Comparison of outcome in patients with acute myocardial infarction aged > 75 years with that in younger patients. Am J Cardiol 1995;75(8):573–6.
66. Riess ML, Stowe DF, Warltier DC. Cardiac pharmacological preconditioning with volatile anesthetics: from bench to bedside? Am J Physiol 2004;286(5):H1603–7.
67. Sniecinski R, Liu H. Reduced efficacy of volatile anesthetic preconditioning with advanced age in isolated rat myocardium. Anesthesiology 2004;100(3):589–97.

68. Lee TM, Su SF, Chou TF, Lee YT, Tsai CH. Loss of preconditioning by attenuated activation of myocardial ATP-sensitive potassium channels in elderly patients undergoing coronary angioplasty. Circulation 2002;105(3):334–40.

69. Dela F, Kjaer M. Resistance training, insulin sensitivity and muscle function in the elderly. Essays Biochem 2006;42:75–88.

70. St-Pierre J, Buckingham JA, Roebuck SJ, Brand MD. Topology of superoxide production from different sites in the mitochondrial electron transport chain. J Biol Chem 2002;277(47): 44784–90.

71. Kudin AP, Debska-Vielhaber G, Kunz WS. Characterization of superoxide production sites in isolated rat brain and skeletal muscle mitochondria. Biomed Pharmacother 2005;59(4): 163–8.

72. Kudin AP, Bimpong-Buta NY, Vielhaber S, Elger CE, Kunz WS. Characterization of superoxide-producing sites in isolated brain mitochondria. J Biol Chem 2004;279(6):4127–35.

73. Panov A, Dikalov S, Shalbuyeva N, Hemendinger R, Greenamyre JT, Rosenfeld J. Species- and tissue-specific relationships between mitochondrial permeability transition and generation of ROS in brain and liver mitochondria of rats and mice. Am J Physiol 2007;292(2): C708–18.

74. Starkov AA, Fiskum G. Regulation of brain mitochondrial H_2O_2 production by membrane potential and NAD(P)H redox state. J Neurochem 2003;86(5):1101–7.

75. Petrosillo G, Di Venosa N, Ruggiero FM et al. Mitochondrial dysfunction associated with cardiac ischemia/reperfusion can be attenuated by oxygen tension control. Role of oxygen-free radicals and cardiolipin. Biochim Biophys Acta 2005;1710(2–3):78–86.

76. Starkov AA, Fiskum G, Chinopoulos C et al. Mitochondrial alpha-ketoglutarate dehydrogenase complex generates reactive oxygen species. J Neurosci 2004;24(36):7779–88.

77. Young TA, Cunningham CC, Bailey SM. Reactive oxygen species production by the mitochondrial respiratory chain in isolated rat hepatocytes and liver mitochondria: studies using myxothiazol. Arch Biochem Biophys 2002;405(1):65–72.

78. Barja G. Mitochondrial free radical production and aging in mammals and birds. Ann N Y Acad Sci 1998;854:224–38.

79. Mansouri A, Muller FL, Liu Y et al. Alterations in mitochondrial function, hydrogen peroxide release and oxidative damage in mouse hind limb skeletal muscle during aging. Mech Ageing Dev 2006;127(3):298–306.

80. Muller FL, Liu Y, Van Remmen H. Complex III releases superoxide to both sides of the inner mitochondrial membrane. J Biol Chem 2004;279(47):49064–73.

81. Turrens JF, Boveris A. Generation of superoxide anion by the NADH dehydrogenase of bovine heart mitochondria. Biochem J 1980;191(2):421–7.

82. Lambert AJ, Brand MD. Superoxide production by NADH:ubiquinone oxidoreductase (complex I) depends on the pH gradient across the mitochondrial inner membrane. Biochem J 2004;382(2):511–7.

83. Boveris A, Costa LE, Poderoso JJ, Carreras MC, Cadenas E. Regulation of mitochondrial respiration by oxygen and nitric oxide. Ann N Y Acad Sci 2000;899:121–35.

84. Zoccarato F, Cavallini L, Bortolami S, Alexandre A. Succinate modulation of H_2O_2 release at NADH: ubiquinone oxidoreductase (complex I) in brain mitochondria. Biochem J 2007;406(1):125–9.

85. Battaglia V, Rossi CA, Colombatto S, Grillo MA, Toninello A. Different behavior of agmatine in liver mitochondria: inducer of oxidative stress or scavenger of reactive oxygen species? Biochim Biophys Acta 2007;1768(5):1147–53.

86. Sohal RS, Svensson I, Brunk UT. Hydrogen peroxide production by liver mitochondria in different species. Mech Ageing Dev 1990;53(3):209–15.

15
Role of the Antioxidant Network in the Prevention of Age-Related Diseases

Mauro Serafini

Summary The "free radical" theory of ageing is based on the fact that the deleterious effects of an excessive production of free radicals produced during aerobic metabolism cause oxidative stress, damaging biomolecules, cell structure, and function. It is widely believed that the release of free radicals, either as a by-product of normal metabolism or associated with inflammatory reactions, can contribute to a number of age-related diseases such as cardiovascular disease and cancer. Several ecological, case-control, and cohort studies indicate that diets rich in plant foods might reduce age-related disease development. Among a number of mechanistic hypotheses, diet-derived antioxidants have been proposed to contribute to explain these findings. However, contrasting results from intervention trials have raised strong concerns about the influence of antioxidants on human health. A vulnerable point of the research on antioxidants is the lack of information on the effect of the whole array of dietary antioxidants in disease prevention, because so far mainly single molecules have been investigated. Epidemiological and experimental evidence is mounting on the potential importance of the so called total antioxidant capacity (TAC). TAC represent a direct measurement of the nonenzymatic antioxidant network considering single antioxidant activity and the synergistic interactions of the redox molecules present in the tested matrix. This chapter describes the results of the main clinical trial involving galenic antioxidants, discussing the importance of properly assessing endogenous and dietary TAC to obtain an integrated and realistic portrait of the complex interactions at the basis of the postulated antiageing effect of antioxidants molecules.

Keywords Total antioxidant capacity, ageing, oxidative stress, redox status, plant foods, diseases.

1 Introduction

It is widely believed that the release of free radicals, either as a by-product of normal metabolism or associated with inflammatory reactions, can contribute to a number of human age-related diseases [1], although it is still unclear whether free radical-induced oxidative stress is the initiating event [2].

From: *Aging Medicine: Oxidative Stress in Aging: From Model Systems to Human Diseases*   269
Edited by: S. Miwa, K.B. Beckman, and F.L. Muller © Humana Press, Totowa, NJ

A large and consistent body of scientific evidence indicates that diets rich in fruit and vegetables provide protection against cardiovascular disease (CVD), several common types of cancer, and other diseases [3, 4]. However, despite the consensus of the evidence about the health effects of vegetables, it still unclear which components are protective and their mechanism of action. On the basis of the premise that diet-derived antioxidants reduce oxidative stress, as suggested by the "antioxidant hypothesis" [5], many clinical trials have tested different redox molecules as a therapy to prevent degenerative diseases. However, contrasting results from intervention trials have raised strong concerns about the influence of antioxidants on human health [6, 7]. A vulnerable point of the research on anti-oxidants is the lack of information on the effect of the whole array of molecules, constituting the antioxidant network, in disease prevention, because so far mainly single molecules have been investigated.

The body's nonenzymatic antioxidant network can be assessed through the measurement of TAC, defined as the moles of radicals neutralized per one liter of tested sample [8]. As displayed in Fig. 15.1, TAC represents the result of many variables, such as redox potentials of the compounds present in the matrix, their cumulative and synergistic interaction, the nature of the oxidizing substrate, and antioxidant localization and solubility. Moreover, TAC considers the cumulative action of all the antioxidants present in the matrix (e.g., plasma, saliva, food extracts, tissues), providing an integrated parameter rather than the simple sum of measurable antioxidants.

This chapter describes results of the main clinical trial involving galenic antioxidants discussing the importance of properly assessing endogenous and dietary TAC to obtain an integrated and realistic portrait of the complex interactions at the basis of the postulated antiageing effect of antioxidants molecules.

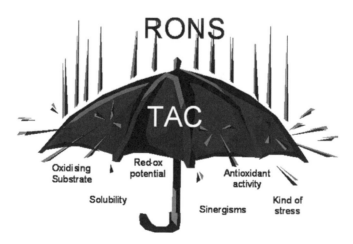

Fig. 15.1 Multiple factors affecting total antioxidant capacity: the TAC concept. RONS = reactive oxygen nitrogen species. Source: [8]

2   Galenic Antioxidant Supplementations and Disease Prevention

2.1   Cardiovascular Diseases

Half of the mortality in Western populations is due to diseases of the cardiovascular system in which the main pathophysiological factor is atherosclerosis. Atherosclerosis is a chronic inflammatory process that involves a complex interplay between circulating cellular and blood elements within the cells of the artery wall [9, 10]. Clinical, genetic, and epidemiological studies demonstrate that oxidative stress, and particularly oxidation of low-density lipoproteins, is a risk factor and plays a significant pathogenetic role for atherosclerotic diseases [11, 12].

The role of antioxidants in CVD prevention has been evaluated in different clinical trials. Rimm et al. [13] examined data from 39,910 U.S. male in the Health Professional's study. Results showed that men in the highest quintile of vitamin E intake had a lower relative risk ([RR] of CVD = 0.60 [95% CI, 0.44–0.81]; RR is the risk of an event [or of developing a disease] relative to exposure; RR is a ratio of the probability of the event occurring in the exposed group versus the control [nonexposed] group) compared with the quintile consuming the lowest amount of vitamin E. The same analyses for β-carotene, showed a relative risk of 0.71 (95% CI, 0.53–0.86) when comparing the highest quintile of intake to the lowest quintile. However, the benefit of β-carotene was only found to be significant in smokers.

In the Nurses' Health Study, 87,245 women aged 34 to 59 were followed for an average of 8 years. Women in the highest quintile of vitamin E intake had a 34% reduction in risk of coronary heart disease events compared with women in the lowest quintile after adjustment for different confounding factors. Intake of β-carotene was not correlated to any reduction in the risk of coronary heart disease [14]. Knekt et al. [15] followed 2,748 men and 2,385 women aged from 30 to 69 years for 18 years. After adjusting for age and other cardiovascular risk factors, vitamin E intake in women showed a significant benefit. People at the highest tertile of intake showed an RR of 0.35 (95% CI, 0.14–0.88) compared with people at the lowest tertile. In men, it only showed a trend toward benefit comparing the highest to the lowest tertile with an RR of 0.68 (95% CI, 0.42–1.11). Vitamin C showed significant benefit in women as well, with an adjusted RR of 0.49 (95% CI, 0.24–0.98) comparing the highest to the lowest tertile of intake. Carotenoids intake was not correlated with a reduction in the risk both for men and women. The Scottish Heart Health Study analyzed 4,036 men and 3,833 women from Scotland aged 40 to 59 years that were followed for 7.7 years [16]. The study showed that, after adjustment for different CVD risk factors, vitamin C and β-carotene intake were found to be associated, only in men, with a significant reduction in the incidence of new Coronary Heart Disease events. Comparing the highest to the lowest quartiles of intake, RR for vitamin C and β-carotene were 0.57 (95% CI, 0.40– 0.83) and 0.68 (95% CI, 0.48–0.96). Vitamin E failed to show any benefit in men. Similarly none of the three antioxidants showed any significant benefit in reducing the endpoints in women after multivariate adjustment.

Three cohort studies [13, 14, 17] enrolling together >150,000 subjects did not show any association between vitamin C intake and CVD mortality.

2.2   Cancer

Carcinogenesis is a complex multistep process that can be regarded as a disease of cells, characterized by an excess of cells beyond the number needed for normal function of the body organ affected [2]. The involvement of oxidative stress is thought to be linked to tumorigenesis at different levels. Oxidative damage to DNA has been shown to lead to DNA single- or double-strand breaks and to chromosomal aberrations. The modifications of DNA bases might result in point mutations, deletions, or gene amplification as a first step of carcinogenesis. Furthermore, free radicals are capable of deactivating detoxifying enzymes responsible for the scavenging of potent carcinogens [2].

The first large intervention trial focused on cancer was the Chinese Cancer Prevention Trial, conducted in Linxian County, an area with a very high incidence in esophageal/gastric cancer. Results showed significant lower cancer morbidity, with an RR of 0.87 for total cancer and 0.79 for stomach cancer, at the end of the 5-year supplementation with a combination of 15 mg of β-carotene plus 30 mg of vitamin E and 50 μg of selenium [18]. The first large-scale European trial to be completed was the Alpha-Tocopherol, Beta Carotene Cancer Prevention (ATBC) [19]. In this case, the supplementation with 20 mg/day β-carotene for 8 years was significantly associated with a higher incidence of lung cancer (18%) and prostate cancer (23%) in male smokers. In contrast, a decrease in the rates of prostate cancer (32%) and mortality (41%) was observed in the α-tocopherol (50 mg/day) supplemented group.

Unexpected results were obtained in the β-Carotene and Retinol Efficacy Trial (CARET trial) that showed a significant increase of lung cancer (28%; RR = 1.36) on asbestos-exposed workers after a 4-year supplementation with 30 mg/day β-carotene and 25,000 IU retinol that constricted to interrupt the trial earlier than planned [20]. No effects were observed on prostate cancer risk. The supplementation with 50 mg/day β-carotene for 12 years in the Physician's Health Study had no effect on lung cancer incidence [21]. However, a decrease in prostate cancer risk was observed in the supplemented people that had low baseline levels of β-carotene. The same dosage was used in the Women's Health Study [22] for 2 years, and no effect was found on cancer risk. Long-term vitamin E supplementation (6 years, 400 IU/day) did not significantly prevent cancer incidence (RR = 0.94) and cancer death (RR = 0.88) in patients from the Heart Outcomes Prevention Evaluation (HOPE) trial [23]. The supplementation with vitamin E (400 IU/day) even increased the all-cause mortality in head and neck cancer patients [24]. The supplementation for 4.5 years with nutritional dosages of selenium (200 μg/day) gave different results; a decrease in lung (RR = 0.54), prostate (RR = 0.37), and colorectal (RR = 0.42) cancer incidence was reported on patients affected with skin disorders [25, 26]. The Supplémentation en Vitamines et en Minéraux Antioxydants (SUVIMAX) study was designed to evaluate the effectiveness of a combination of antioxidant minerals and vitamins at nutritional dosages for 8 years (120 mg of vitamin C, 30 mg of vitamin E, 6 mg of β-carotene, 100 μg of selenium, and 20 mg of zinc, per day) on the mortality by cancer and ischemic heart disease in volunteers of both sexes [27]. Gender differences have been observed in terms of efficacy of the supplementation. A significant protective effect against total cancer incidence was seen in men (RR = 0.69) but not in women (RR = 1.04). A similar trend was observed for all-cause mortality (RR = 0.63 in men vs RR = 1.03 in women). It has been hypothesized that the supplementation may be effective only in men due to their lower baseline status of antioxidants, especially of β-carotene, compared with women [28]. In agreement with this finding, in men with low plasma selenium baseline levels, a preventive effect of selenium supplementation against prostate and colorectal cancer has been observed [25, 26]. Moreover, in the Physician's Health Study [29] a higher effectiveness of supplementation with β-carotene in regard to prostate cancer was found among people in the lowest quartile of baseline plasma β-carotene concentration.

2.3   Evidence from Meta-Analysis

To weigh the evidence, meta-analysis studies of the different clinical trials have been performed. A lack of effect of antioxidant supplementation has been shown in the prevention of lung cancer [30] and CVDs [7]. Bjelakovic et al. [6] reviewed 14 randomized interventions to establish the possible beneficial effect of antioxidant supplements on the incidence of gastrointestinal cancers. A negative effect on overall mortality was described, which was clearly demonstrated for the associations of β-carotene and retinol (RR = 1.29) and β-carotene and vitamin E (RR = 1.10). The potential negative effect of galenic antioxidant supplementation also was observed by Vivekananthan et al. [7], showing that in 138,113 patients assigned to β-carotene or control group, the rates for all-cause mortality, and cardiovascular causes were slightly but significantly higher for β-carotene treatment compared with the control group. With regard to vitamin E, 81,788 patients were included in the all-cause mortality analyses. Results showed that vitamin E did not change the mortality for all-causes (11.3 vs 11.1%) and for cardiovascular mortality (6.0 vs 6.0%) compared with control treatment. Opposite conclusions were reached by Miller et al. [31] who reported, in a meta-analyses on about 136,000 participants enrolled in 19 clinical trials, a statistically significant relationship between vitamin E dosage and all-cause mortality, with increased risk for dosage higher than 150 IU/ d. Recently, Bjelakovic et al. [32] reviewed 68 randomized trials involving β-carotene, vitamin A, ascorbic acid and selenium either singly or combined for a total of 232,606 participants. Data showed that antioxidant supplementation, except for selenium and vitamin C, had no effect on gastrointestinal cancers rates but that it significantly increased all-cause mortality.

The results of the meta-analyses highlight the failure of galenic antioxidant supplementations in reducing age-related diseases, and, even more importantly for public health concerns, introducing the concept of a putative lack of safety arising from long-term supplementation. There are some considerations worth making to give a clearer picture of the phenomenon. Vitamin C, vitamin E, and β-carotene, due to their unquestionable redox properties [33, 34], have gained considerable importance during the last decades. However, the analyses of the composition of fruit and vegetables indicate the existence of hundreds of different antioxidant molecules present in the food matrix that might be involved in their protective effects. Moreover, the long period of supplementations used in most of the clinical trial through the use of only one single antioxidants at doses far exceeding the recommended intake (for their vitamin functions) might be one of the factors at the basis of the negative results. The "antioxidant network" of the body relies on the cooperation between molecules and enzymes for setting physiological and optimal redox potential for oxidative stress prevention. The hypersupplementation with only one (or two) galenic antioxidants for years might have caused strong perturbations to the redox balance, leading to the harmful effect. If antioxidants are essential for metabolic function, it does not necessarily follow that more is better. There is a clear need to investigate the physiology of their interrelationships in the different environment, because the uncontrolled increase of one component might be deleterious. If antioxidants work in concert, the choice of selecting few anti-oxidants at very high doses instead of a mild supplementation within physiological range and respectful of plant food composition might have contributed to the unexpected results.

3 Nonenzymatic Antioxidant Network

Antioxidant defenses of the body are composed of molecular and enzymatic players; however, the composition of the network markedly differs, in terms of concentration and components, in different environments [2]. Protection at the cellular level is mainly guaranteed by enzymes such as superoxide dismutase (SOD), catalase, and glutathione peroxidase, whereas in plasma, nonenzymatic antioxidants are playing the major role. Antioxidant molecules are located in both the hydrophilic and hydrophobic compartments of plasma, and they work in concert to decrease free radical concentrations and to inactivate transition metal ions [2]. Metal-chelating antioxidants such as transferrin, albumin, and ceruloplasmin avoid radical production by inhibiting the Fenton reaction catalyzed by copper or iron. Chain-breaking or free radical-scavenging antioxidants include low-molecular-weight compounds such as uric acid, bilirubin, thiols, vitamin E, ascorbic acid, carotenoids, and coenzyme Q_{10} [2]. α-Tocopherol is widely present within membranes, representing the most abundant lipid-soluble antioxidant. α–Tocopherol can be regenerated from its oxidized form by reduction with vitamin C, but whether this mechanism is actively operative in vivo is still a matter of debate. The lipid-soluble β-carotene is an efficient quencher of singlet oxygen. The water-soluble vitamin C has a broad spectrum of antioxidant activities due to its ability to react with different radicals. Glutathione is synthesized from glutamate, cysteine, and glycine and occurs in millimolar concentration in cells but only in trace amounts in plasma. Bilirubin, the end product of heme catabolism, also has been shown to act as an efficient antioxidant. Urate, a powerful free radical scavenger is the end product of purine metabolism. Cutler [35] suggested that, the loss of the enzyme uricase, which transforms uric acid to allantoin, around the Palaeolithic era, coupled with the loss of the ability to synthesize ascorbic acid, resulted in replacement of ascorbic acid by uric acid as the major endogenous water-soluble antioxidant in biological fluids of humans. The antioxidant potency of polyphenols, secondary metabolites present in plant foods, displays a wide range of in vitro activities ranging from the radical scavenging to the metal chelating. However, the low concentration of phenolics in plasma and the lack of identification of their active metabolites do not allow for firm conclusions about their role as possible component of the in vivo antioxidant network to be drawn.

In terms of the participation of individual components to the network, calculated on the basis of the single individual concentration and respective antioxidant stochiometric coefficient, the main contributor to TAC is uric acid (40–55%), followed by thiol groups (10–24%), ascorbic acid (8–15%), and vitamin E (<10%) [36]. Contribution of single antioxidants to overall TAC leaves unexplained a percentage ranging from 20 to 40% that might be accounted for by unknown components, unmeasured molecules (coenzyme Q_{10}), synergistic interactions, or a combination. The multifunctional properties and the versatility of the antioxidant network highlight the efficiency and the crucial importance of the dynamic interactions between the components of the network in protecting plasma from multi-radical attack driven by different sources.

4 Measurement of Total Antioxidant Capacity In Vivo

In the last decade, a large number of assays for measuring TAC in biological matrixes have been developed [8]. These assays are extremely different one from another; differences exist in the free radical-generating system, molecular target, end point, residence in lipo- and hydrophilic compartment, and physiological relevance. The large number of methodologies makes it difficult to chose an assay with respect to another and to understand the results. It is crucial to know exactly what the different assays are measuring and what kind of information can be obtained adopting one technique instead of another. On the basis of the chemical reaction involved, TAC assays have been classified in two groups: (1) single electron transfer and (2) hydrogen atom transfer mechanisms. Assays involving single electron transfer mechanism include (1) ferric reducing ability of plasma (FRAP) [37] and (2) Trolox equivalent antioxidant capacity (TEAC) [38]. Assays involving hydrogen atom transfer mechanism include (1) total radical-trapping antioxidant parameter (TRAP) [39], (2) fluorescence lipid-oxidation (Fluo-Lip) [40], and (3) oxygen radical antioxidant capacity (ORAC) [41].

4.1   Single Electron Transfer Assays

4.1.1   FRAP Assay

FRAP assay measures the ferric reducing ability of the sample. The method, using iron as redox sensor, determines the reduction of a colorless ferric-tripyridyltriazine complex (TPTZ-Fe^{3+}) to its ferrous-colored form (TPTZ-Fe^{2+}) in the presence of antioxidants. It is a simple and fast spectrophotometric assay carried out at pH 3.6 and 37 °C. Trolox, a water-soluble analog of vitamin E, is used for calibration. FRAP assay measures the reducing power of the sample, including reductants that are not classical biological antioxidants. Lack of measurement of the contribution of thiols groups and the acidic pH represents the main limitation of the method.

4.1.2   TEAC Assay

TEAC assay is based on the formation of ferrylmyoglobin radical (from reaction of metmyoglobin with H_2O_2), which may then react with 2,2′-azino-bis(3-ethylbenzthiazoline-6-sulfonic acid (ABTS) to produce the ABTS$^{•+}$ radical cation. The accumulation of ABTS$^{•+}$ can be inhibited to an extent proportional to the TAC. Biases against this method are the influence of plasma dilution and the nonphysiological conditions of analysis. In fact, the peroxidase process does not mimic the radical formation in vivo and the assay uses a nonphysiological radical [42] that strongly reduce its usefulness in biological fluids.

4.2   Hydrogen Atom Transfer Assays

4.2.1   TRAP Assay

TRAP assay is based on the protection afforded by plasma against the decay of a fluorescent target, the protein R-phycoerythrin, during a chemically generated peroxidation reaction. The constant flux of peroxyl radical is produced by the specific azoinitiator 2,2′-diazobis(2-amidinopropane) hydrochloride (ABAP) at 37 °C and pH 7.4. The antioxidant present in the reactive medium gives a lag time directly related to its amount and activity. The results are quantified by standardizing the lag phase with the phase obtained using a known amount of Trolox, and they are expressed as micromoles of peroxyl radicals trapped by 1 liter of plasma. TRAP assay finds its best application in the measurement of the chain-breaking antioxidant potential of plasma, saliva, urine, and tissues [43], but it is not suitable for measuring antioxidant activity of liposoluble compounds. TRAP assay has been recently adapted to a fluorescent plate reader, lag phase is calculated automatically through a specific program, making the assay suitable for measuring a large number of samples in epidemiological studies [43].

4.2.2 Fluo-Lip Assay

In Fluo-Lip assay, an organic peroxyl radical (generated from ABAP) induces the oxidation of 1-palmitoyl-2-((2-(4-(6-phenyl-trans-1,3,5-exatrienyl) phenyl) ethyl) carbonyl)-sn-glycero-3-phosphocoline incorporated into plasma lipoproteins. The oxidation of the target leads to a decay of its fluorescence, and it is well correlated with the oxidation of natural unsaturated lipids. The antioxidant added to the system gives an increase of lag time, which is proportional to the antioxidant action. Results are expressed as lag phase/min. Similar to the TRAP assay, the Fluo-Lip method is characterized by the simultaneous presence of a prooxidant, an oxidizable substrate (lipid) and the analyzed biological fluid. The two assays integrate each other; TRAP works in a water-soluble environment and Fluo-Lip responds more efficiently to lipophilic antioxidants. However, due to the procedure of serum labeling, vitamin C is destroyed in Fluo-Lip assay, and, probably for this reason, the assay has not found a large application in human studies.

4.2.3 ORAC Assay

ORAC assay is based on the same chemical principle of the TRAP assay, but plasma TAC is measured as the area under the curve (AUC) of fluorescence in time. Because the quantification of plasma TAC is obtained by the measurement of AUC, the main drawback of this method is the interference given by protein of biological fluids [44]. Protein contribution to plasma TAC can be as high as 86% of the total value, a drawback leading to an underestimation of the contribution of main "chain-breaking" antioxidants. Due to these interferences, ORAC assay find its optimal application for the measurement of TAC in food extracts devoid of proteins, with respect to biological fluids.

Currently, there is a strong need to critically assess, on the basis of the results of well tailored in vitro and in vivo studies, the real information that each assay can furnish, to suggest applicability and limitations of the results to specific samples and to drive the choice of one methodology respect to another. However, until this issue is clarified, as an increasing consensus between the international experts and to best fit the generally accepted principle that antioxidant related processes are multifactorial, it is crucial to understand that the "perfect" method does not exist and a battery of tests should be performed on biological samples. We have decided to choose TRAP and FRAP as representative methodologies for the hydrogen atom transfer and single electron transfer mechanism, respectively. Keeping in mind the principal characteristics of each assay, information from the different assays should be merged to have a correct picture of the phenomenon.

5 Dietary Modulation of the Antioxidant Network

Many studies have investigated the ability of diet to modulate plasma TAC after consumption of antioxidant-rich foods in healthy humans [8]. The first among these "acute intervention studies," where TAC was monitored before and after food ingestion, were the work of Serafini et al. [45] and Maxwell et al. [46], which demonstrated that both tea and red wine were able of increasing plasma TAC. This preliminary evidence has been followed by studies showing the ability of different fruit juices, onions, lettuce, chocolate, and honey to boost plasma TAC at different times [8]. Liquids such as wine and tea showed a peak immediately after consumption (30–60 min); and solid foods such as lettuce increase TAC after 2–3 h, with levels returning to baseline in about 4–5 h [39]. The "acute ingestion" protocol represent a reliable experimental model for obtaining information about the potentiality of the selected food items to display an antioxidant activity in vivo or for testing the effect of technology process on the antioxidant properties of foods. However, the homogenous data obtained with acute ingestion studies is not mirrored by long-term dietary intervention trials [47–52; 55–57, 59]. Plasma TAC was shown to increase after consumption of 10 servings of fruit and vegetables per day for 2 weeks [47]. In agreement with this evidence the increase of plasma TAC observed after 7 days of consumption of tomato products with extra virgin olive oil [48] or after 1 mo of supplementation with a plant-based Mediterranean diet [49]. In healthy smoking subjects, plasma TAC increased after consumption of five portions of fruits and vegetables per day for 3 weeks and after consumption of vegetable burger plus a fruit drink for 3 weeks [50, 51]. In the health and nutrition survey carried out in the province of Attica in Greece (ATTICA study), when a large number of subjects was studied (3,042 participants), a correlation between Mediterranean diet score, an index of the adherence of the diet to a standard Mediterranean pattern, and levels of TAC has been revealed [52]. Looking at the single food items, consumption of fruit, vegetables, and olive oil was positively correlated with plasma TAC. In line with this evidence, our group used the Italian TAC database [53, 54] has assessed dietary TAC intake (from vegetables, fruit, oils, pulses, cereals, and alcoholic beverages) in different European Union countries. Dietary information was obtained from the Data Food Networking (DAFNE) study where information from household budget surveys was transformed in daily individual food availability. The TAC food database was adapted to the DAFNE food groups to obtain daily pro capita TAC for each country. Preliminary results (Fig. 15.2) show that Mediterranean countries (Spain, Portugal, Italy, and Greece) had an highest availability in daily dietary TAC (TRAP values ranging from 3,113 to 3,898 μmol of Trolox Equivalent [TE] compared with Northern European countries (Norway, UK, and Ireland; TRAP values ranging from 1,434 to 1,575 μmol TE).

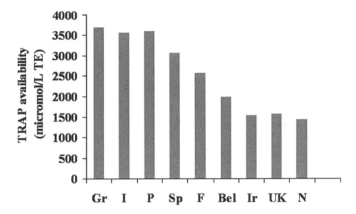

Fig. 15.2 TAC availability (TRAP assay) from overall diet (vegetables, fruit, oils, pulses, cereals, and beverages) in selected European countries. Spain, Sp; Italy, I; Portugal, P; Greece, Gr; France, Fr; Belgium, Bel; United Kingdom, UK; Ireland, Ir; and Norway, N

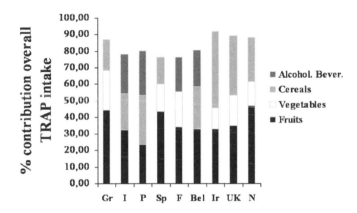

Fig. 15.3 First three contributors (percentage of contribution overall TRAP intake) to dietary TAC (TRAP assay) in selected European countries. Spain, Sp; Italy I; Portugal, P; Greece, Gr; France, Fr; Belgium, Bel; United Kingdom, UK; Ireland, IR; Norway, N

The values for France (2,572 µmol TE) and Belgium (2,000 µmol TE) were intermediate as displayed in Fig. 15.2. Figure 15.3 describes the three main contributors to dietary TAC for the selected countries. Fruit contribution was the highest in all countries except for Portugal, UK, and Ireland, where cereals were the main source of antioxidants. Vegetable contribution was the second highest for Greek and France and the third for Spain, UK, Ireland, and Norway. Italy, Portugal, France, and Belgium display a contribution ranging from 21 to 27% from alcoholic beverages, mainly wine, where Greece shows a very low intake (5%). These results suggest a higher dietary intake of TAC in southern countries compared with northern

European countries, in keeping with the supposed importance of health benefits of the Mediterranean dietary habits.

Lack of change in plasma TAC was observed after consumption of five to seven servings of fruit and vegetables per day, for 2 weeks in 25 healthy nonsmokers [55] or after 1 week of supplementation with 1,500 ml of fruit juice per day [56]. Dragsted et al. [57] conducted an intervention trial on 43 young subjects randomly divided in three groups. The control group was asked to avoid dietary antioxidants for 25 days, where the treatment groups were supplemented with 600 g/day fruit and vegetables and with pills containing the amount of vitamins and minerals contained in 600 g of fruit and vegetables. Results showed no changes in plasma TAC, markers of oxidative damage and antioxidant enzymes after the supplementation with plant foods, pills, or withdrawal of antioxidants from the diet. Serafini et al. [58] pointed out that, due to their ability to cope with light dietary stress, plasma antioxidant defences may need >25 days or specific and stronger dietary stresses, such as a high-fat diet, to be challenged significantly in healthy young subjects.

To further complicate the interrelationships between dietary and endogenous antioxidant network is the recent evidence showing that the improvement of flow mediated dilation, a marker of vascular function, induced by the ingestion of 500 ml of black tea infusion is abolished by milk addition to tea [59]. Findings supported by experiment on endothelial cells showing that the binding complex between milk caseins and tea catechins might be responsible for the lack of effect. The possibility that the in vivo efficiency of dietary antioxidants might be affected by the consumption of different foods is not a new discovery, because the first study to display this phenomenon in vivo was published in 1996 when Serafini et al. [60] reported that the increase in plasma TAC after the consumption of both green and black tea was markedly reduced when the teas were consumed with milk. Strengthening the case, other studies with black tea [61] and chocolate have been shown to lose their ability to increase plasma TAC after milk addition [62]. Inhibitory effects that is in keeping with in vitro evidence in a gastrointestinal model showing that the catechin content, antioxidant capacity and antimutagenic activity of jejunal dialysates obtained were significantly affected by the addition of milk to the teas [63]. The link of phenolics with the proteins of the matrix in which phenolics are ingested might reduce their bioavailability [64] and therefore their potential antioxidant properties in vivo. Contrasting results from this hypothesis came from two studies showing that consumption of a single dose of either lyophilized green or black tea and liquid chocolate together with milk did not abolish the antioxidant effect [65, 66]. However, negligible increase in plasma TAC in the tea study (+3%) and the very low amount of milk (3%) in the chocolate study might have affected results.

Despite the suggested link between diet and endogenous antioxidants network, evidences so far are not conclusive about the identification of the players involved and their interrelationships. Criteria of absorption of dietary redox molecules for optimizing the endogenous antioxidant network according to different conditions (e.g., dietary intake, stress, smoking, physical activity) are largely ignored. Most of the studies have been conducted on a small number of subjects, not using standardized methodologies and without taking into account the existence of homeostatic mechanisms of regulation of plasma TAC. Physiological diversity in the absorption and disposal of antioxidants are variables that might affect the ability of diet to modulate plasma TAC in vivo. The possibility that the modality of ingestion of foods rich in antioxidants can affect their in vivo antioxidant potentiality cannot be ignored and should be taken in account when designing human studies. Only well tailored in vivo studies aiming to consider the effect of these confounders will allow clarifying the potential role of diet in prompting antioxidant defenses.

6   Imbalance of Nonenzymatic Antioxidant Network and Degenerative Diseases

Despite the potential importance of antioxidant interactions, not many studies have investigated the imbalance of the nonenzymatic antioxidant network in patients affected by different diseases. Table 15.1 describes TAC levels in biological fluids of patients with different pathologies or conditions [67–77, 80–87]. Patients affected by Coronary Artery Disease [67], diabetes, menopause [68], anorexia nervosa, AIDS, liver diseases, colitis, Crohn's disease, cystic fibrosis, and psoriasis have significantly lower levels of plasma TAC compared with healthy controls [36]. A recent study has shown, in patients with prehypertension, an inverse association between systolic and diastolic pressure and plasma TAC that was lower compared with normotensive subjects [69]. Moreover a significant decline in plasma levels of TAC with the increase of selected risk factors (e.g., smoking, gender, hypercholesterolemia, diabetes mellitus, and hypertension) has been shown in coronary artery disease patients [70]. On the contrary, other studies have reported a lack of difference in plasma TAC levels in subjects affected by diabetes, Alzheimer's disease, vascular dementia, and Parkinson's disease [36].

Concerning cancer pathology, Table 15.2 shows studies describing levels of TAC in biological fluids of cancer patients. Patients with lung tumours display significantly lower levels of plasma TAC and lower concentrations of thiol groups, ascorbic acid, and vitamin E, respect to healthy control [71]. In colon cancer patients, this decrease in TAC was accompanied by a decrease in total thiol groups and a significant increase in oxidized proteins and lipid peroxides [72]. Significantly lower serum TAC and higher malondialdehyde (MDA) levels were detected in patients with breast cancer [73]. Similar results were observed in a group of patients with cervical intraepithelial neoplasia, who had significant lower plasma levels of TAC and vitamin C with respect to the control group and a higher plasma MDA concentration [74]. In contrast, in another case control study, Ching et al. [75] showed that the highest quartiles of serum concentrations of TAC were associated with significant reductions (about 50%) in breast cancer risk compared with people with the lowest quartiles.

Liu et al. [76] observed a decline in the TAC either in plasma or in malignant pleural effusion supernatant in cancer patients. This lack of an adequate antioxidant system was reflected in an increase in DNA damage, assessed by the Comet assay,

Table 15.1 Selected studies describing TAC levels in biological fluids of patients with different pathologies or conditions

Reference	Biological fluid	TAC method[a]	Pathology or condition	Patients value[b]	Control value
[67]	Plasma	Hydroxyl radical scavenging (mmol/l)	Coronary artery disease	1.14 ± 0.13	1.38 ± 0.20
[68]	Plasma	ABTS (mmol/l)	Menopause	1.14 ± 0.01	1.49 ± 0.02
[69]	Plasma	ImAnOx kit (µmol/l)	Hypertensive	223 ± 30	259 ± 41
			Prehypertensive	242 ± 46	
[80]	Plasma	Crocin assay (µmol/l)	Ulcerative colitis	1.03	1.42
			Chrohn's disease	0.97	
[81]	Plasma	TRAP (µmol/l)	Alzheimer	−24%	100%
[82]	Plasma	Luminol assay (µmol/l)	Alzheimer Vascular dementia	263	292
[83]	Plasma	FRAP (µmol/l)	Steatosis steathopatitis	274 / 21% / 33%	100%
[84]	Serum	ABTS (mmol/l)	Coronary atherosclerosis	1.53 ± 0.20	1.52 ± 0.25
[85]	Plasma	TRAP (µmol/l)	Cystic fibrosis	488 ± 34	580 ± 79
[86]	Plasma	ABTS/peroxidase (µmol/l)	HIV infection	161 ± 97	269 ± 81
[87]	Plasma	ABTS (mmol/l)	Type 2 diabetes	1.70 ± 0.50	2.70 ± 0.50

[a]TEAC, Trolox equivalent antioxidant capacity; FRAP, ferric reducing antioxidant power; TRAP, Total Radical-trapping Antioxidant Parameter; ORAC, oxygen radical antioxidant capacity; ImAnOx, Immunodiagnostik AG, Bensheim, Germany.
[b]Values expressed as percentage are referred to TAC values of control (100%).

Table 15.2 Studies describing levels of TAC in biological fluids of cancer patients compared to their control

Biological fluid	TAC method	Type of cancer	Cancer patients	Healthy subjects
Plasma	TRAP (μmol/l)	Lung	$1,143 \pm 181$ ($n = 57$)	$1,273 \pm 199$ ($n = 76$)
Plasma	Nonproteic antioxidant capacity (%) (inhibition of NADH oxidation)	Colon	55 ($n = 54$)	90 ($n = 20$)
Serum	TEAC (mmol/l)	Breast	2.01 ± 0.01 ($n = 56$)	2.07 ± 0.03 ($n = 18$)
Plasma	TEAC assay (mmol/l)	Cervical intraepithelial neoplasia	1.15 ± 0.17 ($n = 58$)	1.25 ± 0.15 ($n = 86$)
Serum	TEAC (mmol/l)	Breast	Highest vs lowest	($n = 151$) quartile: OR $= 0.47$ (CI 0.28–0.94) ($n = 153$)
Plasma	Reduction of o-phenantroline (U/ml)	Lung, breast, thyroid	8.41 ± 1.78 ($n = 28$)	10.52 ± 1.64 ($n = 33$)
Serum	TEAC (mmol/l)	Tumors at different sites	1.30 ± 0.03 ($n = 82$)	1.10 ± 0.08 ($n = 36$)

on peripheral blood mononuclear cells compared with healthy people. On the contrary, Mantovani et al. [77] found no differences between serum TAC of 82 advanced stage cancer patients and the respective controls. Li et al. [78] investigated the interaction between the MnSOD polymorphism and the baseline plasma levels of a "calculated" TAC (sum of selenium, lycopene, α-tocopherol and γ-tocopherol concentrations), in modifying prostate cancer risk. Results showed that the different genotype make subjects more or less susceptible to differences in baseline plasma antioxidant levels. Among men homozygous for the A allele of MnSOD, people at the lowest quintile of plasma TAC had an RR of 2.5 and 3.1 for total and aggressive prostate cancer, five- and ten-fold higher than people in the highest quintile of calculated TAC (RR = 0.5 and 0.31 for total and aggressive prostate cancer, respectively). This relation was not significant for those men with VV and VA genotype.

The use of TAC in epidemiological studies represents an approach that has been proposed for the first time for investigating the relationship between dietary antioxidant network and gastric cancer risk [79]. Dietary intake of TAC through 12 plant foods was inversely associated with the risk of gastric cancer. Interestingly, in nonsmokers (low radical load) dietary TAC was protective already at the first quartile (the ratio of the odds in favor of getting a disease, if exposed, to the odds in favor of getting a disease if not exposed [OR] = 56) of intake with no further change, in terms of protective effect, for higher TAC levels (OR = 0.48, 0.46, and 0.44). Among smokers, where a possible unbalance of the antioxidant network is likely, a dose–response relationship for increasing levels of TAC was found, suggesting their higher need for dietary TAC. Indirect confirmation of this hypothesis came from a case control study by Ching et al. [54], which showed a significant reduction in breast cancer risk at the second quartile of plasma levels of TAC (OR = 0.52), whereas the risk did not change markedly in the second and third quartile (OR = 0.47 and 0.41), suggesting the existence of threshold levels for the protective effect of endogenous antioxidant network. Recently, in a Prospective study in the Spanish cohort of the European Prospective Investigation into Cancer (EPIC), TAC dietary intake was shown to be associated with reduced mortality in an adult population (n = 41,440), with TRAP assay selected as best indicator of TAC from plant food [88].

7   Conclusions

The potential importance of the antioxidant network in disease prevention is a relatively new topic of research. Although there are still aspects that need to be further investigated, evidence is mounting that TAC measurement might represent a promising tool to monitor the nonenzymatic antioxidant network in vivo and in epidemiological studies. The multivariable nature of antioxidants network make unavoidable, at least for now, the use of a battery of methodologies for measuring the hydrogen atom transfer and the single electron transfer properties of the tested matrix.

The understanding of the role of diet in prompting endogenous antioxidant defenses needs to be clarified due to the indirect evidence suggesting the concept of a homeostatic control of endogenous antioxidant network. The existence of regulatory mechanisms aimed to avoid unnecessary overloading or deficiency of antioxidants that can cause a "reductive/oxidative" unbalance of the redox network highlights the importance of preserving the physiological levels of antioxidants. The concept of antioxidant supplementation needs to be redefined based on the latest results from clinical trials that antioxidant supplementation might be necessary only in, for example, the elderly or in smokers, where oxidative stress may arise.

In conclusion, the road to understanding the preventive actions of antioxidants cannot be focused anymore on the search of the "redox magic bullet" but requires the understanding of the interactions between the different redox machinery to achieve the highest degree of protection against oxidative damage and an optimal preventive action against age-related degenerative diseases.

Acknowledgements We acknowledge the DAFNE Study for providing the data on food availability in different European countries.

References

1. Harman D. Aging. A theory based on free-radical and radiation chemistry. J Gerontol 1956; 11:298–300.
2. Hallowell B, Guttering JMC. In: Hallowell B, Guttering JMC (eds) Free radicals in biology and medicine, 3rd edn, Clarendon Press, Oxford, UK; 1989.
3. Ness AR, Poles JW. Fruit and vegetables, and cardiovascular disease: a review. Int J Epidemiol 1997;26:1–13.
4. Steinmetz KA, Potter JD. Vegetables, fruit, and cancer prevention: a review. J Am Diet Assoc 1996;96:1027–1039.
5. Gey KF. On the antioxidant hypothesis with regard to arteriosclerosis. Biblioth Nutr Dieta 1987;37:53–91.
6. Bjelakovic G, Nikolova D, Simonetti RG, Gluud C. Antioxidant supplements for prevention of gastrointestinal cancers: a systematic review and meta-analysis. Lancet 2004;364:1219–1228.
7. Vivekananthan DP, Penn MS, Sapp SK, Hsu A, Topol EJ. Use of antioxidant vitamins for the prevention of cardiovascular disease: meta-analysis of randomized trials. Lancet 2003;361:2017–2023.
8. Serafini M, Del Rio D. Understanding the association between dietary antioxidants, red-ox status and disease: is the total antioxidant capacity the right tool?. Redox report 2004;9:145–152.
9. Steinberg D, Witzum JL. Lipoproteins and atherogenesis. Current concepts. J Am Med Ass 1990;264:3047–3052.
10. Ross R. Atherosclerosis: an inflammatory disease. N Engl J Med 1999;328:1450–1456.
11. Parthasarathy S, Santanam N, Ramachandran S, Meilach O. Oxidants and antioxidants in atherogenesis: an appraisal. J Lipid Res 1999;40:2143–2157.
12. Witzum JL, Steinberg D. The oxidative modification hypothesis of atherosclerosis: does it hold for humans? Trends Cardiovasc Med 2001;11:93–102.
13. Rimm EB, Stampfer MJ, Ascherio A, Giovannucci E, Colditz GA, Willett WC. Vitamin E consumption and the risk of coronary heart disease in men. N Engl J Med 1993;328:1450–1456.
14. Stampfer MJ, Hennekens CH, Manson JE, Colditz GA, Rosner B, Willett WC. Vitamin E consumption and the risk of coronary heart disease in women. N Eng J Med 1993;328:1444–1449.

15. Knekt P, Reunanen A, Jarvinen R, Heliovaara M, Maatela J, Aromaa A. Antioxidant vitamin intake and coronary mortality in a longitudinal population study. Am J Epidemiol 1994;139:1180–1189.
16. Todd S, Woodward M, Tunstall-Pedoe H, Bolton-Smith C. Dietary antioxidant vitamins and fiber in the aetiology of cardiovascular disease and all-cause mortality: results from the Scottish Heart Health Study. Am J Epidemiol 1999;150:1073–1080.
17. Kushi LH, Folsom AR, Prineas RJ, Mink PJ, Wu Y, Bostick RM. Dietary antioxidant vitamins and death from coronary heart disease in postmenopausal women. N Engl J Med 1996;334:1156–1162.
18. Blot WJ, Li JY, Taylor RR et al. Nutrition intervention trials in Linxian, China. Supplementation with specific vitamin and mineral combinations, cancer incidence and disease-specific mortality in the general population. J Natl Cancer Inst 1993;85:1483–1492.
19. The Alpha-Tocopherol Beta-Carotene Cancer Prevention Study Group. The effect of vitamin E and beta carotene on the incidence of lung cancer and other cancers in male smokers. New Engl J Med 1994;330:1029–1035.
20. Omenn GS, Goodman GE, Thornquist MD et al. Effects of a combination of beta carotene and vitamin A on lung cancer and cardiovascular disease. N Engl J Med 1996;334:1150–1155.
21. Hennekens CH, Buring JE, Manson JE et al. Lack of effect of long-term supplementation with beta carotene on the incidence of malignant neoplasms and cardiovascular disease. N Engl J Med 1996;334:1145–1149.
22. Lee IM, Cook NR, Gaziano JM et al. Vitamin E in the primary prevention of cardiovascular disease and cancer: the Women's Health Study: a randomized controlled trial. J Am Med Ass 1999;694:56–65.
23. Lonn E, Bosch J, Yusuf S et al. Effects of long-term vitamin E supplementation on cardiovascular events and cancer: a randomized controlled trial. J Am Med Assoc 2005;293:1338–1347.
24. Bairati I, Meyer F, Jobin E et al. Antioxidant vitamins supplementation and mortality: a randomized trial in head and neck cancer patients. Int J Cancer 2006;119:2221–2224.
25. Clark LC, Combs GF, Turnbull BW et al. Effects of selenium supplementation for cancer prevention in patients with carcinoma of the skin. A randomized control trial. J Am Med Assoc 1996;276:1957–1963.
26. Reid ME, Duffield-Lillico AJ, Sunga A, Fakih M, Alberts DS, Marshall JR. Selenium supplementation and colorectal adenomas: an analysis of the nutritional prevention of cancer trial. Int J Cancer 2006;118:1777–1781.
27. Hercberg S, Galan P, Preziosi P et al. The SU.VI.MAX Study: a randomised, placebo-controlled trial of the health effects of antioxidant vitamins and minerals. Arch Intern Med 2004;164:2335–2342.
28. Hercberg S, Czernichow S, Galan P. Antioxidant vitamins and minerals in prevention of cancers: lessons from the SUVIMAX study. Br J Nutr 2006;96:S28–S30.
29. Cook NR, Stampfer MJ, Ma J et al. B-carotene supplementation for patients with low baseline levels and decreased risks of total and prostate carcinoma. Cancer 1999;86:1783–1792.
30. Caraballoso M, Sacristan M, Serra C, Bonfill X. Drugs for preventing lung cancer in healthy people. Cochrane Database Syst Rev 2003;2:CD002141.
31. Miller ER, Pastor-Barriuso R, Dalal D, Riemersma RA, Appel L, Guallar E. Meta-analysis: high-dosage vitamin E supplementation may increase all-cause mortality. Ann Inter Med 2004;42:37–46.
32. Bjelakovic G, Nikolova D, Gluud LL, Simonetti GR, Gluud C. Mortality in randomized trials of antioxidant supplement for primary and secondary prevention. Systematic review and meta-analysis. J Am Med Assoc 2007;297:842–857.
33. Benzie JF. Evolution of antioxidant defence mechanisms. Eur J Nutr 2000;39:53–61.
34. Dagenais GR, Marchioli R, Yusuf S, Tognoni G. beta-Carotene, vitamin C, and vitamin E and cardiovascular diseases. Curr Cardiol Rep 2000;2:293–299.
35. Cutler RG. Urate and ascorbate: their possible roles as antioxidants in determining longevity of mammalian species. Arch Gerontol Geriatr 1984;3:321–348.
36. Bartosz G. Total antioxidant capacity. Adv Clin Chem 2003;37:219–93.
37. Benzie IFF, Strain JJ. The ferric reducing ability of plasma (FRAP) as a measure of antioxidant power: the FRAP assay. Anal Biochem 1996;239:70–76.
38. Rice-Evans C, Miller NJ. Total antioxidant status in plasma and body fluids. Methods Enzymol 1994;234:279–293.

39. Serafini M, Bugianesi R, Salucci M, Maiani G, Azzini E. Effect of acute ingestion of fresh and stored lettuce on plasma total antioxidant capacity and antioxidant levels in humans. Br J Nutr 2002;88:615–623.

40. Hofer G, Lichtenberg D, Hermetter A. A new fluorescence method for the continuos determination of surface lipid oxidation in lipoproteins and plasma. Free Radic Res 1995;23:317–327.

41. Cao G, Alessio HM, Cutler RG. Oxygen-radical absorbance capacity assay for antioxidants. Free Radic Biol Med 1993;14:303–311.

42. Strube M, Haenen GRMM, Van Den Berg H, Bast A. Pitfalls in a method for assessment of total antioxidant capacity. Free Radic Res 1997;26:515–521.

43. Serafini M. Total antioxidant capacity of plasma and disease prevention: where are we now? Second International Workshop on Antioxidant Methods, Orlando, Florida (June 2005).

44. Ghiselli A, Serafini M, Ferro-Luzzi A. New approaches for measuring plasma or serum antioxidant capacity: a methodological note. Free Radic Biol Med 1994;16:135–137.

45. Serafini M, Ghiselli A, Ferro-Luzzi A. Red wine, tea and antioxidants. Lancet 1994;344:626.

46. Maxwell S, Cruickshank A, Thorpe G. Red wine and antioxidant activity in serum. Lancet 1994;344:193–194.

47. Cao G, Booth SL, Sadowski JA, Prior RL. Increases in human plasma antioxidant capacity after consumption of controlled diets high in fruit and vegetables. Am J Clin Nutr 1998;68:1081–1087.

48. Lee A, Thurnham DI, Chopra M. Consumption of tomato products with olive oil but not sunflower oil increases the antioxidant activity of plasma. Free Radic Biol Med 2000;29:1051–1055.

49. Leighton F, Cuevas A, Guasch V et al. Plasma polyphenols and antioxidants, oxidative DNA damage and endothelial function in a diet and wine intervention study in humans. Drugs Exp Clin Res 1999;25:133–141.

50. Van den Berg R, van Vliet T, Brockmans WM et al. A vegetable/fruit concentrate with high antioxidant capacity has no effect on biomarkers of antioxidant status in male smokers. J Nutr 2001;131:1714–1722.

51. Roberts WG, Gordon MH, Walker AF. Effects of enhanced consumption of fruit and vegetables on plasma antioxidant status and oxidative resistance of LDL in smokers supplemented with fish oil. Eur J Clin Nutr 2003;57:1303–1310.

52. Pitsavos C, Panagiotakos DB, Tzima N et al. Adherence to the Mediterranean diet is associated with total antioxidant capacity in healthy adults: the ATTICA study. Am J Clin Nutr 2005;82:694–699.

53. Pellegrini N, Serafini M, Colombi B et al. Total antioxidant capacity of plant foods, beverages and oils consumed in Italy assessed by three different in vitro assays. J Nutr 2003;133:2812–2819.

54. Pellegrini N, Serafini M, Salvatore S, Del Rio D, Bianchi M, Brighenti F. Total antioxidant capacity of spices, dried fruits, nuts, pulses, cereals and sweets consumed in Italy assessed by three different in vitro assays. Mol Nutr Food Res 2006;50:1030–1038.

55. Record IR, Dreosti IE, McInerney JK. Changes in plasma antioxidant status following consumption of diets high or low in fruit and vegetables or following dietary supplementation with an antioxidant mixture. Br J Nutr 2001;85:459–464.

56. Young JF, Nielsen SE, Haraldsdottir J et al. Effect of fruit juice intake on urinary quercetin excretion and biomarkers of antioxidative status. Am J Clin Nutr 1999;69:87–94.

57. Dragsted LO, Pedersen A, Hermetter A et al. The 6-a-day study: effects of fruit and vegetables on markers of oxidative stress and antioxidative defence in healthy nonsmokers. Am J Clin Nutr 2004;79:1060–1072.

58. Serafini M, Del Rio D, Crozier A, Benzie IF. Effect of changes in fruit and vegetable intake on plasma antioxidant defenses in humans. Am J Clin Nutr 2005;81:531–532.

59. Lorenz M, Jochmann N, Krosigk A et al. Addition of milk prevents vascular protective effects of tea. Eur Heart J 2007;28:219–223.

60. Serafini M, Ghiselli A, Ferro-Luzzi A. In vivo antioxidant effect of green and black tea in man. Eur J Clin Nutr 1996;50:28–32.

61. Langley Evans SC. Consumption of black tea elicits an increase in plasma antioxidant potential in humans. Int J Food Sci Nutr 2000;51:309–315.

62. Serafini M, Bugianesi R, Maiani M, Valtue a S, De Santis S, Crozier A. Plasma antioxidants from chocolate. Nature 2003,424:1013.

63. Krul C, Schuite AL, Tenfelde A, van Ommen B, Verhagen H, Havenaar R. Antimutagenic activity of green tea and black tea extracts studied in a dynamic in vitro gastrointestinal model. Mutat Res 2001;47:71–85.

64. Charlton AJ, Baxter NJ, Khan ML et al. Polyphenol / peptide binding and precipitation. J Agric Food Chem 2002;50:1593–1601.

65. Leenen R, Rodenburg AJ, Tijburg LB, Wiseman SA. A single dose of tea with or without milk increases plasma antioxidant activity in humans. Eur J Clin Nutr 2000;54:87–92.

66. Schroeter H, Holt RR, Orozco TJ, Schmitz HH, Keen CL. Nutrition: milk and absorption of dietary flavanols. Nature 2003;426:787–788.

67. Demirbag R, Yilmaz R, Abdurrahim K. Relationship between DNA damage, total antioxidant capacity and coronary artery disease. Mutat Res 2005;570:197–203.

68. Leal M, Diaz J, Serrano E, Abellan J, Carbonell LF. Hormone replacement therapy for oxidative stress in postmenopausal women with hot flushes. Obstet Gynecol 2000;95:804–809.

69. Chrysohoou C, Panagiotakos DB, Pitsavos C et al. The association between pre-hypertension status and oxidative stress markers related to atherosclerotic disease: the ATTICA study. Atherosclerosis 2007;192:169–76.

70. Vassalle C, Petrozzi L, Botto N, Andreassi MG, Zucchelli GC. Oxidative stress and its association with coronary artery disease and different atherogenic risk factors. J Intern Med 2004;256:308–315.

71. Erhola M, Nieminen MM, Kellokumpu-Lehtinen P, Metsa-Ketela T, Poussa T, Alho H. Plasma peroxyl radical trapping capacity in lung cancer patients: a case-control study. Free Radic Res 1997;26:439–447.

72. Di Giacomo C, Acquaviva R, Lanteri R, Licata F, Licata A, Vanella A. Nonproteic antioxidant status in plasma of subjects with colon cancer. Exp Biol Med 2003;228:525–528.

73. Sener DE, Gonenc A, Akinci M, Torun M. Lipid peroxidation and total antioxidant status in patients with breast cancer. Cell Biochem Func 2007;25:377–382.

74. Lee GJ, Chung HW, Lee KH, Ahn HS. Antioxidant vitamins and lipid peroxidation in patients with cervical intraepithelial neoplasia. J Korean Med Sci 2005;20:267–272.

75. Ching S, Ingram D, Hahnel R, Beilby J, Rossi E. Serum levels of micronutrients, antioxidants and total antioxidant status predict risk of breast cancer in a case control study. J Nutr 2002;132:303–306.

76. Liu X, Zhao J, Zheng R. DNA damage of tumor-associated lymphocytes and total antioxidant capacity in cancerous patients. Mutat Res 2003;539:1–8.

77. Mantovani G, Maccio A, Madeddu C, Mura L, Gramignano G, Lusso MR et al. Quantitative evaluation of oxidative stress, chronic inflammatory indices and leptin in cancer patients: correlation with stage and performance status. Int J Cancer 2002;98:84–91.

78. Li H, Kantoff PW, Giovannucci E, Leitzmann MF, Gaziano JM, Stampfer MJ et al. Manganese superoxide dismutase polymorphism, prediagnostic antioxidant status, and risk of clinical significant prostate cancer. Cancer Res 2005;65:2498–2504.

79. Serafini M, Bellocco R, Wolk A, Ekstrom AM. Total antioxidant potential of fruit and vegetables and risk of gastric cancer. Gastroenterology 2002;123:985–991.

80. Koutroubakis IE, Malliaraki N, Dimoulios PD, Karmiris K, Castanas E, Kouroumalis EA. Decreased total and corrected antioxidant capacity in patients with inflammatory bowel disease. Dig Dis Sci 2004;49:1433–1437.

81. Repetto MG, Reides CG, Evelson P, Kohan S, de Lustig ES, Llesuy SF. Peripheral markers of oxidative stress in probable Alzheimer patients. Eur J Clin Invest 1999;29:643–649.

82. Foy CJ, Passmore AP, Vahidassr MD, Young IS, Lawson JT. Plasma chain-breaking antioxidants in Alzheimer's disease, vascular dementia and Parkinson's disease. Q J Med 1999;92:39–45.

83. Videla LA, Rodrigo R, Orellana M et al. Oxidative stress-related parameters in the liver of non-alcoholic fatty liver disease patients. Clin Sci (Lond) 2004;106:261–268.

84. Dogru-Abbasoglu S, Kanbagli O, Bulur H et al. Lipid peroxides and antioxidant status in serum of patients with angiographically defined coronary atherosclerosis. Clin Biochem 1999;32:671–672.
85. Langley SC, Brown RK, Kelly FJ. Reduced free-radical-trapping capacity and altered plasma antioxidant status in cystic fibrosis. Pediatr Res 1993;33:247–250.
86. Chanarat N, Chanarat P, Suttajit M, Chiewsilp D. Total antioxidant capacity in plasma of HIV-infected patients. J Med Assoc Thai 1997;80:S116–120.
87. Opara EC, Abdel-Rahman E, Soliman S et al. Depletion of total antioxidant capacity in type 2 diabetes. Metabolism 1999;48:1414–1417.
88. Agudo A, Cabrera L, Amiano P et al. Fruit and vegetable intakes, dietary antioxidant nutrients, and total mortality in Spanish adults: findings from the Spanish cohort of the European Prospective Investigation into Cancer and Nutrition (EPIC-Spain). Am J Clin Nutr 2007;85(6):1634–42.

Section IV
Future

16
Reactive Oxygen Species as Signaling Molecules

Ilsa I. Rovira and Toren Finkel

Summary Although the pathophysiological role of reactive oxygen species (ROS) has received considerable attention, there is a growing realization that oxidants also can play a normal, physiological role within cells. These observations came from studies initially using cultured cells stimulated by a variety of peptide growth factors. In this context, several groups were able to show that after ligand addition, the subsequent production of ROS was a necessary component of downstream signaling. Subsequent experiments have defined the enzymatic source of ligand activated ROS production and some of the relevant molecular targets. Here, we review these observations and discuss the role of ROS in normal growth factor signaling. In addition, we describe how these observations are currently being extended to the production of ROS emanating from the mitochondria, and how mitochondrial ROS also might be involved in certain signaling events. We also discuss how ROS might be involved in the induction of cellular senescence by acting as important intracellular mediators. Finally, we review the role of ROS in stem cells and how oxidants might again act to regulate the biology of these critical cells. Together, it is hoped that these observations will serve as a framework to more fully understand how ROS participate in aging. In particular, these observations provide the starting point to determine whether ROS participate in aging as random, stochastic damaging agents, or whether they function as mediators of critical redox-dependent pathways.

Keywords Reactive oxygen species, signal transduction, tyrosine phosphatases, senescence, stem cells.

1 Reactive Oxygen Species (ROS) in Growth Factor Signaling

Our own interest in the concept of ROS as signaling agents began a little over a decade ago. We were particularly interested in pursuing observations that linked a rise in ROS with the onset of atherosclerosis (see Chapter 10). At that time,

From: *Aging Medicine: Oxidative Stress in Aging: From Model Systems to Human Diseases*   293
Edited by: S. Miwa, K.B. Beckman, and F.L. Muller © Humana Press, Totowa, NJ

it was clear that another reactive oxygen species, namely, nitric oxide (NO), could be produced in a ligand-dependent manner and specifically modulate the activity of a downstream target. For the case of NO, it was known that within the vessel wall, endothelial cells produced the radical species that could then diffuse to the underlying smooth muscle cell layer and modulate the activity of guanylate cyclase [1, 2]. We were intrigued by this general paradigm and wanted to explore the notion that other ROS molecules also could be produced in a ligand-dependent manner and specifically regulate downstream signaling.

As a model, we chose vascular smooth muscle cells (VSMCs) stimulated by the ligand platelet-derived growth factor (PDGF). At that time, there were very few articles describing how to image the production of ROS in living cells. We came across a methodology that used the fluorescent indicator dichlorfluorescein diacetate (DCFDA). This molecule can accumulate within cells, and it fluoresces only when subsequently oxidized. As such, it became a widely used albeit non-qualitative measure of intracellular ROS levels. Although we provided some initial evidence that DCFDA was a useful marker for intracellular hydrogen peroxide levels, subsequent studies by others have questioned to what degree DCFDA is specific for hydrogen peroxide versus other oxidants and whether it is useful as a direct or indirect measure of intracellular ROS levels [3]. Recently, new fluorescent probes have been synthesized that seem to provide the sensitivity and temporal resolution of DCFDA coupled with increased specificity [4].

Using the above-mentioned strategy, we were able to see a burst of hydrogen peroxide production within minutes of PDGF stimulation [5]. Our analysis suggested that ROS generation peaked within minutes of ligand addition and that it returned to baseline within approximately 30–40 min. This time course was similar to what other observers had previously noted for the time course of tyrosine phosphorylation after the addition of PDGF. We were intrigued by this, because it was well known that the addition of exogenous hydrogen peroxide could trigger tyrosine phosphorylation in a ligand-independent manner. We wondered whether the endogenous production of hydrogen peroxide might serve a similar function. To assess this possibility, we increased the scavenging capacity within our VSMCs by increasing the levels of intracellular catalase, an enzyme that specifically degrades hydrogen peroxide. Using this approach we could show that the peak level of hydrogen peroxide after PDGF stimulation was, as expected, inversely proportional to the amount of intracellular catalase activity. Interestingly, in cells where the level of catalase was high, the ability of PDGF to stimulate a burst of tyrosine phosphorylation was severely inhibited. These results led us to conclude that PDGF-stimulated ROS production was required for the subsequent burst of tyrosine phosphorylation. Although our experiments were performed in VSMCs by using PDGF, they were quickly followed by other observations in other cell types by using other ligands [6]. Indeed, it would now seem that multiple different types of cells, and a variety of different ligands, all use some form of ROS-dependent signaling [7–9].

2   NADPH Oxidases

After the observations that growth factors could stimulate the production of ROS within cells and that this increase in ROS was necessary for subsequent signaling, a number of questions arose. What was the source of these oxidants? What was the biochemical pathway between the receptor to the oxidant source? What was the relevant target for ROS within cells, and how could this explain the coupling between ROS and tyrosine phosphorylation? Although many of these questions still remain incompletely understood, significant progress has been made over the ensuing decade.

One critical source of ROS production within cells seems to be a family of enzymes known as the Nox family. This family of enzymes is structurally similar to the classical NADPH oxidase complex found in phagocytic cells. It has been well established that cells such as neutrophils produce large amounts of ROS as part of their role in host defense. Molecular dissection of the phagocytic respiratory burst identified a multisubunit structure that is required for neutrophil ROS production. This complex has both membrane-bound elements (initially termed p22phox and gp91phox) and elements recruited from the cytosol (e.g., p47phox and p67phox). The identity of these subunits was aided greatly by the analysis of rare and unfortunate individuals suffering from the syndrome of chronic granulomatous disease (CGD). Genetic characterization of these individuals has demonstrated that they possess homozygous mutations in one of the four critical NADPH oxidase subunits. Affected individuals rarely survive beyond early adulthood, and their clinical course is manifested by recurrent serious infections. Analysis of phagocytic cells from CGD patients demonstrated that their neutrophils were incapable of producing high levels of ROS (respiratory burst) when challenged with a pathogen [10].

One additional element has been implicated in the function of the neutrophil NADPH oxidase. As opposed to the proteins discussed previously that play a direct role in binding and metabolizing NADPH to produce ROS species, the final element, namely, Rac2, acts more in a regulatory manner. Rac2 is a member of the small GTPase superfamily of proteins, and it is expressed primarily in phagocytic cells, although the highly homologous Rac1 is expressed ubiquitously. For the neutrophil oxidase, Rac2 seems to act as GTP-dependent switch that is required for activation of ROS production when the neutrophil is engaged.

For many years, it was assumed that phagocytic cells were the only cells in the body that made superoxide in a purposeful manner. Indeed, it was generally assumed that expression of the components of the NADPH oxidase were restricted to immune cells. These perceptions have, however, dramatically changed in the past several years. One of the most important reasons for this change is the cloning of novel homologs of the gp91phox subunit. There are now at least six members of this family, with the original gp91phox now more commonly referred to as Nox2. In addition, more recently, homologs of both p47phox and p67phox have been isolated. It would now seem that a variety of cells express Nox family members and that they are capable of producing superoxide in a regulated and purposeful manner [11, 12]. There is also evidence that Rac proteins regulate the activity of these novel Nox family members consistent with older observation that Rac activity was required for the ligand-stimulated production of ROS [13]. The activation of Rac proteins seems to involve the upstream activation of PI3-kinase (PI3-K) (Fig. 16.1). It is generally assumed but not rigorously proven in all cases that the production of ROS is generally lower in the novel Nox systems compared with the original (Nox2)-dependent enzyme. This might naturally reflect the different amounts needed between using superoxide and its derivatives to kill cells (for Nox2) compared with the use of ROS as signaling molecules. In many ways, this is analogous to the production of NO by immune and nonimmune cells. Again, for NO, immune cells acting through NOS2 produce roughly 3 orders of magnitude higher amounts of NO than is normally produced when either NOS1 or NOS3 generate NO to be used as a signaling molecule.

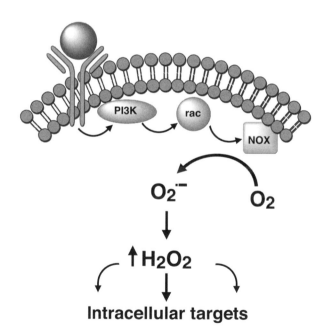

Fig. 16.1 Hydrogen peroxide as a signaling molecule. After the addition of certain ligands (e.g., PDGF and EGF), a series of rapid signaling events unfold. Ligand binding leads to the activation of the lipid kinase PI3-K that in turn activates the small GTPase Rac1. This GTPase acts as a guanine nucleotide-regulated switch to activate the Nox family of cytosolic oxidases. These enzymes generate superoxide that can rapidly dismutate to hydrogen peroxide. The generation of intracellular hydrogen peroxide can then presumably interact with numerous downstream targets

3 Protein Tyrosine Phosphatases

With the realization that ROS levels rapidly increase after growth factor stimulation and that the family of Nox enzymes might be the source of this oxidant burst the question remained, how does a rise in ROS regulate the burst of tyrosine phosphorylation observed after PDGF or epidermal growth factor (EGF) stimulation? More precisely, what are the molecular target(s) of growth factor-induced ROS? One particularly intriguing possibility is that oxidants might regulate the family of intracellular protein tyrosine and dual-specific phosphatases. This possibility was actually suggested initially in the earliest description of the phenomenon of ligand-activated ROS production [5]. Protein tyrosine phosphatases only act on tyrosine-phosphorylated residues, whereas dual-specific phosphatases can act on tyrosine-phosphorylated and serine/threonine-phosphorylated residues. The catalytic mechanism of both tyrosine and dual-specific enzymes both involve a reactive cysteine residue at the catalytic site [9]. In this context, "reactive" means that the proton of the thiol residue in the catalytic cysteine can be oxidized at a physiological pH. Indeed, the pK_a of such reactive cysteines is generally very low (<pH 6.0). As such, the cysteine within the context of the cell exists in both the reduced (SH) and oxidized (S$^-$) thiolate anion form.

Previous studies using purified protein tyrosine phosphatases (PTPs) have clearly shown that the oxidized form of the enzyme is catalytically inactive [14]. However, this inactivation is not irreversible. Indeed, the simple addition of exogenous reducing agents (e.g., dithiothreitol) results in reconstitution of full enzymatic activity. These observations have led to the hypothesis that the burst of ROS after ligand stimulation transiently inactivates intracellular PTPs. As long as ROS are high, the ability of these enzymes to reverse tyrosine phosphorylation is inhibited. As such, the action of activated intracellular tyrosine kinases is essentially unopposed. As ROS levels fall, the PTPs can be reduced, presumably through the activity of the glutaredoxins, thioredoxin networks, or both found within the cell. This reduction back to the SH form shifts the balance away from kinases and back to the tonic inhibition of signaling that PTPs provide (Fig. 16.2).

Careful mass spectrometry studies have led to a more fundamental understanding of the redox chemistry involved in this process. The most carefully studied phosphatase is PTP1B. After growth factor stimulation, increased levels of ROS seem to transiently inactivate the PTP1B phosphatase [15]. The initial oxidation seems to result in the catalytic cysteine being oxidized to a sulfenic (SOH) form. Surprisingly, this oxidized molecule then converts to a rather unusual sulfenic-amide bond (S–N) before further reduction to the fully reduced moiety [16, 17]. For the other PTPs, reduction from the sulfenic form also can be achieved by forming a mixed disulfide with the ubiquitous reductant glutathione (S-GSH) or by forming intermolecular or intramolecular disulfide bonds with other reactive cysteine residues [18]. In this manner, the overall activity of PTPs can be regulated by the reversible oxidation-reduction of the catalytic cysteine residue and provides a relatively simple mechanism through which tyrosine phosphorylation, signaling and redox status can be coupled.

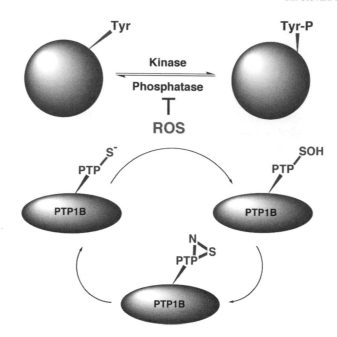

Fig. 16.2 ROS regulate PTPs. The levels of intracellular tyrosine phosphorylation represent the balance between the activity of tyrosine kinases and tyrosine phosphatases. Several lines of evidence suggest that tyrosine phosphatases can be oxidized and inactivated by exogenous or endogenous ROS. This occurs for the case of PTP1B by oxidation of the catalytic cysteine first to a sulfenic form (S–OH) that then resolves through the formation of a sulfenic-amide bond (S–N). These observations suggest a plausible mechanism to couple ROS levels with tyrosine phosphorylation and growth factor signaling

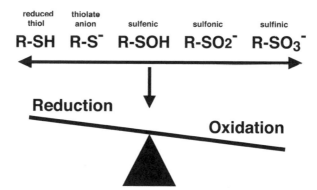

Fig. 16.3 Oxidation state of reactive cysteine residues. Increasing ROS levels results in the oxidation of certain reactive cysteine residues within proteins. The exact identification of a reactive cysteine from an ordinary cysteine is difficult to do based solely on the primary sequence; however, reactive cysteines are usually surrounded by basic amino acids. The sulfenic form is reversible; yet, in the case of phosphatases leads to enzymatic inactivation. Higher oxidation states above sulfenic also can be resolved under certain specific situations in certain proteins

Indeed, the reversibility of cysteine moieties provides a rapid and reversible mechanism for signaling not unlike the more widely studied mechanisms underlying reversible phosphorylation of serine, threonine, or tyrosine (Fig. 16.3).

4 Mitochondrial ROS

Although the preceding discussion centered on the role of ROS generated in the cytosol, there is also growing awareness that ROS generated by mitochondria also participate in signaling pathways. It is generally assumed that oxidants are generated by mitochondria as a by-product of normal aerobic metabolism. Initial estimates in the 1960s and 1970s by using isolated mitochondria and high ambient oxygen concentration have led to the widely quoted number that 1–2% of consumed oxygen are diverted into superoxide. This number is probably a reflection of the experimental and somewhat artificial conditions in which these initial measurements were made. Although the real leakage of mitochondrial electrons under the usual oxygen concentrations seen within cells (i.e., 3–5% oxygen vs 20% laboratory conditions) is difficult to quantify, it is probably 1 or perhaps 2 orders of magnitude lower than the widely quoted value of 1–2%. Nonetheless, even at this lower revised rate, mitochondria generally are thought to be the predominant consumer of molecular oxygen and produce approximately 90% of the ROS within most cells.

Although there is a growing acceptance as discussed that oxidants produced in the cytosol by NOX family members after growth factor stimulation play a physiological role in signal transduction, this concept has not been widely applied to mitochondrial oxidants. Indeed, the bias remains that production of mitochondrial ROS is unregulated and that they are strictly harmful and nonspecific. Here, for the sake of brevity, we review two recent examples that challenge this notion.

The first example centers on a relatively recent report involving tumor necrosis factor (TNF)-α signal transduction. It is well known that TNF-α can induce a number of divergent outcomes in target cells, including both cell growth and cell death. These differences are due in large part to differences in the type of TNF-α receptor that is engaged and what downstream effector pathway is subsequently activated. The activation of nuclear factor-κB (NF-κB) is thought to be one important mechanism to protect cells against TNF-α-induced cell death. In contrast, activation of the c-Jun NH$_2$-terminal kinase (JNK) pathway is usually associated with TNF-α-mediated cell death. Usually, the activation of JNK is transient peaking in the first 20 min or so after ligand addition and returning to baseline within 1 h. In cells or animals in which the NF-κB pathway is inactivated, JNK stimulation is considerably prolonged and the degree of cell death is significantly increased. The molecular basis for the cross talk between TNF-α-stimulated NF-κB and JNK activity remained obscure until it was observed that mitochondrial produced ROS could mediate this interaction [19]. In particular, it was demonstrated that the addition TNF-α stimulated a rise in cellular ROS levels.

This ROS increase could be partially suppressed by increased expression of superoxide dismutase (SOD)2, the manganese-containing form of SOD residing in the mitochondria. The link between mitochondrial ROS release and JNK activity came from the observation that the rise in mitochondrial ROS production was sufficient to oxidize and inactivate various phophatases that regulate the activity of JNK. These investigators specifically demonstrated that mitogen-activated protein kinase phosphatase (MKP)-1, -3, -5, and -7 all became oxidized and inactivated in the setting of TNF-α oxidative stress. Mass spectrometry of in vitro-oxidized MKP-3 revealed that the active site cysteine formed a sulfenic acid intermediate after exposure to hydrogen peroxide. These results suggest two important conclusions. First, the production of mitochondrial oxidants can be regulated by external cytokines; and second, when released, these oxidants can directly target specific intracellular pathways. In this example, the release of mitochondrial oxidants functions to maintain high-intensity JNK signaling that in turn activates programmed cell death pathways. Interestingly, treating cells or animals with an antioxidant prevented JNK activation and inhibited the ability of TNF-α to induce hepatic necrosis.

A second example where the production of mitochondrial oxidants seems to have a specific and important intracellular target centers on the regulation of HIF-1α. The HIF-1α transcription factor is involved in the regulation of >70 separate genes that regulate the response of cells to low oxygen [20]. These gene products provide a coordinated response to the hypoxic environment by encoding for factors that allow for new red blood cell formation, the development of new blood vessels, and augmentation of the cell's intrinsic ability to perform glycolysis. Together, this program allows the cell to maintain energetic stores in the absence of environmental oxygen (e.g., increased glycolysis) and to increase the future delivery of oxygen (e.g., erythropoiesis and angiogenesis). Much of HIF-1α activity is regulated at the level of protein stability. Under well-oxygenated conditions, the half-life of HIF 1α is very short (<5 min), whereas the protein exhibits a considerably longer half life when cell are maintained in a hypoxic environment. The molecular regulation of HIF-1α stability is achieved by the enzymatic action of a recently discovered family of prolyl hydroxylases (PHDs). These enzymes (PHD-1, -2, and -3) catalyze proline hydroxylation of a specific residue with the oxygen-dependent degradation domain of HIF-1α. Hydroxylation of this proline residue stabilizes the interaction of HIF-1α with the von Hippel-Lindau protein and targets HIF-1α for subsequent proteasomal degradation. Because the enzymatic action of the PHDs requires Fe(II), the Krebs' intermediary 2-oxo-glutarate, and most importantly, molecular oxygen, the link between ambient oxygen concentration and HIF-1α stability has a clear potential explanation.

A number of observations suggest that HIF-1α also might be regulated by mitochondrial ROS levels. First, in cells exposed to agents such as t-butyl hydroperoxide, levels of HIF-1α increase. In cells with high ROS levels resulting from genetic manipulation, similar results are observed. For example, one recent study explored the phenotype of mouse embryonic fibroblasts that lacked junD, a member of the activator protein-1 superfamily of transcription factors [21].

Interestingly, the junD$^{-/-}$ cells had higher chronic levels of ROS and elevated HIF-1α levels. Subsequent experiments demonstrated that these effects were mediated by an oxidant induced oxidation of Fe(II) to Fe(III) that in turn inhibited the activity of cellular PHDs.

Although these observations with junD suggest that HIF-1α might be redox regulated, they do not address whether mitochondrial ROS are involved. Nonetheless, several recent studies suggest they might be involved. For example, one study demonstrated that the hypoxic induction of HIF-1α was blocked by using a mitochondrial-targeted antioxidant [22]. Similarly, two groups demonstrated that hypoxic HIF-1α induction was blocked when they used either RNA interference-based methods to knock down components of the electron transport chain or pharmacological inhibitors of mitochondrial activity [23, 24]. These groups also were able to demonstrate the seemingly counterintuitive phenomena of increased mitochondrial ROS levels in the setting of hypoxia—counterintuitive because it is widely held that the production of mitochondrial ROS would demonstrate a linear, or at least a positive, correlation with oxygen concentrations. These results suggest that the relationship between oxygen concentration and mitochondrial ROS levels seems to be more complex and seemingly U-shaped in nature. Nonetheless, these results, and the results involving TNF-α signaling all suggest that the production of mitochondrial oxidants can be regulated and that the release of these oxidants may in turn directly regulate a number of critical intracellular signaling pathways.

5   ROS in Cellular Senescence

If oxidants can contribute to normal growth factor signaling when produced by cytosolic oxidases such as the Nox family of enzymes, and if mitochondrial oxidant can function in an analogous manner during hypoxia or after TNF-α stimulation, what does this mean for the role of ROS in aging? One possibility is that the release of oxidants from the mitochondria secondary to aerobic respiration might be tightly regulated and that once released, ROS might have specific targets within the cell. Redox regulation of these targets would then be expected to influence the overall rate of aging. One particularly intriguing possibility is that gene products linked to life span (e.g., sirtuins, Foxo proteins, and target of rapamycin pathway) might impinge on this system particularly in regulating the release of ROS from the mitochondria.

What are the relevant pathways that mitochondrial oxidants regulate? One possibility is entry into cellular senescence. Senescence represents a permanent growth-arrested state that many, but certainly not all, investigators feel represents a cellular surrogate of overall aging. Initially, most investigators studied the phenomena of replicative senescence. As originally defined by Hayflick and colleagues, primary cells in culture will only divide a finite number of times before they enter a permanent growth-arrested state. This state is known as replicative senescence or the Hayflick limit. At this point, the cells undergo morphological changes, including developing a flattened and elongated appearance. Later, it was realized that these cells also underwent a number of biochemical changes that could lead to their rapid identification. One such change is the appearance of β-galactosidase activity at a neutral pH [25]. This "β-gal" positivity has become a useful marker for both in vitro and in vivo senescence. More recently, the activation of certain pathways, including the DNA damage response and increased levels of p16 protein, have become additional strategies to identify senescent cells [26].

Although senescence was initially viewed as a phenomenon that occurred only with the passaging of primary cells in culture, the ability to more accurately identify the senescent cell made it possible to ask whether there were other triggers of senescence besides the number of previous cell divisions. It quickly became clear that the answer was "yes." Indeed, a host of factors constituting both environmental and genetic stresses seem to trigger entry into senescence [27]. One of the most interesting for our discussion is the observation that moderate levels of exogenous hydrogen peroxide can induce senescence [28]. In these experiments it would seem that senescence is achieved at concentrations of hydrogen peroxide that are below what would normally induce apoptosis. Thus, a continuum of cellular responses could be envisioned after exposure to hydrogen peroxide, ranging from growth to senescence to apoptosis, depending presumably on the integration of the concentration and duration of ROS exposure (Fig. 16.4).

Another trigger for the induction of cellular senescence is the expression of certain oncogenes within the context of a primary cell. The first example of this was

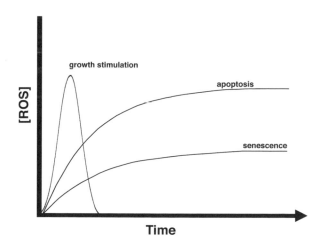

Fig. 16.4 Cell fate and ROS levels. Increased intracellular ROS levels have been linked to growth, death, and senescence. One plausible scenario to explain these various outcomes is that both the concentration and temporal characteristics of the ROS burst is important. Short increases might stimulate growth, whereas prolonged increases might trigger either senescence or apoptosis depending on the absolute levels

the expression of the Ras oncogene in primary fibroblasts derived from either mouse embryos or human skin biopsies [29]. Although expression of tumor oncogenes is generally associated with increased growth and subsequent cellular transformation, such experiments were generally performed in established, immortalized cell lines such as NIH3T3, cells that do not senesce. In contrast to these results, it would seem that expression of a number of oncogenes within normal cells triggers either apoptosis or senescence. This response might in fact have very important consequences for the host cell and for the entire organism. Indeed, for both cases, a cell committed to either cell death (apoptosis) or irreversible growth arrest via senescence cannot divide further and hence cannot pass on the harmful tumor oncogene to subsequent cells. As such, it is now widely held that the induction of cellular senescence by tumor oncogenes represents a fundamental and essential barrier to transformation and a widely used cellular method to block tumor formation [26].

What is the role of ROS in these processes? Early on, it was realized that ambient oxygen concentrations represented an important variable for replicative life span [30]. For example, the number of cellular divisions a primary cell could undergo before becoming senescent varied inversely to the concentration of ambient oxygen. It is common practice to grow cells in tissue culture in an environment that contains room air consisting of approximately 20% oxygen. Although obviously convenient, this concentration is probably 5-fold higher then most cells normally see within the body. Culturing cells in more physiological oxygen concentrations of 5% or even 10% oxygen significantly extends the replicative life span of cells. Conversely, growing cells in higher concentrations (e.g., 50% or greater) reduces their life span [30]. When these initial experiments were performed 30 years ago, the actual basis for replicative senescence was unknown. It is now generally held that replicative senescence is caused by the gradual shortening of telomeres with each cell division. When primary human cells telomeres reach a critical shorten length, a pathway is activated that triggers senescence [31]. The actual length of telomere shortening per each human cell division is approximately 100–200 base pairs, but this number is not fixed. When cells are exposed to higher oxygen concentrations (and hence presumably higher ROS levels), the shortening of telomeres occurs in larger chunks per division [32]. As such, these mechanistic observations provide a reasonable explanation as to why cells senesce at different rates depending on the ambient oxygen concentration.

Oxidants also might play a role in oncogene-induced senescence. For example, we were able to show that in primary human cells engineered to express the Ras oncogene, levels of mitochondrial ROS were increased [33]. These oxidants seemed to be derived from the mitochondria, although the exact reason why Ras expression led to higher levels of intracellular ROS remains obscure. Nonetheless, scavenging of the Ras-induced ROS was sufficient to inhibit the ability of Ras to induce senescence. Recently, several screens have been performed that look for genes whose increased or decreased expression allow for the escape from Ras-induced senescence [34–36]. In many of these cases, the gene products whose altered expression can allow for continuous cell proliferation in the presence of Ras overexpression seems to be involved in either overall metabolism, redox signaling, or both. This has led to the supposition that ROS are important mediators of cellular senescence and that oxidant-mediated modulation of critical redox-dependent pathways are required for entry into senescence [37].

6   ROS and Stem Cells

If oxidants participate in cellular senescence, is this important for organismal aging? Indeed, considerable controversy exists regarding the role, if any, for senescence in overall aging. Although some view the appearance of senescence solely as an in vitro tissue culture phenomenon, others believe that the collection of senescent cells is fundamental to the degenerative changes associated with aging [38]. Thus, the exact contribution of cellular senescence to overall aging remains intensely studied albeit intensely controversial.

The accumulation of senescent cells may have different effects depending on what type of cell is involved. For example, a skin cell that is senescent may be cosmetically important, but perhaps not critical to an organism's overall survival. In contrast, senescence in a postmitotic cardiac myocyte or neuronal cell may have a more important functional consequence, because these cells cannot be easily replaced. The senescence of stem and progenitor cells represents another example of where cellular senescence might have widespread consequences to the organism. Like the heart and brain, stem cells are present at birth and persist throughout the entire life span of the organism. The mobilization of stem and progenitor cells is essential for normal homeostasis and tissue repair, processes whose fidelity declines with age. Clearly, the exhaustion or senescence of the stem and progenitor pool would be expected to have dramatic effects on an organism and perhaps mimic many of the deficiencies associated with advanced age [39].

What then regulates stem and progenitor cell aging? Interestingly, some recent evidence suggests that levels of intracellular ROS might be critical. In a relatively recent manuscript, a group of investigators were probing the phenotype of mice lacking the ataxia-telangiectasia mutated (Atm) gene. These mice exhibit many but not all of the symptoms seen in patients suffering from ataxia-telangiectasia. Like humans with this condition, the mice are cancer prone and many mice develop a lymphoma in the first few months of life. Nonetheless, some animals do not develop tumors and in this context it was observed that cancer-free $Atm^{-/-}$ mice at 6 months started to develop a picture of bone marrow failure [40]. It was quickly discerned that the hematopoietic stem cells (HSC) of the $Atm^{-/-}$ mice had undergone premature exhaustion. In an effort to understand how the absence of Atm led to HSC failure, the authors explored a previously observed phenotype of $Atm^{-/-}$ tissues, namely, that $Atm^{-/-}$ cells demonstrate constitutively high levels of ROS [41]. Consistent with these past observations, the HSC of $Atm^{-/-}$ mice also had higher levels of ROS. More importantly, the premature exhaustion of HSC in this case was rescued by simple antioxidant treatment. This phenomenon is not unique to Atm. Indeed, another very recent study explored the biology of the Foxo family of forkhead transcription factors. Triple-deficient animals lacking FoxO1, FoxO3, and FoxO4 also developed rapid exhaustion of their HSC. These investigators also implicated an ROS-dependent mechanism [42]. Thus, it would seem that stem and progenitor cells might be particularly sensitive to changes in intracellular oxidants. Further studies are needed to understand whether the rise in ROS within the stem cell compartment is leading to exhaustion through the induction of senescence, apoptosis, or both.

7   Summary

The evidence that ROS might contribute to aging is widespread and persuasive. Nonetheless, whether ROS correlate or cause aging remains an open question. Here, we have addressed yet another unresolved issue regarding oxidants and aging, namely, how do ROS influence the rate we age? Although the prejudice has been that oxidants cause random and nonspecific damage that accumulates over time, we have raised an alternative explanation. Namely, we suggest that oxidants may function as intracellular messengers within cells and thereby regulate a number of critical pathways and fate decisions within cells. Under this scenario, the production and release of oxidants from any number of sources, including cytosolic oxidases (e.g., Nox family members) or mitochondrial respiration would be tightly regulated. The release of these oxidants would in turn have specific effects within cells with the discrete redox activation or inactivation of specific target proteins. The result would be that multiple aspects of cellular activity operate under the broad umbrella of redox-dependent regulation. In this context, the challenge would be to understand how the previous evidence for a role of oxidants in aging can be reconciled with this new information. Further understanding regarding what regulates oxidant production, and a more complete delineation of the range of redox targets, hopefully will allow us to better address this question. Until then, we are left with two related and fundamental question. Do ROS make us age? And if so, how?

References

1. Furchgott RF, Zawadzki JV. The obligatory role of endothelial cells in the relaxation of arterial smooth muscle by acetylcholine. Nature 1980; 288:373–6.
2. Lowenstein CJ, Snyder SH. Nitric oxide, a novel biologic messenger. Cell 1992; 70:705–7.
3. Bonini MG, Rota C, Tomasi A, Mason RP. The oxidation of 2′,7′-dichlorofluorescin to reactive oxygen species: a self-fulfilling prophesy? Free Radic Biol Med 2006; 40:968–75.
4. Miller EW, Albers AE, Pralle A, Isacoff EY, Chang CJ. Boronate-based fluorescent probes for imaging cellular hydrogen peroxide. J Am Chem Soc 2005; 127:16652–9.
5. Sundaresan M, Yu ZX, Ferrans VJ, Irani K, Finkel T. Requirement for generation of H_2O_2 for platelet-derived growth factor signal transduction. Science 1995; 270:296–9.

6. Bae YS, Kang SW, Seo MS et al. Epidermal growth factor (EGF)-induced generation of hydrogen peroxide. Role in EGF receptor-mediated tyrosine phosphorylation. J Biol Chem 1997; 272:217–21.

7. Finkel T. Oxidant signals and oxidative stress. Curr Opin Cell Biol 2003; 15:247–54.

8. Rhee SG, Kang SW, Jeong W, Chang TS, Yang KS, Woo HA. Intracellular messenger function of hydrogen peroxide and its regulation by peroxiredoxins. Curr Opin Cell Biol 2005; 17:183–9.

9. Rhee SG. Cell signaling. H_2O_2, a necessary evil for cell signaling. Science 2006; 312:1882–3.

10. Sheppard FR, Kelher MR, Moore EE, McLaughlin NJ, Banerjee A, Silliman CC. Structural organization of the neutrophil NADPH oxidase: phosphorylation and translocation during priming and activation. J Leukoc Biol 2005; 78:1025–42.

11. Lambeth JD. NOX enzymes and the biology of reactive oxygen. Nat Rev Immunol 2004; 4:181–9.

12. Bedard K, Krause KH. The NOX family of ROS-generating NADPH oxidases: physiology and pathophysiology. Physiol Rev 2007; 87:245–313.

13. Sundaresan M, Yu ZX, Ferrans VJ et al. Regulation of reactive-oxygen-species generation in fibroblasts by Rac1. Biochem J 1996; 318(2):379–82.

14. Denu JM, Tanner KG. Specific and reversible inactivation of protein tyrosine phosphatases by hydrogen peroxide: evidence for a sulfenic acid intermediate and implications for redox regulation. Biochemistry 1998; 37:5633–42.

15. Lee SR, Kwon KS, Kim SR, Rhee SG. Reversible inactivation of protein-tyrosine phosphatase 1B in A431 cells stimulated with epidermal growth factor. J Biol Chem 1998; 273:15366–72.

16. Salmeen A, Andersen JN, Myers MP et al. Redox regulation of protein tyrosine phosphatase 1B involves a sulphenyl-amide intermediate. Nature 2003; 423:769–73.

17. van Montfort RL, Congreve M, Tisi D, Carr R, Jhoti H. Oxidation state of the active-site cysteine in protein tyrosine phosphatase 1B. Nature 2003; 423:773–7.

18. Chiarugi P, Buricchi F. Protein tyrosine phosphorylation and reversible oxidation: two cross-talking posttranslation modifications. Antioxid Redox Signal 2007; 9:1–24.

19. Kamata H, Honda S, Maeda S, Chang L, Hirata H, Karin M. Reactive oxygen species promote TNFalpha-induced death and sustained JNK activation by inhibiting MAP kinase phosphatases. Cell 2005; 120:649–61.

20. Semenza GL, Shimoda LA, Prabhakar NR. Regulation of gene expression by HIF-1. Novartis Found Symp 2006; 272:2–8.

21. Gerald D, Berra E, Frapart YM et al. JunD reduces tumor angiogenesis by protecting cells from oxidative stress. Cell 2004; 118:781–94.

22. Sanjuan-Pla A, Cervera AM, Apostolova N et al. A targeted antioxidant reveals the importance of mitochondrial reactive oxygen species in the hypoxic signaling of HIF-1alpha. FEBS Lett 2005; 579:2669–74.

23. Mansfield KD, Guzy RD, Pan Y et al. Mitochondrial dysfunction resulting from loss of cytochrome c impairs cellular oxygen sensing and hypoxic HIF-alpha activation. Cell Metab 2005; 1:393–9.

24. Brunelle JK, Bell EL, Quesada NM et al. Oxygen sensing requires mitochondrial ROS but not oxidative phosphorylation. Cell Metab 2005; 1:409–14.

25. Dimri GP, Lee X, Basile G et al. A biomarker that identifies senescent human cells in culture and in aging skin in vivo. Proc Natl Acad Sci U S A 1995; 92:9363–7.

26. Mooi WJ, Peeper DS. Oncogene-induced cell senescence—halting on the road to cancer. N Engl J Med 2006; 355:1037–46.

27. Ben-Porath I, Weinberg RA. When cells get stressed: an integrative view of cellular senescence. J Clin Invest 2004; 113:8–13.

28. Chen QM, Bartholomew JC, Campisi J, Acosta M, Reagan JD, Ames BN. Molecular analysis of H_2O_2-induced senescent-like growth arrest in normal human fibroblasts: p53 and Rb control G1 arrest but not cell replication. Biochem J 1998; 332:43–50.

29. Serrano M, Lin AW, McCurrach ME, Beach D, Lowe SW. Oncogenic ras provokes premature cell senescence associated with accumulation of p53 and p16INK4a. Cell 1997; 88:593–602.

30. Packer L, Fuehr K. Low oxygen concentration extends the lifespan of cultured human diploid cells. Nature 1977; 267:423–5.
31. Stewart SA, Weinberg RA. Telomeres: cancer to human aging. Annu Rev Cell Dev Biol 2006; 22:531–57.
32. von Zglinicki T, Martin-Ruiz CM. Telomeres as biomarkers for ageing and age-related diseases. Curr Mol Med 2005; 5:197–203.
33. Lee AC, Fenster BE, Ito H et al. Ras proteins induce senescence by altering the intracellular levels of reactive oxygen species. J Biol Chem 1999; 274:7936–40.
34. Wu C, Miloslavskaya I, Demontis S, Maestro R, Galaktionov K. Regulation of cellular response to oncogenic and oxidative stress by Seladin-1. Nature 2004; 432:640–5.
35. Kondoh H, Lleonart ME, Gil J et al. Glycolytic enzymes can modulate cellular life span. Cancer Res 2005; 65:177–85.
36. Nicke B, Bastien J, Khanna SJ et al. Involvement of MINK, a Ste20 family kinase, in Ras oncogene-induced growth arrest in human ovarian surface epithelial cells. Mol Cell 2005; 20:673–85.
37. Colavitti R, Finkel T. Reactive oxygen species as mediators of cellular senescence. IUBMB Life 2005; 57.277–81.
38. Patil CK, Mian IS, Campisi J. The thorny path linking cellular senescence to organismal aging. Mech Ageing Dev 2005; 126:1040–5.
39. Rando TA. Stem cells, ageing and the quest for immortality. Nature 2006; 441:1080–6.
40. Ito K, Hirao A, Arai F et al. Regulation of oxidative stress by ATM is required for self-renewal of haematopoietic stem cells. Nature 2004; 431:997–1002.
41. Barzilai A, Rotman G, Shiloh Y. ATM deficiency and oxidative stress: a new dimension of defective response to DNA damage. DNA Repair (Amst) 2002; 1:3–25.
42. Tothova Z, Kollipara R, Huntly BJ et al. FoxOs are critical mediators of hematopoietic stem cell resistance to physiologic oxidative stress. Cell 2007; 128:325–339.

17
Summary and Outlook

Satomi Miwa, Florian L. Muller, and Kenneth B. Beckman

50 years have passed since Harman first proposed the free radical theory. Many investigators, using different approaches and model systems, have attempted to test it. Is the theory correct? Few investigators would conclude that the most ambitious form of the hypothesis—that oxidative stress is the major *determinant* of life span—has withstood scrutiny. At the same time, there is enough utility in the free radical theory as a framework for experimentation that it has kept the interest of successive generations of researchers. Even when rephrased in more modest terms, i.e. "Oxidative stress contributes to the aging phenotype," the apparent validity of the theory depends on the model organism in question.

Thus, the goal of the first part of this book was to present an assessment of the free radical theory by experts in different model systems. As we have seen, the evidence that oxidative stress contributes to senescence or limits life span is strong in some model organisms, but weak in others.

1 Does Oxidative Stress Limit Life Span?

Passos and von Zglinicki (Chapter 3) make a strong case that mitochondrial reactive oxygen species (ROS) contribute to the senescent phenotype of primary human cells in culture, while Rovira and Finkel (Chapter 16) also point out that oxygen-dependent damage is a key driver of cell senescence itself. Osiewacz and Scheckhuber (Chapter 4) on the fungus *Podospora anserina*, highlight that senescence is largely the result of mitochondria-dependent degenerative processes and that oxidative stress may very well contribute to this phenomenon. Indeed, several genetic modifications affecting the mitochondrial electron transport chain dramatically lengthen the life span of that species. In *Saccharomyces cerevisiae*, Gonidakis and Longo (Chapter 5) conclude that oxidative stress limits chronological life span, which is consistently shortened by knocking out antioxidant enzymes (superoxide dismutase [SOD]), in an oxygen-dependent manner. Furthermore, there is evidence to indicate that reducing oxidative stress is both necessary and sufficient for extended life span. Chronological life span can be extended by pharmacological reduction of the rate of mitochondrial ROS production, or by overexpressing MnSOD. They note, however, that these extensions are

From: *Aging Medicine: Oxidative Stress in Aging: From Model Systems to Human Diseases*   309
Edited by: S. Miwa, K.B. Beckman, and F.L. Muller © Humana Press, Totowa, NJ

modest compared with what is achieved by manipulating nutrient signaling pathways. At the same time, the dramatic life span extension by these pathways is dependent on the presence of MnSOD, indicating that minimizing oxidative stress is necessary, if not sufficient, for life span extension.

Gems and Doonan (Chapter 6) present a solid argument that current data do not support the hypothesis that oxidative stress is limiting to life span in *Caenorhabditis elegans*, although they caution that this conclusion may be "private" to *C. elegans*, because its energy metabolism is unusual. A key observation incompatible with the idea that oxidative stress is life span limiting in *C. elegans* is that manipulating oxygen tension has very little effect on life span, even though this would alter the rates of superoxide production. *Caenorhabditis elegans* is highly unusual among higher metazoans in that it can live and reproduce normally in a 100% oxygen atmosphere; in mammals and *Drosophila*, such a treatment results in high oxidative stress and rapid death. *Caenorhabditis elegans* also seems unusual in that inhibiting the mitochondrial respiratory chain lengthens, rather than shortens, life span.

Mockett et al. (Chapter 7) review a large body of work testing the free radical theory in *Drosophila*. Their overall conclusion is that present data support the idea that oxidative stress contributes to aging in *Drosophila*, but not solidly so. The intervention leading to the greatest life span prolongation to date is the overexpression of glutamate-cysteine ligase (which has increased levels of the antioxidant glutathione), although other antioxidant enzyme overexpression studies have yielded mixed results. It is worth remembering that glutathione is not only an antioxidant but also a critical player in protein folding and detoxification, performing roles that are related but distinct from its antioxidant function. We note that, contrary to *C. elegans*, there is an almost linear decrease in life span with increased oxygen tension in *Drosophila*.

Because of cost and time constraints, much less work has been done on evaluating the free radical theory in mammals. Muller (Chapter 8) reviews the evidence from life span studies of antioxidant knockout mice and concludes that, although subject to some interpretation, the present data are not incompatible with the hypothesis that oxidative stress is limiting to life span.

Thus, there is neither unambiguous support nor refutation of the free radical theory. Why is there such variability between model systems in terms of the degree to which the free radical theory seems predictive? There is no simple relationship to explain this diversity, such as higher versus lower organisms. Consider the example of *Drosophila* and *C. elegans*, both on the protostome evolutionary branch, more closely related to one another than to mammals or yeast. Why does the free radical theory seemingly hold true in *Drosophila*, whereas it comes close to being demonstrably false in *C. elegans*? Although we can give no simple and firm answer, we would echo the sentiment of Gems and Doonan (Chapter 6), that *C. elegans* is unusual for a metazoan in its apparent minimal dependence on the mitochondrial respiratory chain for energy metabolism. Conversely, *Drosophila* is strongly dependent on mitochondrial energy metabolism and uses a tracheal respiratory system that allows rapid and large amounts of oxygen delivery, but at the price of the cells being directly exposed to atmospheric oxygen: this means cells are exposed continuously to 21% O_2, compared with 3% for a typical mammalian cell. Regarding how these results will ultimately translate to humans, results on antioxidant ablation and acute oxygen toxicity in mice would suggest that mammals, in general, behave more like *Drosophila* than *C. elegans*.

A completely different test for the free radical theory was explored in the review by López-Torres and Barja (Chapter 9). Comparisons of vertebrate species with different life spans indicate that species with longer life spans consistently show less oxidative stress than shorter-lived species. This correlation is seemingly not due to differences in antioxidant enzyme levels, but in ROS production by the mitochondrial electron transport chain. It is noteworthy that although many studies have manipulated antioxidant levels, the effect of altering mitochondrial ROS production on life span in mice has not been determined.

2   Role of Oxidative Stress in Pathology

The second part of this book documents the extraordinary explosion of interest in oxidative stress as a agent of pathology. This rapid progress in the field has been possible due to the development of animal (mouse) models of disease, and the fact that interventions and the follow-up to test the role of oxidative stress in pathology do not take as long as those in a direct study of aging.

Experts in various age-related pathologies have provided detailed assessments of the contribution of oxidative stress on development and consequences of the pathologic conditions.

Stuart and Page (Chapter 12) reviewed mouse models deficient in oxidative DNA damage repair, concluding that oxidative stress is a potent driver of DNA mutagenesis and carcinogenesis. Didion et al. (Chapter 13) reviewed the literature on the role of oxidative stress in hypertension, concluding that there is solid evidence implicating oxidative stress in both mice and humans. In mouse models, ablation of selected antioxidant enzymes results in an aggravation of experimentally induced hypertension. Conversely, overexpression of selected antioxidant enzymes or ablation of the superoxide producing vascular NAD(P)H oxidase can attenuate experimentally induced hypertension.

Ballinger (Chapter 10) evaluates the role of oxidative stress in cardiovascular diseases. He presents a paradigm of disease linked to mitochondrial dysfunction. This is supported by extensive correlative evidence in humans and by a key study showing that deficiency of MnSOD dramatically aggravates pathology in a mouse model of atherosclerosis.

Brookes and Hoffman (Chapter 14) discuss the involvement of mitochondrial function in cardiac ischemia reperfusion injury, with detailed accounts of the molecular mechanisms centered on cardiac mitochondria, specifically ROS production. They take a further step in examining how age-related changes in mitochondria (and other related cardioprotective signaling pathways) can affect the pathology. Although direct evidence is lacking, as the authors suggest, this model provides opportunity to examine a potential link between oxidative stress, aging, and pathology (susceptibility to ischemia-reperfusion injury). Although not emphasized in their chapter, ischemia-reperfusion injury can be modulated by antioxidant enzyme modulation.

In diabetes, Abdul-Ghani and DeFronzo (Chapter 11) review in vivo and in vitro cellular studies showing correlations between ROS and vascular complications in diabetes, β-cell failure, and insulin resistance. Moreover, hyperglycemia is linked with overproduction of mitochondrial ROS. However, they raise several issues in relating these studies to human type 2 diabetes, including the lack of consistent effects of antioxidant therapy on diabetic vascular complications.

To conclude, although increasing antioxidant enzyme activities does not typically have positive effects on life span in several model systems, various lines of evidence suggest the beneficial effect of antioxidant treatments against pathologic conditions in mouse models of human age-related disease. But, what evidence do we have that oxidative stress contributes to these pathologies in humans? Despite an abundance of supporting correlative epidemiologic data from observational studies, as noted by Serafini (Chapter 15), major preventive clinical trials of dietary antioxidants in cardiovascular disease and cancer have largely failed, although methodologic constraints might explain null results. Alternatively, it is also possible that oxidative stress is more of a bystander than an active player in the etiology of these diseases.

Remembering that ROS can play a physiologic signaling role (Chapter 16), and that evolution has permitted ROS to be produced constantly with imperfect antioxidant systems, any attempt to test oxidative stress involvement by reducing ROS levels, may have unforeseen consequences.

3   Epilogue

Overall, on its 50th birthday, the status of free radical theory would best be described by the Scottish legal verdict "not proven." The major failure of the theory is that to date, pharmacologic and genetic increases in antioxidant activities have not consistently and dramatically increased life span. At the same time, increasing ROS levels (typically by ablation of antioxidant enzymes) generally results in shortened life span, as the theory predicts. It is possible that oxidative stress is but one of many factors limiting life span, and that in and of itself, reducing oxidative stress is a necessary component but not sufficient for life span extension. Our understanding of the antioxidant system remains incomplete, with new antioxidant genes still being discovered. Thus, we, and many other researchers take the position that although not conclusive, the evidence is sufficiently encouraging to warrant further investigative effort.

> In completing one discovery we never fail to get an imperfect knowledge of others of which we could have no idea before, so that we cannot solve one doubt without creating several new ones.
>
> Joseph Priestley, 1785.

Index